A SOCIAL ARCHAEOLOGY OF
KINSHIP IN IBERIA AND BEYOND

A SOCIAL ARCHAEOLOGY OF KINSHIP IN IBERIA AND BEYOND

Recent Multistranded Approaches from aDNA to Household Archaeology

Antonio Blanco-González &
Eva Alarcón-García (eds)

Published by Sidestone Press, Leiden
www.sidestone.com
E-mail: info @ sidestone.nl
Phone: (+31)(0)71-7370131

Lay-out & cover design: Sidestone Press
Photograph cover: illustration based on a genome sequencing map that also resembles the cross-cultural concentric pattern for a unilineal settlement arranged around a central plaza (image by Tartila | stock.adobe.com)

This publication has been funded by Agencia Estatal de Investigación - Spanish Ministry of Science and Innovation, project ARQPARENT (PID2019-104349GA-I00, MCIN/AEI/10.13039/501100011033).

ISBN 978-94-6426-404-3 (softcover)
ISBN 978-94-6426-405-0 (hardcover)
ISBN 978-94-6426-406-7 (PDF e-book)

DOI: 10.59641/m3p9j0k1l2

Contents

Part I

INTRODUCTORY CONTRIBUTIONS

Introduction: For a Social Archaeology of Kinship

Antonio Blanco-González

Department of Prehistory, Ancient History and
Archaeology, University of Salamanca, Spain,
ablancoglez@usal.es

Eva Alarcón-García

Department of Prehistory and Archaeology,
University of Granada, Spain,
eva@ugr.es

Abstract

This chapter justifies the necessity of a social archaeology of kinship. To do so, the text frames what is understood today by kinship within its intellectual trajectory, giving a brief overview of different disciplinary approaches. Over the previous decades the topic was subject to strong criticism, loss of interest, mistrust or reticence. Nowadays the scholarly outlook is optimistic, the old weaknesses have been contested, interest in the topic is growing steadily, and fresh lines of enquiry have been opened. This paper critically surveys the current confusing panorama and points out unsolved shortcomings that the academic community is yet to face, particularly regarding the mainstream biodeterministic narrative within a context of scientism and neopositivist supremacy. The text then explains the genesis, aims and scope of the volume, framing its contents within the current intellectual landscape, which are arranged into two main research branches: science-based contributions drawing on bioarchaeology (aDNA and isotope analyses) and humanistic-oriented and anthropologically-informed chapters interested in household archaeology. Finally, it comments on the promising prospects for future advancements in this field.

Keywords: *Archaeology of kinship, archaeogenetics, isotope studies, household archaeology, burial archaeology.*

1.1 What does archaeological literature mean when it refers to kinship?

A quick glance at recent archaeological literature shows that the number of contributions with the word kinship in their title has grown in a short matter of time. However, if we delve into such readings, we will soon see that the idea of kinship associated with past human populations covers a varied range of concepts. Indeed, not only does this term refer to different realities, but it is approached from various disciplines, with competing theoretical and conceptual presuppositions and methodological protocols, and mobilising an assorted

In Blanco-González, A. and Alarcón-García, E. (eds.) 2025, *A Social Archaeology of Kinship in Iberia and Beyond. Recent Multistranded Approaches from aDNA to Household Archaeology.* Leiden: Sidestone Press, pp. 9-22.

array of bodies of evidence. All this leads to the elaboration of scientific accounts on kinship that can complement but also oppose each other and even offer incommensurable observations and datasets. Therefore, nowadays, diverse definitions of kinship coexist in scholarly literature, though it is not our objective to offer an all-encompassing one here. In fact, several of the articles collected in this volume explicitly propose their own definition, each one providing richly nuanced details, depending on each researcher's specialisation, bibliographic preferences and research agenda.

Anthropology has been the science of kinship par excellence since its foundation as a discipline (Fox 1967; Schneider 1984). Its interest in kinship has fluctuated throughout its history (see below), but it is essentially concerned with social relationships, analysed through social theory and with methods of inference based on analogy and cross-cultural approach. Archaeology inspired by anthropological methods has drawn on such a social and comparative perspective to address kinship in extinct populations with varying degrees of success (Ensor 2013, 2021; Souvatzi 2017). More recently, some natural scientific disciplines such as physical anthropology have contributed to the study of biological relationships between individuals, a subject which, as its practitioners contend, also falls under the umbrella of kinship (Johnson and Paul 2016). The emergence of human genomics and its application to ancient populations is even more recent and, again, this subfield claims that its goal of tracing genealogical links of consanguinity should also be called kinship (Vai *et al.* 2020; Meller *et al.* 2023).

In view of the above, some archaeologists have proposed to distinguish between "social" and "cultural kinship" approached from the humanities, and "biological kinship" studied by bioarchaeologists and geneticists. However, this divide is false and unnecessary, because the direct and predominant identification of biological progeny with kinship only works with certainty in our society; outside it we cannot extrapolate that equation and social issues encompass most of the casuistry.

This volume aims to contribute to building bridges between all available current visions of ancient kinship from a decidedly social and multiproxy approach. To this end, this book starts by recognising, as the theoretically informed literature does, that kinship is and can only be a social matter irreducible to blood parentage or biological relatedness (Ensor 2013, 2021; Johnson and Paul 2016; Souvatzi 2017; Brück 2021; Brück and Frieman 2021; Alber 2023; Thelen 2023). As Risch and colleagues (2023: 12) put it: "(…) all kinship practices are social constructs", past and present. Kinship is much more than biology; it is not exclusively given by birth or the transference of substances, and beyond the Western-modern obsession with biological progeny, parenthood is a historically contingent and negotiable cultural convention. In a nutshell, a social understanding of kinship involves analysing the crux of interactions between people self-conscious of being relatives, irrespective of the context-dependent criteria used to trace these links of belonging, whether by blood (consanguines), by alliance/marriage (affines) or by any other ways of kin making, under the umbrella of alloparenting (adoptees, godchildren), dependence (patron-clients), etc. In any case, kinship is the basic cross-cutting principle that underpins the livelihood, the physical and intangible heritage, the sociability and the decision-making principles of most societies (Fox 1967; Godelier 2011; Ensor 2013a, 2013b, 2021; Sahlins 2013).

This book compiles approaches to the subject of kinship among past populations using archaeological methods. As editors, we asked different researchers to address the subject from their working protocols, concepts, methods and hypotheses. The final volume is widely heterogeneous and rich in nuanced understandings of what kinship is and how we may approach it. The contents also convey in a clear way the current disparity and some unsolved problems involving confusion and contradiction between conflicting views. To better understand the source of some insufficiencies and misunderstandings in the current landscape, it is essential to briefly review the disciplinary trajectory of the study of kinship.

1.2 A very short overview of the archaeological study of kinship

To understand the role of kinship in archaeology, it is essential to frame its study within the history of humanities at large, with a long tradition, a very rich corpus of anthropological studies and a growing protagonism of natural sciences in the past few decades. The heyday of kinship studies was from the onset of anthropology as a modern discipline –with the foundational contribution of Lewis H. Morgan– up to the mid-twentieth century. By contrast, the period from the 1970s to the 2000s were decades of scholarly distrust on the subject. Such a prolonged negative milieu was led by radical criticism (Needham 1971; Schneider 1984), some of whose challenges were deserved, but which eventually fed interpretive abuses and deeply pervaded prejudices. This move distorted the analysis of kinship and contributed to its ensuing unpopularity, without acknowledging those premises and methods that remained valid. In hindsight, this almost complete banishment of the issue meant throwing the baby out with the bathwater.

The subsequent neglect of archaeologists' theoretical and methodological training on kinship up until today has multiplied the suspicions and uncertainty on this topic. Some scholars have even considered that kinship is irrelevant or inaccessible to archaeology and should be dismissed altogether (cf., Joyce and Gillespie 2000). According to Ensor (2013), this stagnating climate was fuelled by two main "antikinship" misunderstandings: a) the 1970s and 1980s postmodern attacks on kinship, which put improper emphasis on its symbolic dimension, detaching it from subsistence or politics, and regarding it as a mere Western ideological fabrication only existing in the mind of anthropologists (Schneider 1984); and b) the literature on House Societies (*sociétés à maison*) drawing on the notion of *maison* elaborated by Lévi-Strauss (1982), that mistakenly envisaged it as a non-kin-based corporate institution alternative to kinship (Joyce and Gillespie 2000). Ensor (2013: 10-16) has thoroughly responded to such criticism, which was based on oversimplistic and reductionist mischaracterisations. Following these efforts to clarify the situation, it should become clear that: a) kin groupings are very real emic sociological arrangements working in disparate worldwide societies past and present, which can be confidently tackled from diverse approaches; and b) the "house" literature has not realised that Lévi-Strauss (1982) only considered kinship the unilineal descent, whereas *sociétés à maison* are based on cognatic and bilocal kinship practices. After strong rebuttals and clarifications in the 2010s (Godelier 2011; Ensor 2013; Sahlins 2013; Johnson and Paul 2016; Ensor *et al.* 2017) there is no reason to avoid or circumvent this crucial subject anymore. Although there is still some reluctance on its archaeological feasibility, currently an optimistic atmosphere frames the field, and kinship is becoming a buzzword –almost a catchy slogan– that is relentlessly populating the archaeological literature.

1.3 Archaeological trends in the study of past kinship

In the present scholarly panorama, we may glimpse two obvious major strands of archaeological enquiry on kinship, with very unequal attention and numbers of practitioners: one recent and hegemonic trend, biologically-based and focused on the skeletal remains of ancient human populations, and another long-standing, minority anthropologically-oriented tradition working cross-culturally, mainly –but not exclusively– from domestic remains.

On the one hand, an array of bioarchaeological branches has contributed to address kinship. This includes physical anthropology on morpho-metric phenotypic variables and biodistance (e.g., Johnson and Paul 2016; Ensor *et al.* 2017; Ensor 2021). Noteworthy are the biomolecular strands of archaeogenetics and stable isotope analyses –especially dental strontium– that have grown over the last decade (e.g., Vai *et al.* 2020; Bösi and Samida 2021; Pearson *et al.* 2021; Villalba-Mouco *et al.* 2022; Morell-Rovira *et al.* 2024), especially strong when combined (e.g., Sjögren *et al.* 2020; Ensor 2021; Armit *et al.* 2023; Pearson *et al.* 2023). These developments stem from archaeologists, geneticists and chemists working together to infer genomic dynamics and mobility patterns –crucial to unveil marriage and postmarital residence practices– from ancient individuals. As for archaeogenetics, the cutting-edge and high-resolution techniques of genome-wide next generation sequencing (NGS) developed from the mid-2000s are increasingly accurate, quick and cost-effective –in terms of money and sample size– methods to obtain ancient DNA (henceforth aDNA) for modelling biological parentage and ancestry (Reich 2018; Sjögren *et al.* 2020; Vai *et al.* 2020; Bösi and Samida 2021; Kristiansen 2022; Armit *et al.* 2023; Meller *et al.* 2023). Such collaboration has provided impressive and unexpected results. For many archaeologists, conventional wisdom –nurtured by the above-mentioned misunderstandings– has it that this natural science-based avenue is not only mainstream, but the only reliable and "scientific" way to tackle ancient kinship (e.g., Meller *et al.* 2023).

On the other hand, over the last decade there has also been substantial progress in humanistic approaches to kinship drawing on the long tradition of anthropological studies, which is somehow making a strong comeback. This line of enquiry draws on cross-cultural and more theoretically grounded underpinnings through trustworthy and well tested inference strategies which –importantly– strictly resort to archaeological methods. Thus, proxies and indicators of patterns of postmarital residence, filiation/descent and kin group membership have been advocated and statistically revalidated by drawing on household archaeology. These methods pay special attention to the design and spatial arrangement of residential architecture, neighbourhoods and whole settlements, including open spaces, communal facilities and burial grounds (Ensor 2013, 2021; Huebner and Nathan 2017; Souvatzi 2017; Carpenter and Prentiss 2022; Souvatzi *et al.* in press). This avenue of enquiry is to become essential when trying to address kinship in those cultural contexts whose funerary evidence is missing, either due to archaeologically untraceable burial customs –i.e., leading to no human remains– or because of circumstances affecting the preservation of bone, that is, cultural mores –e.g., cremation– or post-depositional vagaries –e.g., osseous dilution in acidic soils.

Strangely enough, in recent compilations bearing the word kinship in their titles, the latter contributions are only quoted in passing (e.g., Brück and Frieman 2021: 50) or rarely feature, perhaps because these volumes mostly deal with sexual intercourse and biological progeny (e.g., Meller *et al.* 2023). By contrast, the all-encompassing book edited by Souvatzi and others (in press) is exceptional. In this line, this volume aims to give equal voice to both research trends and contribute to the breaking down of existing disciplinary barriers, thus opening new pathways of understanding, debate and exchange of experiences and learning. First, however, it is necessary to address some outstanding issues, because is mutual understanding even possible? If current perspectives on kinship do not speak the same language, do they have at least some common grounds? The following section addresses some of the difficulties of the scholarly landscape regarding these questions.

1.4 Problems with the mainstream understanding and narratives on past kinship

As the reader will see, all scientists who today study kinship in past societies are convinced that their concepts and intellectual background are suitable and adequate, their technical strategies are reliable and accurate and that, in short, their object of study is of course kinship. However, it is becoming increasingly clear that the use of the word "kinship" by some biological approaches is restrictive and reductionist, to say the least. Indeed, there seems to be too much conflation of biology and kinship in recent papers drawing on biomolecular research, where the concept of kinship should be used more carefully.

As anthropologically informed scholars have warned repeatedly (Godelier 2011; Sahlins 2013; Ensor 2013, 2021; Johnson and Paul 2016; Ensor *et al.* 2017) blood relatives visible via aDNA only make up a small fraction of the social kin-based groups tackled by anthropology, history or archaeology. The techniques of aDNA are about procreation, that is, tracing biological genealogy between progenitors and their offspring. We should bear in mind that despite the misleading use of the same concept by archaeologists and geneticists, cultural and DNA lineages are two completely different issues (Ensor 2021: 108). The agnatic –i.e., paternally transmitted– Y-haplogroup so-called "patrilineage" and the uterine –i.e., maternally transmitted– "matrilineage" traced via mitochondrial DNA (mtDNA) are genetic abstractions that have very little to do with emic social groupings of people self-identified as relatives. People may arrange in far more variable ways: unilineal descent groups –lineages and clans– but also non-unilineal kindreds –the rare ambilineal ramages or the small bilateral extended groupings. Furthermore, such genetic analytical subunits "reveal only a tiny percentage of an individual's ancestry" (Brück and Frieman 2021: 49). In other words, people related by their Y chromosome –and therefore somehow sharing a genetic line– and the members of their social patrilineage are not the same (cf., Ensor this volume). This is because in unilineal kin groupings –patrilineages and matrilineages, respectively nested within patri- and matriclans– most genealogical/ biological relatives are excluded, whereas membership criteria can be manipulated, and social groupings can opportunistically dissolve or fuse (Ensor 2021: 21). Moreover, from a worldwide and cross-cultural stance, most consanguineous individuals are not considered kin, because it is a cultural category (Fox 1964: 31; Ensor 2013, 2021). Thus, the term kinship –featuring prominently in the titles of contributions drawing on aDNA (e.g., Vai *et al.* 2020; Sjögren *et al.* 2020; Villalba-Mouco *et al.* 2022; Armit *et al.* 2023; Pearson *et al.* 2023)– rarely belongs here, because it is often confounded with sexual procreation, which is a restricted facet of kinship, whose role in social organization is widely variable. Biological progeny relations can be and have been culturally and selectively emphasized, negotiated, concealed or fabricated altogether (Ensor 2021: 11).

The problem of using the same term to convey widely disparate realities also arises in the case of sampled individuals sharing "pedigrees" tracked up to four or more generations along the paternal or maternal lines (e.g., Villalba-Mouco *et al.* 2022; Armit *et al.* 2023) who were hardly contemporaries. In a restrictive sense, centred on social interaction, this putative genealogy should be kept out of the realm of kinship. Most biologically linked relatives past and present never belonged to the same living kin groupings or even never interact(ed) between them. Therefore, they can be considered biological ascendants or descendants yet hardly regarded in terms of kinship relationships.

The bioarchaeology of kinship is purportedly envisaged as revolutionary and a brand-new science of the human past (Reich 2018; Kristiansen 2022). Why is this so? All the above has to do with the present-day climate of increasing hegemony of neo-positivism, science-based functionalism and biodeterministic discourses, which have an overwhelming inertia of their own (Blakey 2020). This hinders critical reflection and relegates social and humanistic contributions to a very minor role (but see Brück and Frieman 2021; Alber 2023; Thelen 2023). Sophisticated description of quantitative variables is mainstream, but this kind of archaeological science rarely questions reality, is often self-protrayed as innocuous and politically deactivated, and thus its historical contribution may be of little relevance. This is more problematic in the current political landscape of the resurgence of the extreme right and discourses based on atemporal essences –e.g., the "race"– and against multiculturalism (Hakenbeck 2019; Blakey 2020). This is another reason why a social archaeology firmly based on humanist foundations is sorely needed.

1.5 Weaknesses and prospects to address kinship today

As we are seeing, despite its century-old disciplinary history, the field of kinship is far from consensual, and many aspects are riddled with conflicting standpoints, internal contractions and interpretive tensions (Godelier 2011; Ensor 2013, 2021; Blakey 2020). All these aspects are present in the contributions collected here. This is something to be expected when dealing with the core of sociability as a subject of study. However, a large part of present difficulties stems from deficiencies attributable to factors external to the development of the discipline itself: lack of knowledge of other approaches, uncommunication between closed scientific communities –despite the constant celebration of interdisciplinarity– and, consequently, reproduction of pitfalls or concerns that have already been solved and answered.

As Ensor (2013, 2021, this volume) and collaborators (Ensor *et al.* 2017) have denounced repeatedly, in absence of some knowledge of the up-to-date anthropological literature on kinship and the lack of training on the subject, most archaeologists address kinship as an easy and unproblematic topic. In this vein, the notions of kinship handled by archaeologists are very elementary, often theoretically outdated and not very refined. The conventional account may differentiate –if ever– between "simple" and "complex" or "nuclear" and "extended" families at most and may even mention lineages or clans (e.g., Huebner and Nathan 2017; Scarre 2018). Yet the relationship between these concepts remain unclear, such broad-brush characterisations conceal as much as they reveal, and non-unilineal groupings are hardly considered. Even some basic cross-cultural assumptions of kinship scholarly, such as the organisation according to two broad kinds of descent patterns –unilineal descent groups and bilateral kindred groupings– is apparently overlooked by most archaeologists and geneticists. This can be seen in recent archaeological literature and is also apparent in several contributions to this volume that only think in terms of unilineal descent (e.g., Chapters 4, 6, 7 or 8). The underlying assumption is that the only possible kinship option is heteronormative, male-dominated and unilineal (Brück 2021). Behind this is the European and evolutionary obsession linking paternity and patrilineality with heritage and economic and political power, as exposed by Thelen (2023: 30-31) while criticising a recent archaeogenetic study on a Bronze Age Argaric site in southeastern Iberia (Villalba-Mouco *et al.* 2022). As we only find what we know how to look for, it is hardly surprising that usually only patrilocality and patrilineality are identified everywhere. This has been refuted by Ensor (2021) for the European Neolithic, something endorsed in recent studies using isotopic evidence (Morell-Rovira *et al.* 2024). In fact, recent wide genetic samplings such as those carried out in aceramic Neolithic Çatalhöyük (Turkey) –where maternally transmitted links have been identified– (Pearson *et al.* 2023) are finding the expected: a past endowed with more variability, that is, less familiar and less like our own world (Brück and Frieman 2021; Thelen 2023). However, it is very infrequent to find in archaeological works any other cultural practices that also define kinship, such as bilocal or neolocal postmarital residence or ambilineal or bilateral filiation (but see Chapters 2, 3, 5, 9, 10 and 11). Far more regrettable is the misuse of technical jargon or the substitution of long-held concepts with new unnecessary and mistaken expressions. This is the case of "male/female exogamy", an unfortunate misnomer recently coined by genetic studies, rightly criticized by Ensor (Chapter 2) and used in several contributions here (Chapters 5, 6 and 8) to actually refer to group exogamy and postmarital male/female mobility.

To overcome such deficiencies, this volume aims to show how archaeology can provide independent lines of evidence and robust inferences on kinship, regardless of whether they are literate or illiterate societies. The task ahead is therefore to re-establish kinship to a central stage as a key matter of debates in humanities and social sciences, including archaeology.

1.6 Genesis, aims and contents of the volume

Building on the above-mentioned concerns, problems and prospects, this collective book aims at contributing to overcome the contradictions, tensions, and limitations of the archaeological record when dealing with kinship in the past. The key objective of the volume is to present an updated overview of current archaeological analyses of kinship from the archaeological record. This collection of papers derives from a two-day workshop (15-16 February 2024) held at the University of Salamanca (Spain) funded by a research project and hosting the most recent advances in the social archaeology of kinship. All the atendees of that meeting also participate in this book, with the exception of Prof. Gonzalo Ruiz Zapatero (Complutense University of Madrid) who was generous enough to act as the final discussant. In addition, a few scholars were subsequently invited (authors of Chapter 5) to include more comprehensive coverage.

The volume is important for it brings together the contributions of leading international scholars from various fields of knowledge tackling ancient kinship from diverse standpoints. Despite the growing number of contributions on this issue, European research has focused intensively on early periods of the later prehistory, especially on the Neolithic (e.g., Ensor 2021; Souvatzi 2017), when kinship was often not as crucial as later in history. By contrast, this volume aimed to pay due attention to case studies on Chalcolithic, Bronze and Iron Ages and Roman agrarian societies in Iberia plus other integrated political organisations and ranked lifestyles elsewhere that required coping with large-scale agrarian labour and heritage management. Even though "it is impossible to know in advance the importance of kinship" (Godelier 2011: 85) such more complex urban societies are the ideal candidates to scrutinize the role of sophisticated kinship practices to emphasise genealogy aimed at guaranteeing the transmission of social power and dynastic material and immaterial wealth. The book delves into key results –often already published yet re-elaborated by their authors for this occasion– from research mainly focused on Iberia, yet drawing also case studies from other regions worldwide (Fig. 1.1).

The Iberian scope is most suitable because the westernmost edge of Europe belongs to a minority research tradition –as defined by Neustupný (1997-8)– which is quickly "moving towards mainstream status" due to its stability, dynamism and sustained efforts to interact with the most influential mainstreams (Ruiz Zapatero 2011: 82) and therefore it can be regarded paradigmatic in the international milieu. Although Spanish funding for research projects is much lower than in other European countries, Spain is home to very dynamic research teams often engaged in international collaborations, something emphatically encouraged by the national research system. Most importantly, archaeological research on kinship in Iberia encapsulates a representative microcosm of the main historiographic trends at the international scene, including weaknesses and inadequacies, as well as strengths and prospects of such an endeavour. Indeed, over a deeply pervaded baseline of culture historical notions, Spanish archaeology is a hub of vibrant theoretical debate, with powerful research groups offering varied and even opposing accounts of the same phenomena, therefore challenging and enriching the research landscape. Within this background, archaeogenetics has recently added an impetuous positivism and biodeterminism (Blakey 2020) giving almost exclusive priority to biomolecular evidence over other more "traditional" ways of inference on the subject.

All those characteristics have been considered when designing the contents of this compilation and are abundantly present in the texts. Thus, the reader will be able to check the similarities in how kinship is conceived and how the results of various sources of information are interpreted by theoretically related teams and find alternative accounts for the available datasets. The contents of the book range from approaches that only conceive of unilineal descent –such as those from genetics (Chapter 4), or others addressing biogenetic datasets from Marxist assumptions (especially Chapters 6 and 7)– to more possibilistic and flexible ones that attach importance to other forms of descent, such as bilateral practices (Chapters 2, 3, 5, 9, 10 and 11). All in all, the selected case studies reflect well the current climate of conceptual disparity, uneasiness, confusion and even improper language use when dealing with this topic. It is therefore an eclectic book, whose heterogeneous contributions draw on assorted theoretical foundations, varied notions of kinship and diverse ways of evidential reasoning in terms of kinship (e.g., variably resorting to soft direct analogy or cross-cultural inference).

The volume is divided into three parts: a short introductory section (Part I) and two blocks devoted respectively to biologically-oriented contributions (Part II) and a few papers delving into household archaeology (Pat III). The very imbalance of both branches

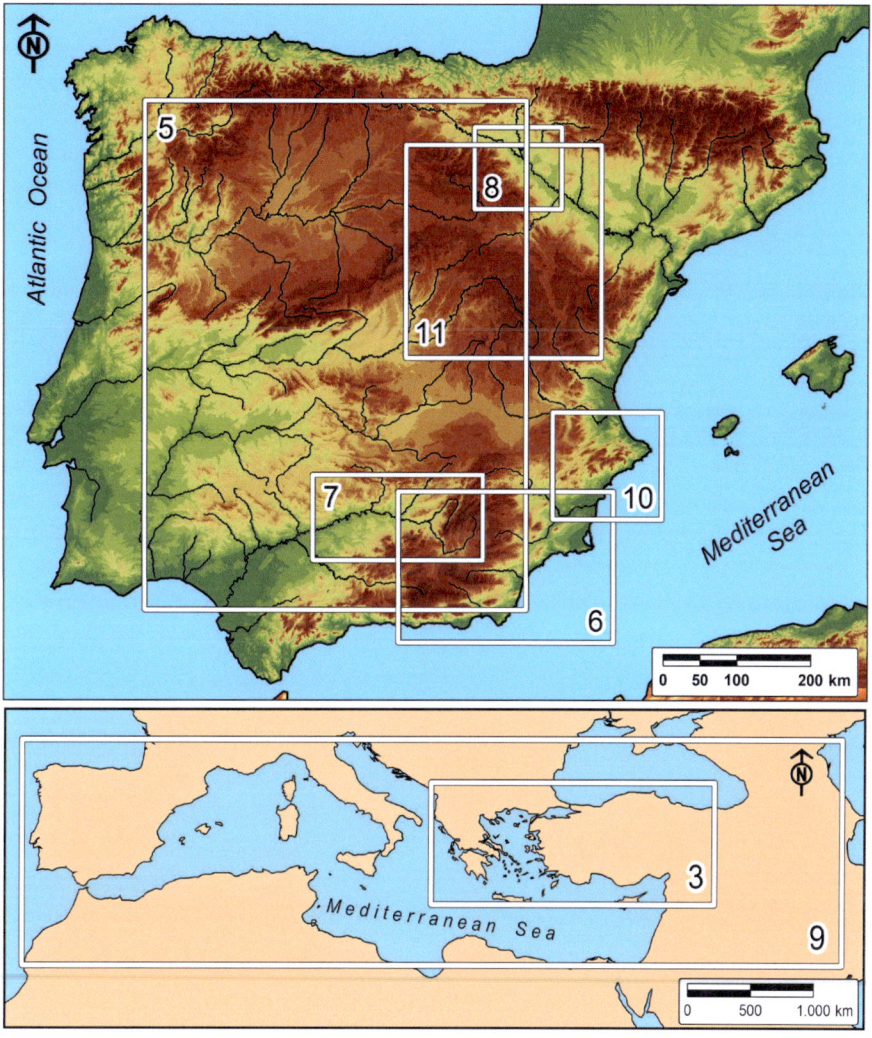

Figure 1.1. Map showing the location of study areas of the contributions, indicated by chapter numbers (by A. Blanco-González. Raw data: Natural Earth).

in the scholarly literature is somehow conveyed by their uneven presence here. The first of the volume's blocks, *Part I. Introductory contributions*, includes the essays of the two keynote papers presented at the original workshop by two internationally renowned scholars with wide expertise in the social analysis of prehistoric kinship. Bradley Ensor (Chapter 2) presents a masterful retrospective on his two-decade experience devoted to inferring social scientific knowledge on kinship from archaeological remains. He vindicates its informative potential, emphasising the need to integrate ethnology, archaeology and bioarchaeology, and reflects on the limitations and problems involved in the challenge of addressing kinship as a dynamic and fluid subject. Particularly instructive and recommendable is his pointing out of common malpractices in current interpretation of kinship. Chapter 3 by Stella Souvatzi continues along the path of critique, reflection and construction of kinship narratives in archaeology. To do so, she examines kinship dynamism and variability in the archaeological record of Greek (Sesklo) and Turkish (Aşikli Höyük and Çatalhöyük) early Neolithic settlements. Her highly textured scrutiny offers a fresh and insightful account on overdue and unconvincing archaeological readings, such as the attempts of interpreting Çatalhöyük without kinship.

Part II. Biomolecular approaches to kinship brings together exemplary cases of characterisation of bioarchaeological bodies of evidence –archaeogenetic and isotope– to track biological relationships variably related to past kinship. Rosa Fregel (Chapter 4) presents in a didactic way the current international protocol for the recovery of aDNA and discusses its explanatory viability. She underlines how the rigour of the procedure is to enhance the chances for obtaining positive results. The text also posits the need for this data to be contextual rather than abstract and concludes that only by addressing these aspects, paleogenomic analyses can shed light on kinship and social organisation in past societies. In Chapter 5, Marta Cintas-Peña and Ana Herrero-Corral combine aDNA and strontium isotopes from a rich corpus of published datasets to address so-called "genetic kinship" and postmarital residence patterns as interdependent variables. Their overview underlines problems in the available evidence –e.g., the lack of sex and age determination in sampled individuals– and concludes acknowledging the occurrence of patrilocal and bilocal practices during the third millennium BCE in Chalcolithic Iberia. Among the tensions within this contribution is the habitual conflation of post-mortem and postmarital mobility, a key issue often assumed yet rarely properly discussed. The following paper by Eva Celdrán and collaborators (Chapter 6) tackles kinship and gender identities in the extensively open-area excavated site of La Almoloya (Murcia, Spain), which has produced the largest aDNA sample in prehistoric Iberia, tackling the Early Bronze Age Argaric society (2200-1550 BCE) in the Iberian Southeast. The text, soundly backed by Marxist underpinnings, points to the identification of unilineal descent and patrilocality, which are fully in keeping with the excavators' class-based and state-like hypothesis. Interestingly, results from both genetics and household archaeology may be interpreted in other ways and this may prompt fruitful debate. In Chapter 7, Carmen Rísquez and Arturo Ruiz depart from the premise that the spatial distribution of tombs within their target monumental tomb can be interpreted in terms of kinship relationships. Based on such an approach, they use bioarchaeology, burial furnishings and the spatial arrangement of tombs in the Early Iron Age tumulus of Cerrillo Blanco (Porcuna, Jaén) to read the articulation and legitimation of power at a crucial moment in the forging of protohistoric Iberian societies. These researchers also draw on Marxist social theory to analyse family compositions and embrace a unilineal interpretation to account for the presence of hierarchical "extended" and "nuclear families", thus adhering to the agnatic-clientele model posited for several nearby necropolises. Concluding this section, Roberto Risch and coauthors (Chapter 8) draw on already published and ground-breaking datasets to reflect on the pros and cons of archaeogenetics (aDNA) in the identification of "biological kinship" and the performative complexity of funerary practices. They are interested in why during the Early Iron Age in the Ebro Valley (Spain), funerary rituals for newborn children within the dwellings were so common. Drawing on the high-resolution genetic information from Alto de la Cruz and Las Eretas villages, these scholars suggest that such intramural burials responded to a most inusual casuistry –twins, trisomies, Down and Edwards syndromes– due to selective deposition. They eventually conclude that the target sample was extracted from a pool larger than expected and suggest a complex mix of residence and descent practices hardly compatible from cross-cultural standards. Such contentious reading will hopefully provoke responses attentive to the rich combination of burial and household records.

The final section, *Part III. Addressing kinship from household archaeology*, covers a handful of articles with a special focus on the domestic realm. They resort to the spatial arrangement of settlement layouts and domestic subunits to glimpse how and in what way kinship practices were channeled and materialised and vice versa. The three chapters also impinge upon the reintegration of the House Society model in the discussion in terms of kinship of Mediterranean cases, where such an historical institution was pervasive and widespread (García Moreno 2022). Thus, these authors pay attention to unveiling how everyday social practices such as cohabitation, commensality, co-working and maintenance chores can become effective ways of kin making. Instead of a complete characterization of the rich variability to be expected in the ancient societies analysed in this volume (e.g., Chapter 3), Antonio Blanco (Chapter 9) traces a concrete way of life and co-residence: the composite virilocal aggregate. His text adopts a comparative and transcultural perspective to study well-known examples worldwide of this ubiquitous collective social agent –though neither exclusive nor necessarily predominant– in patriarchal societies and uses bilocal and bilateral cases as counterexamples to emphasize commonalities. On his part, Ignasi Grau (Chapter 10) suggests a highly detailed archaeological look at residential buildings of the Late Iron Age communities of South-eastern Iberia to offer an in-depth collation of unilineal/patrilocal and bilateral/bilocal kinship patterns patent in domestic archaeology. In so doing, he claims for the heuristic potential of the household cycles and dynamics and their contextual analysis as a reliable and highly informative strategy to evaluate social organization. Lastly, Chapter 11 by Jesús Bermejo choses another richly documented case to conduct an alternative, anthropologically oriented analysis of a Late Roman *villa* in Central Iberia. His household archaeology study is exemplary in including assorted lines of evidence –e.g., architecture, heraldry, literary and epigraphic data– paying attention to symbolic capital and combined with an unusual bottom-up social perspective within this stagnant and undertheorized subfield of enquiry. This approach allows the author to envisage a social scenario marked by social resilience and the central role of heterarchical and bilateral kinship models, in contrast to the classical decline-driven interpretations of the end of the Roman Empire.

In short, this book has aimed at highlighting the prospects and revalorising the strengths of both lines of work on the subject: the quantitative, science-based, deterministic and burial-centred approach and the more humanistic, qualitative and relativistic one. This duality of approaches is indeed conveyed by the cover image of the book: an illustration based on a genome sequencing map which also recalls the cross-cultural spatial pattern of residential groups practicing unilineal descent organised around a large central plaza (cf., Ensor 2013, 2021, Chapters 2 and 3). The main objective of the volume is to contribute to bridging the current divide while promoting much-needed interdisciplinary dialogue and fostering mutual learning among kinship studies in archaeology. The topic is attracting growing interest and has been enriched with very substantial recent pieces (Ensor 2013, 2021; Ensor *et al.* 2017; Meller *et al.* 2023; Souvatzi *et al.* in press), unthinkable only a decade ago. However, it demands much-needed intellectual elaboration and reliable and theoretically supported approaches to avoid habitual misunderstanding and faulty reasoning (Ensor 2013, 2021; Brück 2021; Thelen 2023). We do hope that this selection of texts will be stimulating for the international community of archaeologists, and that mainstream and other minority academic traditions –*sensu* Neustupný (1997-8)– beyond Iberia will be challenged. In this way, the volume could fulfill its purpose of raising awareness of the challenges facing us in the archaeological study of kinship, but also of the barely glimpsed pathways and possibilities that remain to be explored. Indeed, the future of the study of kinship relies in more integrated multistranded cooperation sharing well-supported, sophisticated and theoretically and anthropologically informed underpinnings while mobilising an array of independent bodies of evidence.

Acknowledgements

This book and the workshop have been made possible through funding from the Spanish Ministry of Science and Innovation, via the project *Archaeology of kinship through the household in the Iberian Northern Plateau (1800-400 BCE) - ARQPARENT* (PID2019-104349GA-I00, MCIN/AEI/10.13039/501100011033) whose principal investigators were the co-editors. We are very grateful to all contributors for their openness to dialogue and exchange standpoints, challenges and experiences, and especially for their commitment and patience with this editorial endeavour. Charles Bashore Acero was crucial in proofreading the English version of all the chapters. We would also like to thank the Sidestone team for their superb work.

References

Alber, E. (2023): Multiple relations: Towards and anthropology of parenting. In Meller, H.H., Krause, J., Haak, W. and Risch, R. (eds.), *Kinship, sex, and biological relatedness. The contribution of archaeogenetics to the understanding of social and biological relations.* Tagungen des Landesmuseums für Vorgeschichte Halle, 28. Heidelberg: Propylaeum, pp. 35-41. DOI:10.11588/propylaeum.1280.c17992.

Armit, I., Fischer, C.E., Koon, H., Nicholls, R., Olalde, I., Rohland, N., Buckberry, J., Montgomery, J., Mason, Ph., Črešnar, M., Büster, L. and Reich, D. (2023): Kinship practices in Early Iron Age South-east Europe: Genetic and isotopic analysis of burials from the Dolge njive barrow cemetery, Dolenjska, Slovenia. *Antiquity* 97(392), pp. 403-418. DOI:10.15184/aqy.2023.2.

Blakey, M.L. (2020): On the biodeterministic imagination. *Archaeological Dialogues* 27, pp. 1-16. DOI:10.1017/S1380203820000021.

Bösi, E. and Samida, S. (eds.) (2021): *Special Topic: Next generation sequencing. Challenges for science and society. TATuP Zeitschrift für Technikfolgenabschätzung in Theorie und Praxis. Journal for Technology Assessment in Theory and Practice* 30(2), pp. 11-52. https://www.tatup.de/index.php/tatup/issue/view/170.

Brück, J. (2021): Ancient DNA, kinship and relational identities in Bronze Age Britain. *Antiquity* 95(379), pp. 228-237. DOI:10.15184/aqy.2020.216.

Brück, J. and Frieman, C.J. (2021): Making kin: The archaeology and genetics of human relationships. In Bösi, E. and Samida S. (eds.), *Special Topic Next generation sequencing. Challenges for science and society. TATuP Zeitschrift für Technikfolgenabschätzung in Theorie und Praxis. Journal for Technology Assessment in Theory and Practice* 30(2), pp. 47-52. DOI:10.14512/tatup.30.2.47.

Carpenter, L.C. and Prentiss, A.M. (eds.) (2022): *Archaeology of Households, Kinship, and Social Change.* Oxford: Routledge.

Ensor, B.E. (2011): Kinship theory in archaeology: From critiques to the study of transformations. *American Antiquity* 76(2), pp. 203-227. DOI: 10.7183/0002-7316.76.2.203.

Ensor, B.E. (2013): *The Archaeology of Kinship: Advancing Interpretation and Contributions to Theory.* Arizona: The University of Arizona Press.

Ensor, B.E. (2021): *The Not Very Patrilocal European Neolithic. Strontium, aDNA, and Archaeological Kinship Analyses.* Oxford: Archaeopress Archaeology.

Ensor, B.E., Irish, J.D. and Keegan, W.F. (2017): The bioarchaeology of kinship: Proposed revisions to assumptions guiding interpretation. *Current Anthropology* 58(6), pp. 739-761. DOI: 10.1086/694584.

Fox, R.A. (1967): *Kinship and Marriage. An Anthropological Perspective.* Cambridge: Cambridge University Press.

García Moreno, J.C. (ed.) (2022): *From House Societies to States. Early Political Organisation, from Antiquity to the Middle Age*s. Oxford: Oxbow Books.

Godelier, M. (2011): *The Metamorphoses of Kinship.* London: Verso.

Hakenbeck, S.E. (2019): Genetics, archaeology and the far right: An unholy trinity. *World Archaeology* 51, pp. 517-527. DOI: 10.1080/00438243.2019.1617189.

Huebner, S.R. and Nathan, G. (eds.) (2017): *Mediterranean Families in Antiquity: Households, Extended Families, and Domestic Space.* Oxford: Wiley-Blackwell.

Johnson, K.M. and Paul, K.S. (2016): Bioarchaeology and Kinship: Integrating Theory, Social Relatedness, and Biology in Ancient Family Research. *Journal of Archaeological Research* 24(1), pp. 75-123. DOI: 10.1007/s10814-015-9086-z.

Joyce, R.A. and Gillespie, S.D. (eds.) (2000): *Beyond Kinship: Social and Material Reproduction in House Societies.* Philadelphia: University of Pennsylvania Press.

Kristiansen, K. (2022): *Archaeology and the genetic revolution in European prehistory.* Cambridge: Cambridge University Press.

Lévi-Strauss, C. (1982): *The Way of the Masks.* Seattle: University of Chicago.

Meller, H., Krause, J., Haak, W. and Risch, R. (eds.) (2023): *Kinship, sex, and biological relatedness. The contribution of archaeogenetics to the understanding of social and biological relations.* Tagungen des Landesmuseums für Vorgeschichte Halle, 28. Heidelberg: Propylaeum. DOI: 10.11588/propylaeum.1280.c18044.

Needham, R. (ed.) (1971): *Rethinking Marriage and Kinship.* London: Routledge.

Neustupný, E. (1997-98): Mainstreams and minorities in archaeology. *Archaeologia Polona* 35-36, pp. 13-24.

Pearson, J., Lamb, A. and Evans, J. (2021): Multi-isotope evidence of diet (carbon and nitrogen) and mobility (strontium) at Neolithic Çatalhöyük. In Hodder, I. (ed.), *Peopling the Landscape of Çatalhöyük: Reports from the 2009–2017 Seasons.* London: British Institute at Ankara, pp. 217-244.

Pearson, J., Evans, J., Lamb, A., Baird, D., Hodder, I., Marciniak, A., Larsen, C.S., Knüsel, C.J., Haddow, S.D., Pilloud, M.A., Bogaard, A., Fairbairn, A., Plug, J., Mazzucato, C., Mustafaoğlu, G., Feldman, M., Somel, M. and Fernández-Domínguez, E. (2023): Mobility and kinship in the world's first village societies. *Proceedings of the National Academy of Sciences USA* 120(4), p. e2209480119. DOI: 10.1073/pnas.2209480119.

Reich, D. (2018): *Who we are and how we got here. Ancient DNA and the new science of the human past.* Oxford: Oxford University Press.

Risch, R., Haak, W., Krause, J. and Meller, H. (2023): Kinship, sex, and biological relatedness. The contribution of archaeogenetics to the understanding of social and biological relations. In Meller, H., Krause, J., Haak, W. and Risch, R. (eds.), *Kinship, sex, and biological relatedness. The contribution of archaeogenetics to the understanding of social and biological relations.* Tagungen des Landesmuseums für Vorgeschichte Halle, 28. Heidelberg: Propylaeum, pp. 9-25. DOI: 10.11588/propylaeum.1280.c18044.

Ruiz Zapatero, G. (2011): Settlement and Landscape in Iron Age Europe: Archaeological Mainstreams and Minorities. In Moore, T. and Armada, X.L. (eds.), *Atlantic Europe in the First Millennium BC: Crossing the Divide.* Oxford: Oxford University Press, pp. 81-108. DOI: 10.1093/acprof:osobl/9780199567959.003.0002.

Sahlins, M. (2013): *What Kinship Is - And Is Not.* Chicago: University of Chicago Press.

Scarre, C. (ed.) (2018): *The Human Past. World Prehistory and the Development of Human Societies.* Fourth edition. London: Thames and Hudson.

Schneider, D. (1984): *A Critique of the Theory of Kinship.* Chicago: University of Chicago Press.

Sjögren, K.G., Olalde, I., Carver, S., Allentoft, M.E., Knowles, T., Kroonen G., Pike, A.W., Schröter, P., Brown, K.A., Brown, K.R., Harrison, R.J., Bertemes, F., Reich, D., Kristiansen, K. and Heyd. V. (2020): Kinship and social organization in Copper Age Europe. A cross-disciplinary analysis of archaeology, DNA, isotopes, and anthropology from two Bell Beaker cemeteries. *PLoS One* 15(11), p. e0241278. DOI: 10.1371/journal.pone.0241278.

Souvatzi, S. (2017): Kinship and Social Archaeology. *Cross-Cultural Research* 51(2), pp. 172-195. DOI: 10.1177/1069397117691028.

Souvatzi, S., Bickle, P. and Cvecek, S. (eds.) (in press): *Prehistoric Kinship: Contemporary Perspectives in Archaeology and Bioarchaeology.* Cambridge: Cambridge University Press.

Thelen, T. (2023): Kinship: Old problems and new prospects in the conversation between archaeology and social anthropology. In Meller, H., Krause, J., Haak, W. and Risch, R. (eds.), *Kinship, sex, and biological relatedness. The contribution of archaeogenetics to the understanding of social and biological relations.* Tagungen des Landesmuseums für Vorgeschichte Halle, 28. Heidelberg: Propylaeum, pp. 29-34. DOI: 10.11588/propylaeum.1280.c17991.

Vai, S., Amorim, C.E., Lari, M. and Caramelli, D. (2020): Kinship Determination in Archaeological Contexts Through DNA Analysis. *Frontiers in Ecology and Evolution* 8, p. 83. DOI: 10.3389/fevo.2020.00083.

Villalba-Mouco, V., Oliart, C., Rihuete-Herrada, C., Rohrlach, A.B., Fregeiro, M.I., Childebayeva, A., Ringbauer, H., Olalde, I., Celdrán Beltrán, E., Puello-Mora, C., Valério, M., Krause, J., Lull, V., Micó, R., Risch, R. and Haak, W. (2022): Kinship practices in the early state El Argar society from Bronze Age Iberia. *Scientific Reports* 12, p. 22415. DOI: 10.1038/s41598-022-25975-9.

Observations on Methods from Two Decades of Interdisciplinary and Transatlantic Prehistoric Kinship Research

Bradley E. Ensor

Department of Sociology, Anthropology and Criminology,
Eastern Michigan University, United States of America,
bensor@emich.edu

Abstract

Ethnographic research since the 1960s illustrates how kinship is central to political economy, corporate group organization, gender relations, agency, identities, mobility, and exchange. If prehistorians wish to address these themes, kinship is essential. The problem is how to interpret kinship in prehistory –an issue some see as impossible while others have approached it by various theoretical perspectives and fields with different materials. This chapter reviews the various materials, theoretical genres, and methods used for interpretations on prehistoric kinship. It is written from a critical perspective after two decades of engaging with ethnology (analyses of ethnographic data); archaeology (analyses of material culture); bioarchaeology (analyses of human skeletal/dental phenotypic and isotopic data by archaeological contexts); and paleogenetics (analyses of ancient DNA) in the Americas and Europe. Although much progress has been achieved in each of these fields, the major challenges today have less to do with limitations in archeological or biological materials but instead are unfamiliarity with kinship and kinship research and disparate theoretical perspectives. For Spain, there is a third major challenge: austerity's impacts on the kinds of data most useful for prehistoric kinship research.

Keywords: *prehistory, archaeology of kinship, inference methods, bioarchaeology, social anthropology.*

2.1 Kinship in ethnographic research

Because there has been so much confusion about kinship emanating from "house theory" and postmodernism (see Ensor 2011, 2013a: 10-16), an overview is warranted on the ethnographic perspectives on the subject that make it valuable for prehistorians' questions over the past

In Blanco-González, A. and Alarcón-García, E. (eds.) 2025, *A Social Archaeology of Kinship in Iberia and Beyond. Recent Multistranded Approaches from aDNA to Household Archaeology.*
Leiden: Sidestone Press, pp. 23-52.

few decades. Kinship among differentially situated agents in any community is the specific combination of strategies for corporate groups, residence, marriage, and kin terminology. Since the 1960s research on these dimensions of relationality was seldom a practice in classification or genealogical reductionism (Ensor 2011). Instead, kinship research changes along with major theoretical shifts; there is no "kinship theory" (Ensor 2017a: 2). For today's questions on political economy, corporate groups, agency, and identities, the presentations of kinship in ethnographic literature over the past half century should provide guidance. "Kinship in terms of social relations among variably situated actors engaged in the practice of social reproduction within broader political economic contexts have become central to contemporary anthropology" (Peletz 1995: 366). Corporateness, conflict, agency, emotion, communities of practice, cultural landscapes, and the making and reproducing of identities, social memories, and histories are alive in ethnographic descriptions of kinship since the 1970s, decades before these topics became trendy in archaeology.

Beginning in the 1960s, ethnographers increasingly used the concept of corporate kin groups to describe what it means to belong to a lineage, clan, or other descent group. The notion of corporateness worked its way into introductory textbooks on the subject (e.g., Keesing 1975). Corporate group membership, through different filiation or affiliation criteria (Scheffler 2001), includes rights to collectively owned estates and resources with which to make a living that were provided by the group's ancestors and must be maintained for the group's future generations. Membership provides mutual support among comembers, the most important layer of identity, and a collective history and ancestral spirituality. Residential groups are usually not the same as the corporate groups but reproduce the latter through postmarital residence. Marital rules or preferences distribute kindred networks, alliances, material exchanges, and biological relations across corporate groups. Kin terminology is the manipulable linguistic classification of relationships and identities –how people express their relationality with others regardless of genealogy.

Kinship is political economy; it is impacted by political economy. Put simply, through kinship social labor is "locked up," or "embedded in particular relations between people" (Wolf 1982: 91). "To pursue the relations of production to their heart only to find structures of kinship is by now predictable" (Modjeska 1982: 51). Corporate kin group membership provides access to collectively owned resources with which to produce and exchange. Their makeup and division of property define the social relations of production. Marital alliance systems may be reciprocal or competitive –the latter kinds promoting the need for surplus production to attract marriages– and define responsibilities for material exchanges, cooperation, and assistance that come with each marriage (e.g., Ensor 2013a: 197-217, 2017b). In class-differentiated societies, the social relations of production between the classes determine the economic conditions of each class, and therefore the kinds of kinship strategies taken within each class. For example, nobility have corporate estate-owning groups of one kind or another, commoners owning resources –e.g., peasants– but with tributary obligations to nobility may have different corporate kin groups, and those without resources who labor on nobility estates –e.g., serfs– are more likely to form non-corporate neolocal residences combined with bilateral descent (e.g., Ensor 2013b).

All dimensions of kinship have a dialectic interface with gender relations (e.g., Stone and King 2019). Historical domains of kinship allow for the transcendence of "domestic" versus "political" distinctions for gender (Tsing and Yanagisako 1983). Descent group membership and residence practices are commonly through gender. Gender status and conditions are manipulated through descent and residence. Social labor and reproduction are typically for the perpetuation of the corporate groups, influencing both gender and age status; children belong to their corporate group, not to one or both parents (Ensor 2013a: 20-22). Even under the combination of patrilineal descent and patrilocality, women have roles and support through their own patrilineal group unless there is membership transfer upon marriage to their husband's group to control unrelated wives. A similar impact on gender status occurs within matrilineal descent groups when men institutionalize avunculocality (Ensor 2021a; Ensor *et al.* 2017). Within kinship practices, we find the strategies and responses of gendered agents (Stone and King 2019).

Kinship, political economy, gender, and identity are reproduced through practice. Kinship varies within communities as differentially situated agents strategically respond to their conflicts and aspirations. Kinship changes as agents manipulate membership criteria, residence strategies, and marriage practices in response to conflicting relations, endogenous crises, or exogenous factors (Ensor 2013a, 2013b, 2017a, 2017b, 2020: 191-198, 2021a: 143-144).

On the question of biological relations, it was always apparent that genealogical networks cast by marriages across corporate groups are recognized but vary in their importance and uses. On one end of the spectrum, corporate groups exclude most of an individual's genealogical relations who contribute little to that person's life and have no role in their identity. The only important people are comembers, most of whom lack biological relatedness. On the other end is neolocality and bilateral descent in the absence of corporate groups –prevalent in the global north and wherever people depend on wages– where, ideologically, genealogical and affinal networks are the only "kin" relations available to them, though the same roles are often assumed by "fictive kin". Although one school of British anthropologists unfortunately defined kinship as genealogical relatedness and kin groups as "political organization" (e.g., Evans-Pritchard 1940), the question on this matter among anthropologists since the mid-twentieth century has not been whether kinship is biological relatedness; but rather, how socially-constructed kin groups include and exclude biological and affinal relations and how genealogical networks are used or deemphasized (e.g., Gjessing 1956; Lévi-Strauss 1956; Fortes 1958, 1959: 149, 1969; Fox 1967).

All this and more were known for at least 60 years in social anthropology. This perspective on kinship addresses contemporary questions far better than the less sophisticated, and poorly informed, reinventions of kinship. The question for prehistorians of Iberia –and beyond– is how to interpret the different dimensions of kinship in prehistory as the first step to address these themes that remain important to archaeology.

2.2 Ethnological approaches to prehistoric kinship

The longest lasting, most influential approaches to prehistoric kinship are ethnological. After the rejection of speculative and ethnocentric nineteenth century unilineal evolution and the period of cognitive anthropology (e.g., Benedict 1934), historical particularism (e.g., Boas 1920), and structural functionalism (e.g., Radcliffe-Brown 1952) in the early half of the twentieth century, functionalist-materialist neoevolution became the major theoretical orientation in anthropology until the 1970s (e.g., Service 1962; Steward 1963; Fried 1967). Using the corpus of ethnographic works, ethnologists could develop social typologies. A simple-to-complex evolution in types was assumed and individual ethnographic communities could be used to infer the cause of transformation from one type to another.

Additionally, the origins of culture among early hominins were assumed to be reflected in primate and simple types for human hunter-gatherers.

The functionalist neoevolutionism of the 1960s continues today in kin terminology research (e.g., Godelier *et al.* 1998; Kronenfeld 2004; Trautmann and Whiteley 2012). Formalist kin terminology analyses use a preconceived neoevolutionary unilinear sequence from simple to complex abstract types (Fig. 2.1). Scholars then use ethnographic kin terminology models to identify a culture's placement within the preconceived evolutionary sequence. Structural functionalist assumptions are used to assume social organization and marriage rules for each abstract type of kin terminology. Because the ethnographic models have elements from multiple abstractions, those attributes considered archaic according to the position in the preconceived sequence are used as "evidence" on which elements must precede the evolution of an abstract type (Fig. 2.1). At the same time, the placement along the evolutionary ladder is used to interpret the specific ethnographic culture's prehistoric kinship practices (Ensor 2021b).

For many reasons, neoevolutionism waned by the 1980s (e.g., Peregrine 1996, 2001b). However, kin terminology neoevolutionism continues today with many of the same problems (Ensor 2013b, 2016, 2017a, 2021b). One major issue is the "primitivizing" of nonwestern peoples –the assumption that their histories are governed by evolutionary processes rather than their creativity in responding to circumstances. The theoretical perspective calls for abstract types. One type, Allen's (2008, 2012) "tetradic" kinship receiving much attention, is a logic-based fictitious model. There is no evidence for the preconceived transformations from one type to another. The linear evolution assumes all societies pass through the same stages. It ignores the social and economic factors that are known to cause changes in kinship. The ethnographic models are normative –one homogeneous depiction of cultural-linguistic groupings based on few informants– when variation among communities with different political economic histories within such groupings are well known (e.g., Moore 1988). The ethnographic models are synchronic; yet are assumed to contain evidence for diachronic change. Some elements are assumed to be earlier remnants while others are supposed to have been adopted later –in accordance with the preconceived type of sequence– without observing the actual changes over time. Additionally, most were collected after dramatic impacts from colonialism, depopulation, and capitalism (McKnight 2004; Ensor 2013b, 2016). Frequently, the ethnographic models confuse the altered ethnographic present with the traditional past (Ensor 2021b: 27, 31-32). Many of the structural functionalist leaps from kin terms to marriage or descent may not be acceptable for generalization (e.g., Goodenough 1970). None of the datasets date to prehistory; the entire approach relies on modern ethnographic information from nineteenth-twentieth century interpreted in accordance with the preconceived theory on how cultures change.

A second ethnological approach adopts phylogenetics (cladistics) (Fig. 2.2). Phylogenetics was the first form of historical linguistics introduced at the turn of the nineteenthcentury to explain how assumed human races and their cultures diverged and evolved –or not– from a common ancestral population (Campbell and Poser 2008). The similarity and differences in linguistic elements observed in the present are used to classify populations into clades, each sharing a protolanguage. Sometimes archaeologists are asked to use pottery types as indicators of the distinct bio-linguistic-cultural populations. Recently, paleogenetics has used haplogroups for the same ends. Today, kin terminology, and sometimes residence are treated the same, albeit with more sophisticated analytical tools that perform

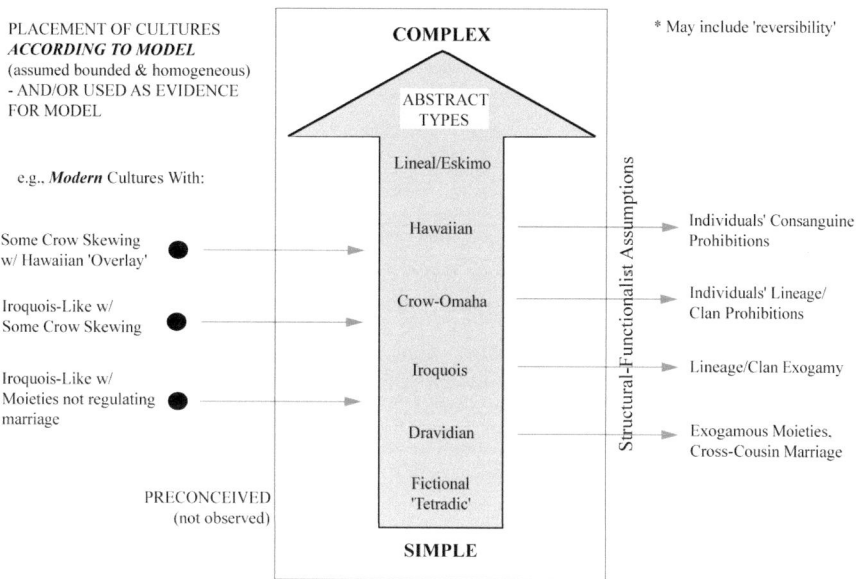

Figure 2.1. The neoevolutionary approach to kinship in deep time.

cluster analyses. Similarities and differences in kin term elements or semantics from cultural-linguistic groupings are used to generate clades to identify a common "genetic ancestry" sharing a "protokinship" that analysts claim is a "reconstruction" of kinship and change within a region in deep time –i.e., tens of thousands of years– (e.g., Jones 2003; Fortunato 2010; Fortunato and Jordan 2011; Jones and Milicik 2011; Ehret 2012; Korotayev *et al.* 2019; Dziebel 2021). In other words, the approach assumes distinct biological-linguistic-cultural populations, each with one homogeneous kin term system, that remain isolated and immutable for thousands of years after divergence.

The phylogenetic approach has been intensively critiqued in social anthropology (e.g., Moore 1994; Sims-Williams 1998; McKnight 2004; Wildcat *et al.* 2004), biological anthropology (e.g., Armelagos and VanGerven 2003; Steele and Kandler 2010), linguistic anthropology (e.g., Bateman *et al.* 1990; Clendon 2006; Campbell and Poser 2008), and archaeology (e.g., Trigger 2006: 213-241; Ensor 2017a, 2021a). One of the greatest problems is the essentializing equation of biology, culture, language, and kinship. The phylogenetic model ignores the more common "horizontal spread" of genes, culture, and linguistics across populations, and the common fusing of culturally and linguistically diverse groups (e.g., Moore 1994; Sims-Williams 1998). Phylogenetics also ignores similar kinship strategies among people with similar socioeconomic contexts or ecological adaptations across the defined populations (e.g., Szołtysek 2008, 2012, 2015; Wheeler *et al.* 2012; Szołtysek *et al.* 2020). It ignores variation in practices among people living in the same populations; even those with different socioeconomic contexts in the same local communities (e.g., Kertzer and Brettel 1987). As with neoevolutionism, phylogenetics also relies on modern data interpreted in accordance with the theory, questionable normative and synchronic kinship models for each population, and structural functionalist generalizations (Ensor 2017a, 2021a: 94-102).

PHYLOGENETIC/CLADISTIC MODEL

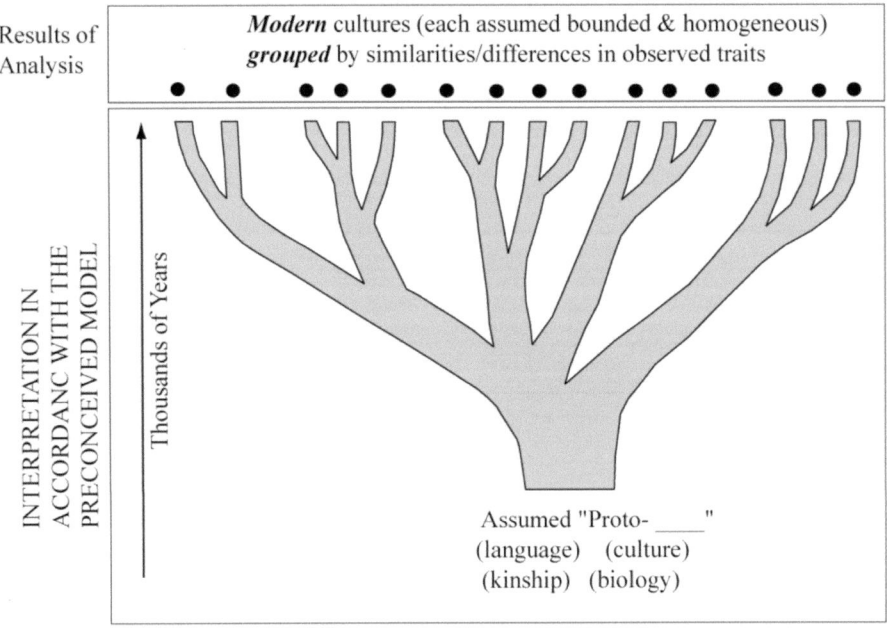

Figure 2.2. The phylogenetic approach to kinship in deep time.

Functionalist cross-cultural analyses opened another avenue for interpretation of prehistoric kinship through associations with archaeological evidence for non-kinship practices (e.g., Peregrine 1996). For example, Driver and Massey's (1957) hypothesis that matrilocality and patrilocality can be explained by the localization of women's or men's gender roles were tested through cross-cultural analyses several times with some support (e.g., Ember and Ember 1971; Korotayev 2003). Bilocality has been cross-culturally associated with depopulation, migration, gender equality in inheritance, and resource insecurity (Murdock 1949; Eggan 1966; Ember and Ember 1972; Pasternak 1976). Warfare among communities within societies was associated with patrilocality, warfare between societies was associated with matrilocality, and the presence of warfare with unilineal descent groups –as opposed to bilateral descent– (Ember *et al.* 1974).

Such statistically confirmed global associations are sometimes used for interpreting prehistoric kinship practices. The major problem with this indirect approach is, once again, the normative data for the cultural units of analysis (Ensor 2016). Additionally, although one set of circumstances may be associated with a kinship practice, others may also be associated with the same kinship practice –as with bilocality– (Ensor 2016). Another problem is the use of these synchronic associations for diachronic explanation, whereby the associated variable is treated as a causative agent for a kinship practice (Ensor 2016). There is also considerable misunderstanding among prehistorians over which factors are associated with kinship practices. For example, a common assumption in the Central European bioarchaeological literature is that patrilocality is associated with agriculture, patrilineal descent, or nuclear families (e.g., Bentley *et al.*

2012; Bentley *et al.* 2013) when, in fact, none of these are correlated with patrilocality (Ensor 2021a: 81-85).

Given the wide-ranging problems with both the neoevolutionary and phylogenetics theoretical perspectives and their data, neither can accurately inform about prehistoric kinship. The indirect associations between practices may be helpful for framing hypotheses for specific prehistoric communities. However, when prehistorians consume ethnological interpretations without question, assuming they are more accurate, they perpetuate the problems rather than use their data to test the models or reveal new insights. This phenomenon –known as "ethnological tyranny" (Wobst 1978; Keegan and Maclachlan 1989; Ensor 2013a, 2016, 2017a, 2021b)– requires solutions for identifying kinship practices using datasets that date to the prehistoric periods in question (Ensor 2013a: 272-298, 2016, 2017a).

2.3 Archaeological approaches

In general, archaeologists have been reluctant to interpret kinship. Most interpretations are atheoretical, inductive, and vague, such as claims that perhaps kinship was involved in social groupings or nonspecific interpretations are made for possible families or households. Rather than a critique of that reluctance to address kinship, this section reviews the methodological trends guiding the interpretation of kinship practices.

Culture historical archaeology sometimes used direct historical analogy to infer kinship practices among prehistoric peoples by projecting normative characterizations of the earliest historically known practices of a population onto its prehistoric cultural ancestors, linked through material trait continuities (e.g., Steward 1937, 1942; Rouse 1992). This was another ethnological approach that assumed bounded populations, internal homogeneity, and long-term continuity over hundreds or thousands of years. Functionalism introduced interpretations of kinship practices based on presumed associations with archaeologically identified subsistence strategies (e.g., Haury 1956; Gjessing 1975), but those associations were later disproven through cross-cultural analyses (as above). Post-processualist "House" advocates misrepresented "kinship" as passive and a product of Western genealogical bias. To replace kinship, lists of traits defined by Lévi-Strauss and exemplified in selective ethnographic analogies were used to interpret "Houses" that were claimed to be the only means to understand corporate organization. However, most archaeological approaches to kinship developed from processual archaeology's questions on social organization and to make archaeology more anthropological. These include artifact distribution, cemetery structure, and community pattern approaches. Though the artifact-based approaches carry questionable assumptions, the community pattern approach is the more resilient and addresses questions on agency, practice, and identity, yet requires substantial site excavation and dating of structures.

2.3.1 Artifact-based approaches

Artifact-based approaches include 1960s "ceramic sociology" that used few selected ethnographic analogies to assume that, in the absence of pottery specialization, women make pottery and copy their mother's decorative motifs. Matrilocality –questionably assumed to be associated with matrilineal descent– would concentrate motifs within residential areas whereas patrilocality –questionably assumed to be associated with patrilineal descent– would distribute motifs across residential areas (e.g., Longacre 1964, 1966; Deetz 1965; Hill 1966; McPherron 1967; Whallon 1968). These efforts were soon criticized: the use of pottery from post-occupation depositional contexts in dwellings were questioned, cross-cultural research found no association between women and pottery production in the absence of specialization (e.g., Murdock and Provost 1973), and descent cannot be predicted from residence (Allen and Richardson 1971). Recent versions engender other artifacts, production techniques, or activities to interpret residence through intra- and inter-settlement distributions (e.g., Sanger *et al.* 2020). However, like the 1960s "ceramic sociology" such claims are also based on few selected ethnographic analogies that cross-cultural research has disproven or on direct historical analogy that assumes no change to gender roles for hundreds or thousands of years. One contrasting approach is to use artifacts associated with institutions to view how the social organization in the direct historical analog developed in prehistory through diachronic archaeological analyses (e.g., Ware 2018). Whereas such uses of direct historical analogy may be useful for prehistoric periods separated only by a century or two from the analog if continuity can be established, it is hazardous for inferences across longer spans of time or if discontinuity is apparent due to cultural change. Presumably for this reason, selective ethnographic analogy is more commonly used for artifact-based approaches in Europe.

Another artifact-based approach uses the assumption that pottery and other objects with distinct styles, raw stone materials, or even food remains, can represent distinct kin groups or moieties (e.g., Binford 1972; Gibson 1973; Van de Velde 2007; Bogaard *et al.* 2011). However, intermarrying groups exchange significant quantities of materials. If symbols or raw materials distinguish kin groups, they may be more useful for modeling marriage networks across kin group estates (e.g., Pechtl 2015; Souvatzi 2017). Despite the problematic assumptions in artifact-based approaches to kinship, they may be the most practical under austerity in Spain if archaeologists rely mostly on data from survey and limited excavations. Artifact approaches could at least produce hypotheses on residence, descent, and/or marriage patterns.

2.3.2 Cemetery structure

Cemetery structure, on its own, can reveal corporate groups and subgroupings but little more. 1970s-1980s cross-cultural studies suggested that unilineal descent groups are associated with distinct demarcated cemeteries (e.g., Saxe 1970; Goldstein 1981). However, Christopher Carr's (1995) more thorough analysis revealed that any corporate groups –not just unilineal descent groups– have demarcated cemeteries. That study also confirmed that burial clusters within cemeteries indicate subgroupings like extended or conjugal families. This approach requires funding for large-scale cemetery excavations but under austerity older projects with extensive cemetery excavations might be revisited. However, as revealed by Carr (1995: 165-182) the distribution of burials does not indicate the kinds of corporate groups involved. Some recent Central European literature assumes that male-child coburials indicate patrilineal descent (e.g., Le Roy *et al.* 2016). However, male-child or female-child coburial could conceivably occur with any kinship configuration. Both cemetery structure and coburial patterns need contextualizing within kinship practices; the identification of the latter requires other methods such as bioarchaeological and genetic data from the same cemeteries, or archaeological data from associated settlements.

2.3.3 Trait lists

The House Society concept was defined by Waterman (1920) and later reiterated by Spott and Kroeber (1942: 166) to indicate how some communities have corporate groups not based on unilineal descent that function the same as lineages. Lévi-Strauss (1982: 163-187, 1987) used the model as a neoevolutionary type for the transition between kin-based and non-kin-based social organization. He presented it at a time when cognatic social organization was beginning to be explored more thoroughly in anthropology (Ensor 2011, 2013a: 13-19, 2013b). "Houses" are bilocal residential groups combined with bilateral descent, whereby any relations may be negotiated to gain membership to the corporate estate. Exemplifying postprocessualism's sensationalism, unfamiliarity with kinship, and focus on the symbolic at the turn of the millennium, early "house" advocates in archaeology argued the House model must replace kinship to learn about corporate organization. Unaware that corporate organization, agency, and ideology had been central to ethnographic treatments of kinship since the 1960s, "house" literature first sought to find problems in kinship –universalizing Schneider's (1984) critique of "kinship theory" to claim it was based on Western biological relations and no culture practices that ethnographers described– to justify a replacement with Lévi-Strauss' House Societies (e.g., Gillespie 2000a, 2000b; Joyce 2000). The early literature had a difficult time defining what was a "House"; it instead focused on what it was not –a non-kinship-based corporate organization. Descriptions relied heavily on selective ethnographic analogies (e.g., Joyce and Gillespie 2000). By the end of the first decade, the literature became "house-centric" interpreting what a "House" does –the corporate characteristics of unspecified groups– rather than what a "House" is or is not –the relations upon which the corporate groups are created and maintained– (e.g., Beck 2007).

Using Lévi-Strauss as a guide, the "house" literature typically cites lists of characteristics that archaeologists could use to identify "Houses" in antiquity. The most common traits used to infer "Houses" were collectivity, longevity, heirlooms, and moral personhood; thought the latter was usually an assumption based on the material expressions of the others. Selective ethnographic analogies were used to illustrate these characteristics. The archaeological application that followed those descriptions commonly employed evidence for longevity in settlement location or the rebuilding of dwellings –neither of which conforms with any specific form of social organization– to interpret a House Society (e.g., Marshall 2000; Tringham 2000; Chesson 2007; Craig 2007; Joyce 2007). Nevertheless, the same lists of traits are equally applicable to corporate descent groups (Ensor 2011, 2013b).

Before waning, the "house" literature became more receptive to the reality of corporate descent groups –that "House" was only one strategy to form corporate groups– and began assuming that any corporate descent group may be envisioned as a "House" (e.g., Coupland *et al.* 2009) or by attempting to distinguish what material characteristics constitute a lineage versus a "House". For example, González Ruibal (2006: 146) specifically lists nine traits assumed to indicate House Societies, the last two of which may be more difficult to identify materially: hierarchy; unclear or mixed descent; symbolic relevance of dwellings; dwelling investments as an arena of competition; the existence of titles or symbols associated with dwellings; inherited heirlooms or symbols of rank; female and male lines are manipulated for, with women's contribution to, house wealth; territorial collectives; and text references to houses as social units. The problem with such lists, of course, is that all these traits are also compatible with corporate descent groups. Even the manipulation of female and male lines is not exclusive to bilocality and bilateral descent (aka "Houses") –it is also a membership principle for large corporate ambilineal descent groups. Another concern is the absence of archaeological evidence for some traits, which cannot be equated with evidence for absence. Lists of traits are unlikely to adequately identify "Houses" –small corporate bilocal residential groups using bilateral descent– or to distinguish those from corporate descent groups.

2.3.4 Community patterns

Unlike ethnographic analogy, which selects cases to exemplify a presumed association between material patterns and social or ideological practices, cross-cultural research is designed to test such claims using representative samples of ethnographic communities from different cultural-linguistic groupings around the world, controlling for intracommunity variability and impacts from colonialism and capitalism, and evaluating interobserver recording of ethnographic data (Ember and Ember 2009). Weak correlations between the examined variables results in a dismissal of the association. Strong correlations, in contrast, indicate a global pattern or rule of associations that provide strength to archaeologists' inferences. Deviating ethnographic examples do not overrule the global generalization; the latter provides greater confidence in accuracy than selective ethnographic analogy that may or may not reflect the pattern or rule (Ember and Ember 1995; Peregrine 1996).

Cross-cultural testing and retesting have revealed very strong global associations between dwellings and residential group compositions and between community patterns for corporate descent groups versus bilateral descent. A major advantage to this approach is that it avoids normative inferences on prehistoric communities; it enables the identification of community variation and change, which are necessary for narratives on agency (Ensor 2013a, 2013b, 2013c, 2016, 2017a, 2020: 191-198, 2021a: 104-197). Moreover, the correlations are not influenced by high-level theories; they are independent and can be used to archaeologically test different high-order theories on how kinship changes (Ensor 2017a). However, the community pattern approach to residence and descent is suited only to sedentary communities, regardless of subsistence strategy; it is less reliable for inferences on mobile foraging communities (Ensor 2021a: 122).

Some residential group practices are safely inferred from two lines of evidence: 1) the "living floor area" of dwellings, and 2) the spatial arrangements of dwellings for residential groups with multiple dwellings. Common mistakes are to use total dwelling size (e.g., Hrnčíř *et al.* 2020) and to treat multiple rooms within a dwelling as separate dwellings (e.g., Schillaci and Stojanowski 2002). "Living floor area" is defined as the internal space remaining within a dwelling after subtracting any internal cooking and storage areas. The data come from ethnographic illustrations linked to occupants' residential practices (Ember 1973). Dwellings with living floor areas above 80 m^2 are strongly correlated around the world with matrilocal dwellings –large structures housing multiple sisters' conjugal families (Ember 1973; Ensor 2021a: 117-122). Internal partitions for each sister's conjugal family may exist. Those should not be treated as individual dwellings but rather summed for the total living floor area under the roof (Ember 1973; Peregrine and Ember 2002). In contrast, all non-matrilocal residential groups use dwellings with living floor areas less than 60 m^2 (Ensor 2021a: 117-122). Each dwelling is for a single conjugal family. These associations were first tested by M. Ember (1973) and retested in four additional studies (Divale 1977; Brown 1987; Porčić 2010; Hrnčíř *et al.* 2020). The Hrnčíř *et al.* (2020) study was the only one to use pre-recorded data. When removing questionable ethnographic cases –e.g., foragers or noble palaces– the correlations between the living floor area cutoffs for matrilocal versus non-matrilocal practices have even greater statistical strength (Ensor 2021a: 117-122).

The spatial arrangement of conjugal family dwellings –living floor areas less than 60 m^2– is the basis for differentiating non-matrilocal residential strategies. The data are from ethnographic illustrations or descriptions of dwelling arrangements linked to occupants' residential practices (Ensor 2021a: 123-126). For example, the multiple conjugal family dwellings for patrilocal residential groups –multiple brothers' conjugal family dwellings– are nearly universally formally arranged around, and with entries facing, a small plaza space. In contrast, the multiple dwellings for bilocal residential groups –with unpredictable, internally diverse residential negotiations– form a haphazardly arranged cluster (Ensor 2021a: 123-126).

Unspecified corporate descent groups versus bilateral descent can be distinguished with great confidence through settlement layouts. These data also come from ethnographic illustrations and descriptions of settlements (Ensor 2021a: 126-131). Chang's (1958) early use of cross-cultural research, which is impressively sophisticated even by contemporary standards, was the first to demonstrate a strong statistical association between unilineal descent groups and either 1) residential groups surrounding a plaza or central ceremonial structure for single descent group settlements or 2) segments –distinct spatial concentrations of residential groups– for multi-unilineal descent group villages. The pattern for bilateral descent, in contrast, involves non-formal distributions of residential groups: scattered haphazardly across a settlement or more broadly across the landscape. Adding additional ethnographic cases, while controlling for overlap among cultural/linguistic groupings, for a much larger sample size confirms the differences between settlements with descent groups and bilateral descent (Ensor 2021a: 126-131).

The specific descent strategies can be inferred from the combination with residential groups (Fig. 2.3). Although we cannot predict descent from residence, the demonstration of a corporate unilineal descent group having matrilocal residential groups indicates a matrilineal descent group. Alternatively, the settlement layout for a unilineal descent group combined with patrilocal groups indicates a patrilineal descent group. These associations are supported through previous cross-cultural analyses (e.g., Pasternak 1976). Meanwhile, bilateral descent can be combined with any residential strategy: e.g., bilocal, matrilocal, neolocal, patrilocal, and virilocal residential groups (Ensor 2021a: 132-135) (Fig. 2.3).

Additional hypotheses awaiting cross-cultural analysis are currently based on ethnographic analogies (Ensor 2013a: 155-158, 2021a: 132-135). For example, the descent group pattern combined with bilocal residential groups may indicate ambilineages – whereby individuals negotiate exclusive group membership through either patrilineal or matrilineal relations, which is not the same as bilateral where all genealogical relations (cross and parallel) of each spouse are used to gain rights among multiple residential groups. If individual conjugal family dwellings surround a plaza, they may signal virilocality with a patrilineage or avunculocality with a matrilineage (Ensor 2013a: 156-157, 2021a: 132-135). Stem families –whereby only one heir's (any gender) conjugal family replaces the parents of a neolocal residence (e.g., through primogeniture or ultimogeniture)– may be indicated by rebuilding the dwelling across generations but without growth into an extended residential group (e.g., Ensor 2021a: 162-163; Blanco González this volume). The implications of the latter are that additional children join or establish residential groups elsewhere, which in some cases may explain settlement spatial growth or migration.

The community pattern approach detects change and variation in the residential and corporate membership strategies taken, which is necessary for inferences on agency.

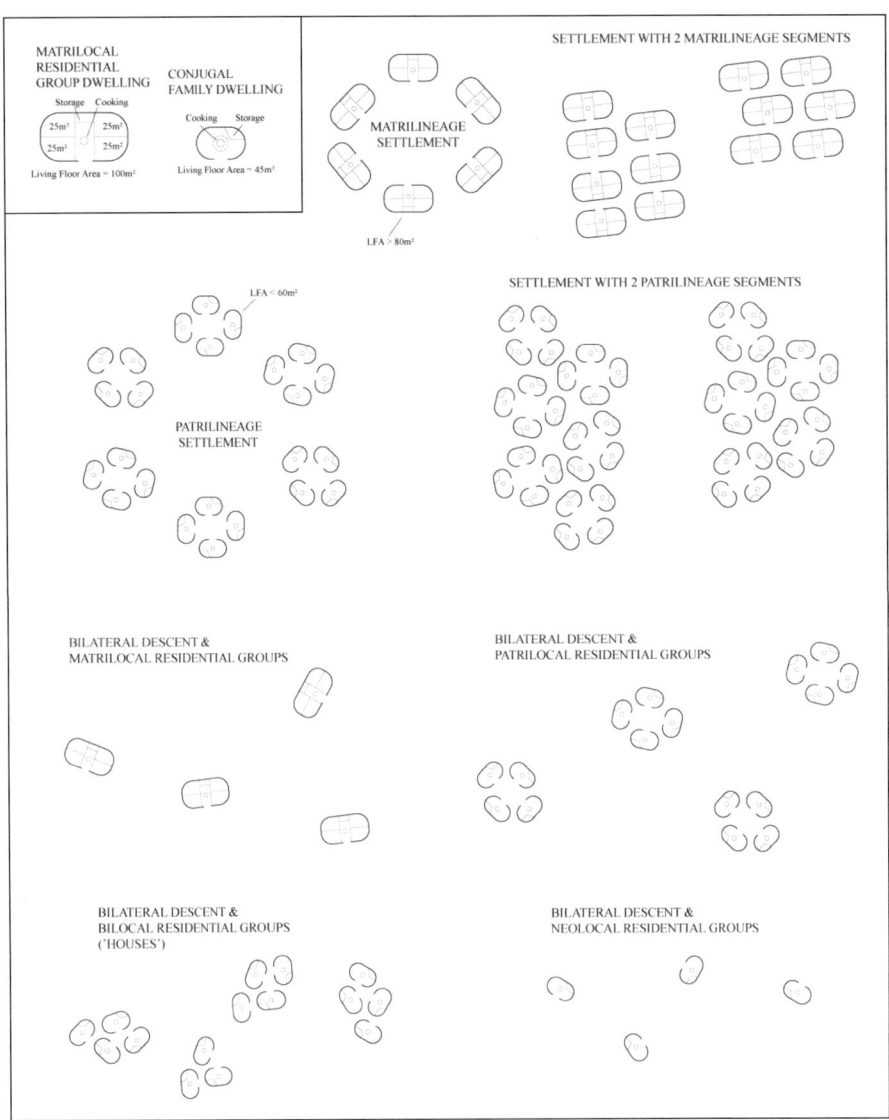

Figure 2.3. Community patterns by descent and residence.

Elsewhere, it enabled the observation of several transformations in individual descent groups across hundreds of years through the manipulation of membership and residential strategies: in some cases, including the strategies to develop corporate groups by founding ancestors (e.g., Ensor 2013a, 2013c). Moreover, different corporate group strategies are often evident within settlements of non-class differentiated, illustrating agency in strategies and identities (e.g., Ensor 2013a, 2021a: 155-197), and, as should be expected, among differently situated classes in tributary political economies (e.g., Ensor 2013b, 2020: 178-198).

Although the cross-cultural tests reveal strong statistical correlations, providing more confidence than found in most correlational studies in archaeology, and although having the distinct advantages of being independent of high-level theory and enabling the interpretation of change and variation, the archaeological data needed demand high

|

levels of funding. Excavations are needed across much, if not all, of each settlement. Archaeologists need sufficient excavation evidence to distinguish dwellings from other structures, and to calculate living floor areas –unless the total interior sizes are less than 60 m². Dwellings need to be dated to observe synchronic variation and diachronic change. These demands pose challenges under austerity where archaeologists must make do with survey and small-scale excavations, and reliance on ceramic chronologies and few absolute dates for structures. For example, my analyses of reports available online for the Iberian Neolithic led to limited interpretations on residence –mostly non-matrilocal– and less on descent due to the limited scale of excavations, though matrilocality and bilateral descent could be inferred from Chalcolithic Camino de las Yeseras (Madrid) (Ensor 2021a: 148-155). Elliptical ditch settlements in Southern Iberia –like Perdigões (Valera 2018) and Mamoa 2 do Leandro in Portugal (Valera and Antunes 2008)– might suggest descent groups but research designs for kinship analysis would require more extensive excavations to shed light on dwelling sizes and their spatial distributions, like those for Neolithic Greece (Souvatzi 2017, this volume). In contrast, more extensive excavations at Iberian Iron Age settlements enable more successful analyses on residential and corporate group strategies (e.g., Blanco González this volume; Grau Mira this volume). For all periods, there may be older reports on extensive excavations useful for analyzing community patterns. For new site investigations at settlements with above-ground masonry architecture, low-cost mapping projects that can distinguish residential and non-residential structures without extensive excavations can be useful, particularly if those structures can be placed within chronological frameworks.

2.4 Bioarchaeological approaches

Bioarchaeology was originally a processualist perspective developed within biological/ physical anthropology that considers the archaeological contexts of human skeletal data for interpretation (Buikstra 1977) as opposed to culture historical uses of biological data to interpret population interaction and migration over time. It has since adapted to agency perspectives. Bioarchaeologists also developed approaches to interpret kinship practices using prehistoric data. Bioarchaeology has also struggled with biological versus non-biological conceptualizations of kinship when using biological data. Today, there are two major sources of bioarchaeological data used to interpret descent and postmarital residence: phenotypic traits and strontium isotope ratios. Postmodernist perspectives may use these data sources for interpretations on non-Western identities to replace a biological perspective on kinship (e.g., Johnson and Paul 2016). Although research traditionally maintains two problematic assumptions –that descent groups should be homogeneous and that spouses are buried together– revisions to improve interpretations have been offered (Ensor *et al.* 2017). One limitation is the difficulty in identifying variation and change, leading to normative presentations of descent and residence practices.

2.4.1 Morphometric phenotypic trait approaches

Methods using cranial and dental morphological scores and/or metric trait measurements were developed in the 1980s to address questions on descent and residence. After scoring (nonmetric traits) or measuring (metric) individuals from a cemetery, or from multiple cemeteries or sites, biodistance using multivariate statistical analyses enables the identification of patterns in the degrees of phenotypic similarities and differences. These have been used to test for descent groups and postmarital residence. For non-specialists it may be worth pointing out that phenotypic traits are the assumed expression of genotypes derived from the inherited alleles –from a Mendelian perspective– from both parents.

When used for interpreting descent from intra-cemetery biodistance analyses, research traditionally assumes that unilineal descent groups should be relatively homogeneous (e.g., Howell and Kintigh 1996; Stojanowski and Schillaci 2006: 53-64). The underlying assumptions are that 1) groups like lineages grow intergenerationally out of a single nuclear family and the unilineal principle of membership remains constant, and 2) spouses are members of the same lineage and therefore buried in the same cemetery. If these were the case, then a cemetery for a lineage should demonstrate limited –non-statistically significant– intra-cemetery biodistance. Alternatively, a Multi-Dimensional Scaling diagram should illustrate a tight clustering of all individuals from the cemetery. Additionally, limited intra-cemetery biodistance between adult males and females is assumed to reflect lineage endogamy (e.g., Stefan 1999; Schillaci and Stojanowski 2003: 10). Conversely, there should be significant inter-cemetery differences in biodistance.

From an ethnological perspective, there are several problems with these assumptions: 1) lineages and clans are more likely to form from the fusing of multiple groups rather than from growing out of single founding nuclear families; 2) descent groups are known to change their membership principles over time, causing more internal heterogeneity; 3) siblings –not spouses– are more likely to be buried in the same descent group cemetery; and 4) group exogamy –necessary to maintain unilineal membership criteria– prevents homogeneity within descent groups (Keegan 2009; Ensor *et al.* 2017). For these reasons, unilineal descent groups should be internally heterogeneous rather than homogeneous. Within a lineage, there should be multiple intra-lineage biodistance clusters of biologically close individuals –each cluster is different from the others despite lineage comembership. Moreover, there should be close biodistances among some clusters of different lineages due to exogamy and the return of husbands/fathers –if matrilineal– or wives/mothers –if patrilineal– to their respective lineages for burial, unless wives' memberships are transferred to that of their husband's patrilineage –a strategy to socially control wives for men's empowerment. Considering these factors, revised models for interpreting descent groups versus bilateral descent from phenotypic data have been offered (Ensor *et al.* 2017). Figure 2.4 illustrates those revised expectations for unilineal descent groups. Although unilineal descent group cemeteries may be identified in this manner, one problem is that matrilineal and patrilineal descent cannot be distinguished in the absence of wife membership transfers by patrilineages (Ensor *et al.* 2017).

The use of phenotypic biodistance for interpreting postmarital residence dates to the early development of bioarchaeology. Stemming from the traditional assumption that spouses are universally buried in the same cemetery, intra- and inter- site/cemetery biodistance by sex is used to infer which moved after marriage (e.g., Lane and Sublett 1972; Spence 1974; Konigsberg 1988; Schillaci and Stojanowski 2003; Tomczak and Powell 2003). If males exhibited greater biodistance with one another and with females, then matrilocality was interpreted. Conversely if females exhibited greater biodistance with one another and with males, then patrilocality was interpreted. If there were significant differences among both sexes, then bilocality would be interpreted. Others examined

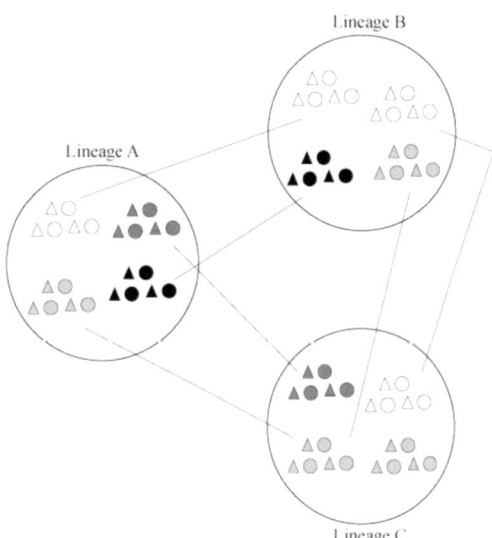

Figure 2.4. Hypothetical biodistance (or molecular genetic relations) within and among unilineal descent group cemeteries. Shadings indicate heterogeneity among biodistance clusters within lineages. Lines indicate gene flow and homogeneity among inter-lineage biodistance clusters through exogamy to maintain the unilineal membership principle.

females and males across settlements. For example, if females across settlements had no statistical differences in biodistance, then patrilocality was interpreted, or if males across settlements had close biodistance, then matrilocality was interpreted (e.g., Aguiar and Alvez Neves 1991; Hubbe *et al.* 2009). As indicated above, however, spouses are not universally buried together: not with matrilineal descent groups; only in some communities with patrilineal descent groups –if wives' membership is transferred, and that too is variable within communities; and only predictably are they buried in the same cemetery when there is bilateral descent (Ensor *et al.* 2017). For this reason, the revised models for interpretation proposed by Ensor and colleagues (2017) emphasize models by combination of descent, residence, and marital practices.

2.4.2 Strontium isotope ratios

By the turn of the millennium, dental and bone isotope ratios had become common new sources of data on prehistoric diet and mobility. Specifically, strontium isotope ratios (^{87}Sr/^{86}Sr) from dental crowns –representing childhood, or young adults in the case of third molars– or bone –representing the years before death– are used to model geographic mobility. Strontium ratios suggest the type of landform in a region where people lived; they do not indicate specific locations where people lived (Pollard 2011). For instance, people living at different, and perhaps distant, settlements but with similar geological contexts may have a similar range in ratios despite having migrated among those settlements. Only those who lived at one type of landform and later migrated to another having a different "local" range in ratios, where they were buried, stand out as having migrated. Nevertheless, strontium isotope ratios are a common basis for interpreting postmarital residence. For example, in Central European research, patrilocality is routinely interpreted if more males than females have ratios falling within the geological range for where they were buried (e.g., Haak *et al.* 2008; Bentley *et al.* 2012; Bentley 2013).

Several problems with the prevailing models for interpretation of strontium isotope ratios in Central Europe are outlined elsewhere (Ensor 2021a). These studies assume that spouses are buried together by not considering descent –as above– and typically employ a simplistic matrilocal versus patrilocal dichotomy. For these reasons, Ensor (2021a: 29-34) presents ethnologically informed revised models for interpretation. Male homogeneity in local ratios and female heterogeneity reflect bilateral descent and patrilocality, corporate patrilineal descent groups with wives' membership transfers, or corporate matrilineal descent groups and avunculocality. Both male and female homogeneity and local ratios reflect either matrilineal descent groups with matrilocality or patrilineal descent groups with patrilocality without wives' membership transfers. The same could also be found with any descent and residence practices if marriages were consistently among people of the same landform type –local or not. Female local homogeneity and male heterogeneity should be exclusive to bilateral descent with matrilocality. Variability in local and nonlocal ratios among both sexes reflects bilateral descent combined with bilocality. By not having considered the influence of descent, there are often biases in the Central European interpretations. For example, the data presented in some of those studies concluding patrilocality suggest a significant percentage of both females and males –sometimes equal or higher percentages of males– with nonlocal ratios, which would not occur if patrilocality was a norm, and which is instead consistent with bilateral descent combined with bilocality (Ensor 2021a: 41-62). Even lower percentages of males with nonlocal ratios may merely reflect a commonly found patrilocal bias in bilocality (Fox 1967; Keesing 1975).

One challenge, relevant to Neolithic-Chalcolithic Iberia, is when there are low numbers of burials present at settlements. Cintas Peña and García Sanjuán (2022) attempt to resolve this issue for Chalcolithic Southern Iberia by pooling the sex-based strontium results from numerous settlements spanning hundreds of years. This forces the authors to assume regional social homogeneity in practices across those settlements, and over time. Despite that caveat, their interpretation of bilocality with a patrilocal bias exemplifies a shift toward a more ethnologically informed perspective than the prevalent matrilocal-patrilocal dichotomy and provides a methodological solution for other regions and time periods where few burials are present at sites.

2.4.3 Additional challenges for bioarchaeological approaches

In contrast to the archaeological community pattern approach, which enables the observation of variation and change, a major limitation in sampling for phenotypic trait and strontium isotope analyses is the need to approach kinship through a normative perspective. Variation in practices within settlements is difficult to interpret when needing to pool all females and males from each cemetery to match the pooled results to a single model for interpretation. Likewise, with few exceptions (e.g., Hubbe *et al.* 2009), most studies present a synchronic perspective by needing to pool females and males buried over long spans of time to match with a model for interpretation –there may be too few individuals dating to each phase for a phase-by-phase analysis to test for change. To some, the necessary normative perspective may appear to exemplify a neoevolutionary typologizing for classification. However, as Cintas Peña and García Sanjuán (2022) demonstrate, interpretation may still conform with a contemporary theoretical perspective.

In Spain and elsewhere under austerity, funding limitations influence the ability to conduct bioarchaeological analyses. First, the bioarchaeological approaches may require extensive cemetery excavations to yield a significant sample of individuals. However, older projects' collections may be revisited for analyses. There may be limited funding available for strontium analyses. However, cumulative small samples can be used for pooling (e.g., Cintas Peña and García Sanjuán 2022). Phenotypic biodistance approaches provide the more affordable of the two approaches and should not be overshadowed by costly ancient DNA (aDNA).

2.5 Paleogenetics

The significant revolutionary advances in aDNA research are unfortunately overshadowed by negative theoretical baggage consistently rejected by anthropology since the early twentieth century. These include the reintroduction of late nineteenth to early twentieth century racial essentializing: the assumption of genetically bounded populations, each with innate, unchanging cultural characteristics, which has been thoroughly critiqued elsewhere (e.g., Furholt 2018, 2020; Frieman and Hofmann 2019; Hackenbeck 2019; Battle-Baptiste 2020; Blakey 2020; Crellin and Harris 2020; Ensor 2021a: 94-102, 2021b; Lee 2020). That theoretical divide and most problems involve a poor understanding of anthropology, in general, and an even more impoverished understanding of the basics of kinship research (Ensor 2021a, 2021b). For instance, geneticists commonly reference "female exogamy". "Female exogamy" literally means men, women, and other genders marry outside their "female". Groups –like lineages or clans– not sexes, are exogamous. Moreover, if a woman married a man outside her group, then her husband simultaneously married a woman outside his group –that man also practiced group exogamy. The concept also confuses marriage –e.g., what groups are exogamous– with postmarital residence –e.g., where people live after marriage– which are completely different dimensions of kinship practices. Other problems abound. For example, a recent article using numerous individual lines of genetic inheritance makes the claim that no consanguinity was detected (Gnecchi Ruscone *et al.* 2024). Consanguinity refers to genealogical or "blood" relations –i.e., genetic relations. The literature also confuses genetic lines of inheritance with lineages; thus, interpreting that the society had matrilineal or patrilineal descent when only one or few individual multigenerational lines of inheritance are detected. Haplogroup admixture from females is assumed to indicate patrilocality when adoption of women, or men, has no bearing on descent and residence practices. There is also a dubious assumption that male-child coburial indicates patrilineal descent. On the positive side, the abundant critiques of a biological perspective on kinship have by now forced most palaegeneticists to entertain a vague "social kinship" versus "biological kinship" dichotomy. Of course, even that is problematic for not recognizing that all kinship practices are social –even the biologically based ideology of the West is a social construction– and that corporate unilineal descent groups have genetically unrelated comembers with the same obligations as those having genetic relations –as above, in reference to phenotypic approaches. Concluding "social kinship" does not tell us what kinship practices were used. For these reasons, Ensor (2021a) introduced ethnographically informed models for interpreting descent and residence using aDNA, each based on the expectations for the social distributions of mtDNA, Y-chromosome, and autosomal relations.

This section focuses on another problem: sampling issues. Although Ensor (2021a) found that sample sizes from European cemeteries, tumuli, or cave sites were not the only problem, the detection of haplotypes (mtDNA and/or Y-chromosome) significantly influenced the ability to interpret kinship practices. Haplogroups alone are generally insufficient for interpreting kinship. Most importantly, children, infants, and unsexed adults are less useful, whereas sex-identified adults, ideally in similar proportions, are most useful for kinship analysis. To understand the problems in sampling issues, we need to understand the theory and perspectives that have guided kinship research in paleogenetics.

Unlike bioarchaeology, the recent applications of paleogenetics in Europe –and the sampling methods– are strongly influenced by phylogenetic theory. I believe this stems from the first applications using haplogroups in essentializing narratives imagining unchanging Mesolithic-foraging and Southwest Asian-farming races (e.g., Cavalli-Sforza and Minch 1997; Bramanti *et al.* 2009). The phylogenetic perspective only requires few individuals of any age or sex from a few cemeteries to interpret racial isolation or admixture. The small samples of adults and infants or children are also useful if we maintain a biological definition of kinship and seek nuclear family relations, or individual lines of genetic inheritance, though often the evidence contradicts that biological definition. But such sampling for phylogenetic purposes is inappropriate for kinship analyses.

Ideally, we need adult females and males from each cemetery –those who lived long enough to have married and had children and who themselves grew up to marry and have children– and in greater numbers from each cemetery to identify the relevant patterns in the haplotype and autosomal relations for different descent and residence configurations. Kinship analysis requires understanding the compositions of corporate groups (Ensor 2021a). For example, the expectation for a matrilineal descent group's cemetery involves multiple nongenetically related sets of adult women and men sharing haplotypes across generations (women, as well as their brothers and mothers belong to the same corporate matrilineal group); any sharing of Y-chromosome haplotypes should be rare (only among sets of biological brothers); and there should be no fathers to the men and women present –since fathers would be buried in their own matrilineal group's cemetery– (e.g., Fig. 2.4, applied to genetic relations). Patrilineal descent group cemeteries should exhibit the opposite. Patrilineal descent groups lacking wives' membership transfers and patrilocality combined with bilocality would appear the same: in either case, spouses are buried together. Matrilocal residential groups with bilateral descent should have related women but unrelated men. Bilateral descent combined with bilocality should have some genetically related women and some genetically related men and other men and women lacking genetic relations to comembers (Ensor 2021a). As can be seen from these distinctions, the interpretation of descent and residence using aDNA should ideally be based on sampling from numerous sex-identified adults within a given cemetery –following the example set by bioarchaeology.

Using these models for inference, Ensor (2021a) found variability and change in few selected cemeteries across Iberia. Early Neolithic Avellanar Cave, Cataluña (Lacan *et al.* 2011; Gibaja *et al.* 2018) best matches a patrilineal descent group. Burials at Middle Neolithic Can Gambús in the same region near the Mediterranean coast (Gomes *et al.* 2017) was either for a matrilineal descent group or for a matrilocal group with bilateral descent. The sample size was too small to distinguish between those. Though restricted to haplogroups and subhaplogroups, the data from the Alto de Reinoso barrow (Alt *et al.* 2016) also suggest either a matrilineal descent group or a matrilocal group with bilateral descent. Each of these inferences differs from Cintas Peña's and García Sanjuán's (2022) interpretations of bilocality in the Chalcolithic of Southern Iberia based on strontium isotope ratios. Perhaps there was a chronological trend from patrilineal to matrilineal or matrilocal with bilateral to bilocal –with bilateral descent– with a patrilocal bias from the Early Neolithic to the

Chalcolithic for the whole of Iberia. Or perhaps more numerous applications will reveal mosaics in practices in each region and period.

Obviously, the costs of aDNA sampling for numerous sex-identified adults may be prohibitive under austerity measures, in which case phenotypic trait approaches in bioarchaeology would provide better approach. Phenotypic biodistance research also carries less negative baggage. On the other hand, because the sensationalism of paleogenetics has attracted greater international attention and funding than other approaches, and with the development of testing facilities in Spain, aDNA research may become feasible even under austerity. The greatest challenge for Spain, as elsewhere, is better understanding, through cumulative ethnographic knowledge, how kinship variably distributes genetic relationships to avoid the unproductive outcomes currently characterizing paleogenetic research across much of Europe.

2.6 Summary and considerations for Iberian prehistoric kinship research

Ethnological approaches by necessity use modern data and therefore carry evolutionary assumptions about how to use and treat those data. However, both the phylogenetic and neoevolutionary assumptions and interpretation are unacceptable to contemporary anthropology. These problems, and to avoid ethnological tyranny, highlight the need for archaeological, bioarchaeological, and paleogenetic approaches to prehistoric kinship.

The greatest challenges to prehistoric kinship research in Spain and elsewhere are unfortunate conceptual limitations. After its long period of absence in higher education, prehistorians need greater familiarity with the subject of kinship (Cveček 2024). In archaeology, vague references to "families", "households", "house", or "possible lineages/clans" do not inform on the compositions of residential groups, corporate kin groups, or marital alliances from which to infer gender dynamics, practice and social reproduction, and agents' manipulations of relationships and identities. In bioarchaeology, simplistic assumptions of descent group homogeneity or placement of spouses in the same cemetery illustrate a need to better understand various kin group membership compositions, postmortem location, and marriage systems. Solutions are also needed to replace normative interpretation resulting from the pooling of individuals. Paleogenetics is the least informed area on kinship and relies heavily on disproven theoretical assumptions, which have negatively impacted its potential for useful contributions. At best, it has so far provided a vague "biological" versus "social" dichotomy; but that also tells us little about corporate group memberships, residential dynamics, or marriage systems, let alone how people practiced and manipulated relationalities.

Archaeological approaches are the most diverse. Direct historical analogy and the later disproven early functionalist associations with subsistence strategies developed from culture historical and early functionalist questions. The postmodernist House perspective relies on trait lists to distinguish "Houses" from descent groups using selective ethnographic analogies, but the traits claimed to represent houses apply equally to any corporate descent groups. Most methods developed out of processual archaeology may be applied to various theoretical perspectives. Artifact-based approaches assume relations between gender and styles or materials to interpret postmarital residence or by assuming group-based stylistic affinities. However, ethnographic analogy and direct historical analogy to impose gender roles in prehistory can be hazardous and the distribution of stylistic attributes may better reflect marriage networks. Cross-cultural testing –avoiding the hazards of ethnographic analogy, direct historical analogy, and theoretical biases– has been used for inferences on cemetery structure and community patterns. Demarcated cemeteries are demonstrated to be associated with corporate groups, and intra-cemetery subgroupings with conjugal or extended families, but these alone do not reveal the kinds of corporate groups or residential strategies used. Community pattern associations with kinship practices are favored by the author for withstanding repeated testing, theoretical independence, and, unlike most other approaches, avoid normative interpretations on practices by illuminating variation and change within and among settlements, which are needed for interpreting agency.

Bioarchaeology –developed initially for processualist questions but valuable for practice and agency perspectives today– propelled biological anthropology beyond culture historical assessments of population biological "affinities" for interpreting interaction and migration. Using multiple phenotypic traits, biodistance studies enabled avenues for interpreting postmarital residence and descent. The major problems were the assumptions that spouses are universally buried in the same cemeteries and that unilineal descent groups should be internally homogeneous. However, those problems can be addressed through revised models for interpreting intra- and intercemetery biodistance patterns. More recent strontium isotope studies initiated with the same assumption that spouses should be buried together, yet that too can be remedied with revised models for expectations for varied descent and residence combinations. Perhaps a greater challenge for bioarchaeological approaches is overcoming the normative characterizations that are a product of pooling individuals for the multivariate statistical methods.

The sensational revolutionary paleogenetics is unfortunately matched by the most impoverished understanding of kinship. Although paleogenetic interpretations have been of little use, if not harmful, the data may still be valuable if assessed through a twentieth-century understanding of how descent and residence combinations distribute biological data. Emphasized here is the need for different sampling to address kinship. Although haplogroups among small samples of infants, children and few adults may address misguided questions from phylogenetics or uninformative biological versus social dichotomizing, they are inappropriate for kinship analyses. More refined haplotyping and sampling more adult females and males within and among cemeteries are needed –as practiced in bioarchaeology– for comparison with informed models on how descent, marriage, and residence combinations distribute biological relations.

Given the impacts on funding under austerity, serious attention needs to be given to which methods are the most cost-effective. For archaeological survey projects with limited excavations, artifact-based approaches could produce hypotheses. In the absence of funding for large-scale excavations with abundant radiometric dating, mapping of settlements where structure walls are visible from the ground surface and/or the use of older reports on extensive excavations may provide the best avenues for the community pattern approach –the most proven method with the most rewarding outcomes for contemporary theoretical perspectives. Of the biological lines of data, dental morphometric multi-trait analyses using collections from appropriate older projects provide the least costly but demonstrably most productive method. These approaches, together with greater awareness of kinship and its history of research, may provide the best avenues for prehistoric kinship research under severe funding limitations.

References

Aguiar, G.S. and Alvez Neves, W. (1991): Postmarital residence pattern and within-sex genetic diversity among the Urubu-Ka'apor Indians, Brazilian Amazon. *Human Biology* 63, pp. 467-488.

Allen, N.J. (2012): Tetradic theory and Omaha systems. In Trautmann T.R. and Whiteley, P.M. (eds.), *Crow-Omaha: New Light on a Classic Problem of Kinship Analysis*. Tucson: University of Arizona Press, pp. 51-66.

Allen, N.J. (2008): Tetradic theory and the origin of human kinship systems. In Allen, N.J., Callan, H., Dunbar, R. and James, W. (eds.), *Early human kinship: From sex to social reproduction*. Oxford: Blackwell, pp. 96-112.

Allen, W.L. and Richardson, J.B. (1971): The reconstruction of kinship from archaeological data: The concepts, the methods, and the feasibility. *American Antiquity* 36, pp. 41-53.

Alt, K.W., Zesch, S., Garrido Pena, R., Knipper, C., Szécsényi Nagy, A., Roth, C., Tejedor Rodríguez, C., Held, P., García Martínez de Lagrán, I., Navitainuck, D., Arcusa Magallón, H. and Rojo Guerra, M.A. (2016): A community in life and death: The late Neolithic megalithic tomb at Alto de Reinoso (Burgos, Spain). *PLoS One* 11(1), p. e0146176. DOI: 10.1371/journal.pone.0146176.

Armelagos, G.J. and Van Gerven, D.P. (2003): A Century of Skeletal Biology and Paleopathology: Contrasts, Contradictions, and Conflicts. *American Anthropologist* 105, pp. 51-62. DOI: 10.1525/aa.2003.105.1.53.

Bateman, R., Goddard, I., O'Grady, R., Funk, V.A., Mooi, R., Dress, W.J. and Cannell, P. (1990): The feasibility of reconciling human phylogeny and the history of language. *Current Anthropology* 31, pp. 1-24. DOI: 10.1086/203800.

Battle-Baptiste, W. (2020): The strange afterlife of biodeterministic imagination. *Archaeological Dialogues* 27, pp. 27-35. DOI: 10.1017/S1380203820000069.

Beck, R.A. (ed.) (2007): *The durable house: House society models in archaeology*. Carbondale: Southern Illinois University. DOI: 10.1017/S0002731600048782.

Benedict, R. (1934): *Patterns of culture*. Boston: Houghton Mifflin.

Bentley, R.A. (2013): Mobility and the diversity of Early Neolithic lives: Isotopic evidence from skeletons. *Journal of Anthropological Archaeology* 32, pp. 303-312. DOI: 10.1016/j.jaa.2012.01.009.

Bentley, R.A., Bickle, P., Francken, M., Gerling, C., Hamilton, J., Hedges, R., Stephan, E., Wahl, J. and Whittle, A. (2013): Baden-Württemberg. In Bickle, P. and Whittle, A. (eds.), *The First Farmers of Central Europe: Diversity in LBK Lifeways*. Oxford: Oxbow Books, pp. 251-290.

Bentley, R.A., Bickle, P., Fibiger, L., Nowell, G.M., Dale, C.W., Hedges, R.E.M., Hamilton, J., Wahl, M., Francken, G., Grupe, E., Lenneis, M., Arbogast, R.M., Hofmann, D. and Whittle, A. (2012): Community differentiation and kinship among Europe's first farmers. *Proceedings of the National Academy of Sciences USA* 109, pp. 9326-9330. DOI: 10.1073/pnas.1113710109.

Binford, L.R. (1972): Galley Pond Mound. In Binford, L.R. (ed.), *An archaeological perspective*. New York: Seminar Press, pp. 390-420.

Blakey, M.L. (2020): On the biodeterministic imagination. *Archeological Dialogues* 27, pp. 1-16. DOI: 10.1017/S1380203820000021.

Boas, F. (1920): The methods of ethnology. *American Anthropologist* 22, pp. 311-321.

Bogaard, A., Krause, R. and Strien, H.C. (2011): Towards a social geography of cultivation and plant use in an early farming community: Vaihingen an der Enz, South-West Germany. *Antiquity* 85, pp. 395-416. DOI: 10.1017/S0003598X00067831.

Bramanti, B., Thomas, M.G., Haak, W., Unterlaender, M., Jores, P., Tambets, K., Antanaitis-Jacobs, I., Haidle, M.N., Jankauskas, R., Kind, C.J., Lueth, F., Terberger, T., Hiller, J., Matsumura, S., Forster, P. and Burger, J. (2009): Genetic Discontinuity Between Local Hunter-Gatherers and Central Europe's First Farmers. *Science* 326, pp. 137-140. DOI: 10.1126/science.1176869.

Brown, B.M. (1987): Population estimation from floor area: A restudy of 'Naroll's Constant'. *Cross-Cultural Research* 21, pp. 1-49. DOI: 10.1177/106939718702100101.

Buikstra, J.E. (1977): Biocultural dimensions of archaeological study: A regional perspective. In Blakely, R.L. (ed.), *Biocultural Adaptation in Prehistoric America.* Athens: University of Georgia Press, pp. 67-84.

Campbell, L. and Poser, W.J. (2008): *Language classification: History and method.* Cambridge: Cambridge University Press. DOI: 10.1017/CBO9780511486906.

Cavalli-Sforza, L.L. and Minch, E. (1997): Paleolithic and Neolithic lineages in the European mitochondrial gene pool. *American Journal of Human Genetics* 61(1), pp. 247-254. DOI: 10.1016/S0002-9297(07)64303-1.

Chang, K.C. (1958): Study of the Neolithic social grouping: Examples from the New World. *American Anthropologist* 60, pp. 298-334. DOI: 10.1525/aa.1958.60.2.02a00080.

Chesson, M.S. (2007): House, town, field, and wadi: Landscapes of the Early Bronze Age southern Levant. In Beck, R.A. (ed.), *The durable house: House society models in archaeology.* Carbondale: Southern Illinois University, pp. 317-343.

Cintas Peña, M. and García Sanjuán, L. (2022): Women, residential patterns and early social complexity: From theory to practice in Copper Age Iberia. *Journal of Anthropological Archaeology* 67, p. 101422. DOI: 10.1016/j.jaa.2022.101422.

Clendon, M. (2006): Reassessing Australia's linguistic prehistory. *Current Anthropology* 47, pp. 39-61. DOI: 10.1086/497671.

Coupland, G., Clark, T. and Palmer, A. (2009): Hierarchy, communalism, and the spatial order of Northwest Coast plank houses: a comparative study. *American Antiquity* 74, pp. 77-106. DOI: 10.2307/25470539.

Craig, D.B. (2007): Courtyard groups and the emergence of house estates in early Hohokam society. In Beck, R.A. (ed.), *The durable house: House society models in archaeology.* Carbondale: Southern Illinois University, pp. 446-463

Crellin, R.J. and Harris, O.J.T. (2020): Beyond binaries. Interrogating ancient DNA. *Archaeological Dialogues* 27, pp. 37-56.

Cveček, S. (2024): Why kinship still needs anthropologists in the 21st century. *Anthropology Today* 40(1), pp. 3-7.

Deetz, J. (1965): *The dynamics of stylistic change in Arikara ceramics.* Champaign: University of Illinois Press.

Divale, W.T. (1977): Living floors and marital residence: A replication. *Behavior Science Research* 12, pp. 109-115.

Driver, H.E. and Massey, W.C. (1957): *Comparative Studies of North American Indians.* Philadelphia: American Philosophical Society.

Dziebel, G. (2021): Crow-Omaha Kinship: Revitalizing a Problem or Generating a Solution. *Kinship* 1(2), pp. 1-23.

Eggan, F. (1966): *The American Indian*. Chicago: Aldine.

Ehret, C. (2012): Deep-Time Historical Contexts of Crow and Omaha Systems: Perspectives from Africa. In Trautmann, T.R. and Whiteley, P.M. (eds), *Crow-Omaha: New Light on a Classic Problem of Kinship Analysis*. Tucson: University of Arizona Press, pp. 173-202.

Ember, C.R. and Ember, M. (1972): The conditions favoring multilocal residence. *Southwestern Journal of Anthropology* 28, pp. 382-400.

Ember, C.R. and Ember, M. (2009): *Cross-Cultural Research Methods*. Second edition. Lanham: AltaMira Press.

Ember, C.R., Ember, M. and Pasternak, B. (1974): On the development of unilineal descent. *Journal of Anthropological Research* 30, pp. 69-94.

Ember, M. (1973): An archaeological indicator of matrilocal versus patrilocal residence. *American Antiquity* 38, pp. 177-182.

Ember, M. and Ember, C.R. (1971): The conditions favoring matrilocal versus patrilocal residence. *American Anthropologist* 73, pp. 571-594.

Ember, M. and Ember, C.R. (1995): Worldwide cross-cultural studies and their relevance for archaeology. *Journal of Archaeological Research* 3, pp. 87-111. DOI: 10.1007/BF02231488.

Ensor, B.E. (2011): Kinship theory in archaeology: From critiques to the study of transformations. *American Antiquity* 76(2), pp. 203-227. DOI: 10.7183/0002-7316.76.2.203.

Ensor, B.E. (2013a): *The Archaeology of Kinship: Advancing Interpretation and Contributions to Theory*. Tucson: University of Arizona Press.

Ensor, B.E. (2013b): *Crafting Prehispanic Maya Kinship*. Tuscaloosa: University of Alabama Press.

Ensor, B.E. (2013c): Kinship and social organization in the Prehispanic Caribbean. In Keegan, W.F., Hoffman, C.C. and Rodríguez-Ramos, R. (eds.), *The Oxford Handbook of Caribbean Archaeology*. Oxford: Oxford University Press, pp. 84-96.

Ensor, B.E. (2016): Ethnological Problems and the Production of Archaeological Kinship Research. *Structure and Dynamics* 9(2), pp. 80-109. DOI: 10.5070/SD992032329.

Ensor, B.E. (2017a): Testing Ethnological Theories on Prehistoric Kinship. *Cross-Cultural Research* 51(3), pp. 1-29. DOI: 10.1177/10693971176976.

Ensor, B.E. (2017b): Kin-mode contradictions, crises, and transformations in the Archaic Lower Mississippi Valley. In Rosenswig, R.M. and Cunningham, J.J. (eds.), *Modes of Production and Archaeology*. Gainesville: University Press of Florida, pp. 123-143.

Ensor, B.E. (2020): *Oysters in the Land of Cacao: Archaeology, Material Culture, and Societies at Islas de Los Cerros and the Western Chontalpa*. Tucson: University of Arizona Press.

Ensor, B.E. (2021a): *The not very patrilocal European neolithic: Strontium, aDNA, and archaeological kinship analyses*. Oxford: Archaeopress.

Ensor, B.E. (2021b): Crow-Omaha and the future of kin term research. *Kinship* 1(2), pp. 24-67.

Ensor, B.E. (2021c): Making aDNA useful for kinship analysis. *Antiquity* 95(379), pp. 241-243. DOI: 10.15184/aqy.2020.234.

Ensor, B.E., Irish, J.D. and Keegan, W.F. (2017): The Bioarchaeology of Kinship: Proposed Revisions to Assumptions Guiding Interpretation. *Current Anthropology* 58(6), pp. 739-761. DOI: 10.1086/694584.

Evans-Pritchard, E.E. (1940): *The Nuer.* New York: Oxford University Press.

Fortes, M. (1958): Introduction. In Goody, J. (ed.), *The Developmental Cycle in Domestic Groups.* Cambridge: Cambridge University Press, pp. 1-14.

Fortes, M. (1959): Primitive kinship. *Scientific American* 200(6), pp. 146-158.

Fortes, M. (1969): *Kinship and the social order: The legacy of Lewis Henry Morgan.* Chicago: Aldine.

Fortunato, L. (2011): Reconstructing the History of Residence Strategies in Indo-European-Speaking Societies: Neo-, Uxori-, and Virilocality. *Human Biology* 83(1), pp. 107-128. DOI: 10.3378/027.083.0107.

Fortunato, L. and Jordan, F. (2010): Your Place or Mine? A Phylogenetic Comparative Analysis of Marital Residence in Indo-European and Austronesian Societies. *Philosophical Transactions of the Royal Socitey B: Biological Sciences* 365, pp. 3913-3922.

Fox, R. (1967): *Kinship and Marriage: An Anthropological Perspective.* Cambridge: Cambridge University Press.

Fried, M. (1967): *The evolution of political society.* New York: Random House.

Frieman, C.J. and Hofmann, D. (2019): Present pasts in the archaeology of genetics, identity, and migration in Europe: a critical essay. *World Archaeology* 51, pp. 528-545. DOI: 10.1080/00438243.2019.1627907.

Furholt, M. (2018): Massive migrations? The impact of recent aDNA studies on our view of third millennium Europe. *European Journal of Archaeology* 21(2), pp. 159-191. DOI: 10.1017/eaa.2017.43.

Gibaja, J.F., Morell, B., López Onaindía, D., Zemour, A., Bosch, À., Tarrús, J., and Eulàlia-Subirà, M. (2018): Nuevos datos cronológicos sobre la cueva sepulcral neolítica de l'Avellaner (Les Planes d'Hostoles, Girona). *Munibe Antropologia-Arkeologia* 69. DOI: 10.21630/maa.2018.69.01.

Gibson, J.L. (1973): *Social systems at Poverty Point: An analysis of intersite and intrasite variability.* PhD dissertation. Dallas: Southern Methodist University.

Gillespie, S.D. (2000a): Rethinking ancient Maya social organization: Replacing "lineage" with "house". *American Anthropologist* 102, pp. 467-484.

Gillespie, S.D. (2000b): Beyond kinship: An introduction. In Joyce, R.A. and Gillespie, S.D. (eds.), *Beyond kinship: Social and material reproduction in house societies*: Philadelphia: University of Pennsylvania Press, pp. 1-21.

Gjessing, G. (1956): *Socio-culture: Interdisciplinary essays on society and culture.* Oslo: Universitetes Etnografiske Museum.

Gjessing, G. (1975): Socio-archaeology. *Current Anthropology* 16, pp. 323-341.

Gnecchi Ruscone, G.A., Rácz, Z., Samu, L., Szeniczey, T., Faragó, N., Knipper, C., Friedrich, R., Zlámová, D., Traverso, L., Liccardo, S., Wabnitz, S., Popli, D., Wang, K., Radzeviciute, R., Gulyás, B., Koncz, I., Balogh, C., Lezsák, G.M., Mácsai, V., Burnbury, M.M.E., Spekker, O., Le Roux, P., Szécsényi-Nagy, A., Gusztáv-Mende, B., Colleran, H., Hajdu, T., Geary, P., Pohl, W., Vida, T., Krauss, J., and Hofmanová, Z. (2024): Network of large pedigrees reveals social practices of Avar communities. *Nature* 629, pp. 376-383. DOI: 10.1038/s41586-024-07312-4.

Godelier, M., Trautmann, T.R. and Tjon Sie Fat, F.E. (eds.) (1998): *Transformations of Kinship.* Washington: Smithsonian Institution Press.

Goldstein, L.G. (1981): One-dimensional archaeology and multi-dimensional people: Spatial organization and mortuary analysis. In Chapman, R., Kinnes, I. and Randsborg, K. (eds.), *The Archaeology of Death.* Cambridge: Cambridge University Press, pp. 53-70.

Gomes, C., Gibaja, J.F., Roig, J., Buxó, I., Baeza Richer, C., López Matayoshi, C., López Parra, A.M., Paloma Díez, S., Subirà, M.E. and Arroyo Pardo, E. (2017): Biological kinship analysis in extremely critical samples: The case of a Spanish Neolithic necropolis. *Forensic Science International: Genetics Supplement Series* 6, pp. e421-e422. DOI: 10.1016/j.fsigss.2017.09.154.

González Ruibal, A. (2006): House societies vs. kinship-based societies: An archaeological case from Iron Age Europe. *Journal of Anthropological Archaeology* 25(1), pp. 144-173. DOI: 10.1016/j.jaa.2005.09.002.

Goodenough, W. (1970): *Description and comparison in cultural anthropology*. Chicago: Aldine.

Haak, W., Brandt, G., Jong, H.N., Meyer, C., Ganslmeier, R., Heyd, V., Hawkesworth, C., Pike, A.W.G., Meller, H., and Alt, K.W. (2008): Ancient DNA, strontium isotopes, and osteological analyses shed light on social and kinship organization of the later Stone Age. *Proceedings of the National Academy of Sciences USA* 105(47), pp. 18226-18231. DOI: 10.1073/pnas.080759210.

Hackenbeck, S.E. (2019): Genetics, archaeology and the far right: An unholy trinity. *World Archaeology* 51(4), pp. 517-527. DOI: 10.1080/00438243.2019.1617189.

Haury, E. (1956): Speculation on prehistoric settlement patterns in the Southwest. In Willey, G.R. (ed.), *Prehistoric settlement patterns in the New World*. New York: Viking Fund, pp. 3-10.

Hill, J.N. (1966): A prehistoric community in eastern Arizona. *Southwestern Journal of Anthropology* 22, pp. 9-30.

Howell, T.L. and Kintigh, K.W. (1996): Archaeological identification of kin groups using mortuary and biological data: An example from the American Southwest. *American Antiquity* 61(3), pp. 537-554. DOI: 10.2307/281839.

Hrnčiř, V., Duda, P., Šaffa, G., Květina, P. and Zrzavy, J. (2020): Identifying post-marital residence patterns in prehistory: A phylogenetic comparative analysis of dwelling size. *PLoS One* 15(2), p. e0229363. DOI: 10.1371/journal.pone.0229363.

Hubbe, M., Alves Neves, W., Castro de Oliveira, E. and Strauss, A. (2009): Postmarital residence practice in Southern Brazilian coastal groups: Continuity and change. *Latin American Antiquity* 20(2), pp. 267-278. DOI: 10.1017/S1045663500002637.

Johnson, K.M. and Paul, K.S. (2016): Bioarchaeology and Kinship: Integrating Theory, Social Relatedness, and Biology in Ancient Family Research. *Journal of Archaeological Research* 24, pp. 75-123. DOI: 10.1007/s10814-015-9086-z.

Jones, D. (2003): Kinship and Deep History: Exploring Connections between Culture Areas, Genes, and Languages. *American Anthropologist* 105, pp. 501-514.

Jones, D. and Milicik, B. (eds.) (2011): *Kinship, Language, and Prehistory: Per Hage and the Renaissance in Kinship Studies*. Salt Lake City: University of Utah Press.

Joyce, R.A. (2000): Heirlooms and houses: Materiality and social memory. In Joyce, R.A. and Gillespie, S.D. (eds.), *Beyond kinship: Social and material reproduction in house societies*. Philadelphia: University of Pennsylvania Press, pp. 189-212.

Joyce, R.A. (2007): Building houses: The materialization of lasting identity in formative Mesoamerica. In Beck, R.A. (ed.), *The durable house: House society models in archaeology*. Carbondale: Southern Illinois University, pp. 53-72.

Joyce, R.A. and Gillespie, S.D. (eds.) (2000): *Beyond kinship: Social and material reproduction in house societies*. Philadelphia: University of Pennsylvania Press.

Keegan, W.F. (2009): Central plaza burials in Saladoid Puerto Rico: An alternative perspective. *Latin American Antiquity* 20(2), pp. 375-385. DOI:10.1017/S1045663500002686.

Keegan, W.F. and Maclachlan, M.D. (1989): The evolution of avunculocal chiefdoms: A reconstruction of Taino kinship and politics. *American Anthropologist* 91, pp. 613-630.

Keesing, R.M. (1975): *Kin Groups and Social Structure.* New York: Holt, Rinehart and Winston Inc.

Kertzer, D.I. and Brettell, C. (1987): Advances in Italian and Iberian Family History. *Journal of Family History* 12, pp. 87-120.

Konigsberg, L.W. (1988): Migration models of prehistoric postmarital residence. *American Journal of Physical Anthropology* 77, pp. 471-482.

Korotayev, A. (2003): Form of marriage, sexual division of labor, and postmarital residence in cross-cultural perspective. *Journal of Anthropological Research* 59(1), pp. 69-89.

Korotayev, A., Borinskaya, S., Starostin, G. and Meshcherina, K. (2019): Evolution of Eurasian and African Family Systems, Cross-Cultural Research, Comparative Linguistics, and Deep History. *Social Evolution and History* 18, pp. 286-312. DOI:10.30884/seh/2019.02.15.

Kronenfeld, D.B. (2004): Definitions of cross versus parallel: Implications for a new typology (an appreciation of A. Kimball Romney). *Cross-Cultural Research* 38(3), pp. 294-269. DOI:10.1177/1069397104264276.

Lacan, M., Keyser, C., Ricaut, F.-X., Brucato, N., Tarrús, J., Bosch, A., Guilaine, J., Crubézy, E. and Ludes, B. (2011): Ancient DNA suggests the leading role played by men in the Neolithic dissemination. *Proceedings of the National Academy of Sciences USA* 108(45), pp. 18255-18259. DOI:10.1073/pnas.1113061108.

Lane, R.A. and Sublett, A.J. (1972): Osteology of social organization: Residence pattern. *American Antiquity* 37, pp. 186-201.

Le Roy, M., Rivollat, M., Mendisco, F., Pemonge, M.H., Coutelier, C., Couture, C., Tillier, A-M., Rottier, S. and Deguilloux, M.F. (2016): Distinct ancestries for similar funerary practices? A GIS analysis comparing funerary, osteological and aDNA data from the Middle Neolithic necropolis Gurgy "Les Noisats" (Yonne, France). *Journal of Archaeological Science* 73, pp. 45-54. DOI:10.1016/j.jas.2016.07.003.

Lee, N. (2020): Here we go again. The need to contest and refute biological determinism in archaeology. *Archaeological Dialogues* 27(1), pp. 20-22. DOI:10.1017/S1380203820000045.

Lévi-Strauss, C. (1956): The family. In Shapiro, H.L. (ed.), *Man, culture, and society.* New York: Oxford University Press, pp. 261-285.

Lévi-Strauss, C. (1982): *The way of the masks.* Seattle: University of Washington Press.

Lévi-Strauss, C. (1987): *Anthropology and myth: Lectures, 1951-1982.* Oxford: Basil Blackwell.

Longacre, W.A. (1964): Archaeology as anthropology: A case study. *Science* 144, pp. 1454-1455.

Longacre, W.A. (1966): Changing patterns of social integration: A prehistoric example from the American southwest. *American Anthropologist* 68, pp. 94-102.

Marshall, Y. (2000): Transformations of Nuu-cha-nulth houses. In Joyce, R.A. and Gillespie, S.D. (eds.), *Beyond kinship: Social and material reproduction in house societies.* Philadelphia: University of Pennsylvania Press, pp. 73-102.

McKnight, D. (2004): *Going the Whiteman's way: Kinship and marriage among Australian Aborigines.* Ashgate: Aldershot.

McPherron, A. (1967): Pottery style clustering, marital residence, and cultural adaptations of an Algonkian-Iroquoian Border. In Tooker, E. (ed.), *Proceedings of the 1965 conference on Iroquois research*. Albany: New York State Museum, pp. 101-107.

Modjeska, N. (1982): Production and inequality: Perspectives from central New Guinea. In Strathern, A. (eds.), *Inequality in New Guinea Highland Societies*. Cambridge: Cambridge University Press, pp. 50-108.

Moore, J.H. (1988): The dialectics of Cheyenne kinship: Variability and change. *Ethnology* 27, pp. 253-269.

Moore, J.H. (1994): Putting anthropology back together again: The ethnogenetic critique of cladistic theory. *American Anthropologist* 96, pp. 925-948.

Murdock, G.P. (1949): *Social structure*. New York: Macmillan.

Murdock, G.P. and Provost, C. (1973): Factors in the division of labor by sex: A cross-cultural analysis. *Ethnology* 12, pp. 379-392.

Pasternak, B. (1976): *Introduction to kinship and social organization*. Englewood Cliffs: Prentice Hall.

Pechtl, J. (2015): Linearbandkeramik pottery and society. In Fowler, C., Harding, J. and Hofmann, D. (eds.), *The Oxford Handbook of Neolithic Europe*. Oxford: Oxford University Press, pp. 555-572.

Peletz, M.G. (1995): Kinship studies in late twentieth-century anthropology. *Annual Review of Anthropology* 24, pp. 343-372. DOI: 10.1146/annurev.an.24.100195.002015.

Peregrine, P.N. (1996): Ethnology versus ethnographic analogy: A common confusion in archaeological interpretation. *Cross-Cultural Research* 30(4), pp. 316-329. DOI: 10.1177/106939719603000403.

Peregrine, P.N. (2001a): Matrilocality, corporate strategy, and the organization of production in the Chacoan world. *American Antiquity* 66(1), pp. 36-46. DOI: 10.2307/2694316.

Peregrine, P.N. (2001b): Cross-cultural comparative approaches in archaeology. *Annual Review of Anthropology* 30, pp. 1-18. DOI: 10.1146/annurev.anthro.30.1.1.

Pollard, A.M. (2011): Isotopes and impact: A cautionary note. *Antiquity* 85(328), pp. 631-638. DOI: 10.1017/S0003598X00068034.

Porčić, M. (2010): House Floor Area as a Correlate of Marital Residence Pattern: A Logistic Regression Approach. *Cross-Cultural Research* 44(4), pp. 405-424. DOI: 10.1177/1069397110378839.

Radcliffe-Brown, A.R. (1952): *Structure and function in primitive society: Essays and addresses*. London: Cohen and West.

Rouse, I. (1992): *The Tainos: Rise and Decline of the People Who Greeted Columbus*. New Haven: Yale University Press.

Sanger, M., Bourcy, S., Ogden, Q.M. and Troutman, M. (2020): Postmarital residence patterns in the Late Archaic coastal Southeast USA: Similarities in stone tools revealed by geometric morphometrics. *Journal of Archaeological Method and Theory* 27, pp. 327-359. DOI: 10.1007/s10816-019-09435-9.

Saxe, A.A. (1970): *Social dimensions of mortuary practices in a Mesolithic population from Wadi Halfa, Sudan*. PhD dissertation. Ann Arbor: University of Michigan.

Scheffler, H.W. (2001): *Filiation and affiliation*. Boulder: Westview Press.

Schillaci, M.A. and Stojanowski, C.M. (2002): A reassessment of matrilocality in Chacoan Culture. *American Antiquity* 67(2), pp. 343-356. DOI: 10.2307/2694571.

Schillaci, M.A. and Stojanowski, C.M. (2003): Postmarital residence and biological variation at Pueblo Bonito. *American Journal of Physical Anthropology* 120(1), pp. 1-15. DOI:10.1002/ajpa.10147.

Schneider, D.M. (1984): *A critique of the theory of kinship.* Chicago: University of Chicago Press.

Service, E.R. (1962): *Primitive social organization: An evolutionary perspective.* New York: Random House.

Sims-Williams, P. (1998): Genetics, Linguistics, and Prehistory: Thinking Big and Thinking Straight. *Antiquity* 72(277), pp. 505-527. DOI:10.1017/S0003598X00086932.

Souvatzi, S. (2017): Kinship and Social Archaeology. *Cross-Cultural Research* 51(2), pp. 172-195. DOI:10.1177/1069397117691028.

Spence, M.W. (1974): Residential practices and the distribution of skeletal traits in Teotihuacan, Mexico. *Man* 9, pp. 262-273.

Steele, J. and Kandler, A. (2010): Language trees ≠ gene trees. *Theory in Biosciences* 129(2-3), pp. 223-233. DOI:10.1007/s12064-010-0096-6.

Stefan, V.H. (1999): Craniometric variation and homogeneity in prehistoric/protohistoric Rapa Nui (Eastern Island) regional populations. *American Journal of Physical Anthropology* 110, pp. 407-419.

Steward, J.H. (1937): Ecological aspects of Southwestern society. *Anthropos* 32, pp. 87-104.

Steward, J.H. (1942): The direct historical approach to archaeology. *American Antiquity* 7, pp. 337-343.

Steward, J.H. (1963): *Theory of culture change: The methodology of multilinear evolution.* Urbana: University of Illinois Press.

Stojanowski, C.M. and Schillaci, M.A. (2006): Phenotypic approaches for understanding patterns of intracemetery biological variation. *Yearbook of Physical Anthropology* 131(43), pp. 49-88. DOI:10.1002/ajpa.20517.

Stone, L. and King, D. (2019): *Kinship and gender: An introduction.* Sixth edition. New York: Routledge.

Szołtysek, M. (2008): Rethinking Eastern Europe: Household Formation Patterns in the Polish-Lithuanian Commonwealth and European Family Systems. *Continuity and Change* 23(3), pp. 389-427. DOI:10.1017/S0268416008006929.

Szołtysek, M. (2012): Spatial Construction of European Family and Household Systems: Promising Path or Blind Alley? An Eastern European Perspective. *Continuity and Change* 27(1), pp. 11-52. DOI:10.1017/S0268416012000057.

Szołtysek, M. (2015): *Rethinking East-Central Europe: Family systems and co-residence in the Polish Lithuanian Commonwealth.* Bern: Peter Lang Press.

Szołtysek, M., Ogórek, B. and Gruber, S. (2020): Global and Local Correlations of Hahnal's Household Formation Markers in Historical Europe: A Cautionary Tale. *Population Studies* 75(1), pp. 67-89. DOI:10.1080/00324728.2020.1832252.

Tomzack, P.D. and Powell, J.F. (2003): Postmarital residence practices in the Windover population: sex-based dental variation as an indicator of patrilocality. *American Antiquity* 68(1), pp. 93-108. DOI:10.2307/3557034.

Trautmann, T.R. and Whiteley, P.M. (eds.) (2012): *Crow-Omaha: New Light on a Classic Problem of Kinship Analysis.* Tucson: University of Arizona Press.

Trigger, B.G. (2006): *A History of Archaeological Thought.* Cambridge: Cambridge University Press.

Tringham, R. (2000): The continuous house: A view from the deep past. In Joyce, R.A. and Gillespie, S.D. (eds.), *Beyond kinship: Social and material reproduction in house societies*: Philadelphia: University of Pennsylvania Press, pp. 115-134.

Tsing, A.L. and Yanagisako, S.J. (1983): Feminism and Kinship Theory. *Current Anthropology* 24(4), pp. 511-516. DOI:10.1086/203039.

Valera, A.C. (ed.) (2018): *Os Perdigões neolíticos: Génese e desenvolvimento (de meados do 4o aos inícios do 3o milénio AC)*. Lisboa: Era Arqueologia.

Valera, A.C. and Antunes, S. (2008): A Mamoa 2 do Leandro (Maia, Porto): Intervenção de minimização no ámbito do alargamento da A3. *Apontamentos de Arqueologia e Patrimonio* 2008(3), pp. 7-17.

Van de Velde, P. (2007): On the Neolithic pottery from the site. In Van de Velde, P. (ed.), *Excavations at Geleen Janskamperveld 1990/1991*. Leiden: Leiden University, pp. 99-125.

Ware, J. (2018): Kinship and community in the northern Southwest: Chaco and beyond. *American Antiquity* 83(4), pp. 639-658. DOI:10.1017/aaq.2018.48.

Waterman, T.T. (1920): Yurok Geography. *University of California Publications in American Archaeology and Ethnology* 16, pp. 177-324.

Whallon, R.E. (1968): Investigations of late prehistoric social organization in New York. In Binford, S.R. and Binford, L.R. (eds.), *New Perspectives in Archaeology*. Chicago: Aldine, pp. 223-244.

Wheeler, W.C., Whiteley, P.M. and Powers, T. (2012): Phylogenetic Analysis of Sociocultural Data: Identifying Transformation Vectors for Kinship Systems. In Trautmann, T.R. and Whiteley, P.M. (eds.), *Crow-Omaha: New Light on a Classic Problem of Kinship Analysis*. Tucson: University of Arizona Press, pp. 109-131.

Wildcat, D., Sumi, I. and Deloria Jr., V. (2004): Commentary: A Response to Doug Jones. *American Anthropologist* 106, p. 641.

Wobst, H.M. (1978): The archaeo-ethnology of hunter-gatherers or the tyranny of the ethnographic record in archaeology. *American Antiquity* 43(2), pp. 303-309. DOI:10.2307/279256.

Wolf, E.R. (1982): *Europe and the People without History*. Berkeley: University of California Press.

3

Kinship, Social Space and History in Prehistory

Stella Souvatzi

Department of History, Archaeology and Social
Anthropology, University of Thessaly, Greece,
ssouvatzi@uth.gr

Abstract

This chapter derives from my keynote speech at the conference and has two interrelated aims. The first aim is to take a step towards a theoretical framework for social kinship research in archaeology; the second is to show the potential of bringing kinship as a dynamic process into archaeological analysis, using examples of settlements from Neolithic Greece and Turkey. Accordingly, the first part of the chapter discusses critically the current state of research, focussing principally on the social definition of kinship and pointing out a series of wider theoretical problems that impede an understanding of its dynamics, and suggests alternative perspectives. The second part examines kinship as a major factor of the spatial and social organisation of Neolithic communities and how it provided a dynamic potential for connections and continuous transformations through everyday lived space as well as at a larger context.

Keywords: *social relations, archaeological theory, Neolithic, Greece and Turkey, habitation patterns.*

3.1 Introduction

Kinship is a most significant organising principle of human grouping, social production, relationality and identity cross-culturally and diachronically. Kinship also structures time and incorporates transitional processes, continuity and changes of membership and partnerships, shifts in internal and external relationships, and interactions between kinship groups and the broader society. Kinship has both political and moral dimensions. Kinship is fundamental in social analysis and theoretical discourse across a range of disciplines, including anthropology, sociology and social history. Archaeology shares many interests with this discussion, and the many theoretical and practical justifications for the significance of kinship in the related social disciplines are also highly relevant to archaeology.

However, kinship has rarely been actively incorporated within archaeological practice, especially in the archaeology of the Old World, including social archaeology (see Souvatzi

In Blanco-González, A. and Alarcón-García, E. (eds.) 2025, *A Social Archaeology of Kinship in Iberia and Beyond. Recent Multistranded Approaches from aDNA to Household Archaeology.* Leiden: Sidestone Press, pp. 53-70.

et al. in press). Despite abstract references to family, lineage, ancestry, genealogy and so on, most archaeological studies do not actually show how exactly kinship can play a role in connections and transformations, nor do they make suggestions regarding specific ways of detecting such aspects. This is surprising, given the growing interest in the social analysis of action and everyday life, as well as the fact that kinship has material, spatial and temporal dimensions that make it particularly appropriate for archaeological study. The recent fascination with bioarchaeological data, primarily from burial sites (e.g., Ensor 2021: 63-79; Fowler 2022; Cummings and Fowler 2023; Meller *et al.* 2023), addresses only one of the many aspects of kinship, biological relatedness, and therefore provides "only a partial window on what are complex patterns of behavior" (Bickle *et al.* in press; Johnson in press). Studies using other lines of evidence such as settlement and economic systems, material things and activity patterns that specifically address questions of kinship as a major factor in settlement organisation and history remain few (e.g., Souvatzi 2017; Hrnčíř *et al.* 2020; Ensor 2021). At the same time, much of the archaeology that uses these more "traditional" methods –i.e., settlement and material data– has largely continued to view kinship through stereotypical assumptions and earlier functionalist equations of house typologies with family typologies or has simply side-lined kinship in favour of other social categories and analytical units.

In my opinion, this phenomenon has deeper root. I believe that archaeological practice in general –as opposed, perhaps, to theory– is marked by a series of contradictions and by uncritical adoption of inherited models that impede an understanding of kinship as a social process.

3.2 Theoretical limitations and potential

The first contradiction relates to the approaches to space and time, and especially, to the habitual treatment of space and time as two separate categories rather than as interrelated. While the dialectical relationship of space and time has been convincingly argued for in several theoretical publications, it has rarely been applied in actual empirical analysis, in which the two concepts are usually examined separately, with most archaeological works focusing either on space or on time (see Souvatzi 2012a; Souvatzi and Hadji 2014). However, space and time both are fundamental in the formation and transformation of social relations. Kinship, like all social relations and all historical processes, is intrinsically both temporally and spatially contingent.

Mainstream space-based archaeological arguments about kinship, and particularly, American-influenced processual approaches are characterised by a continuing presence of functionalism. This is evident, among other things, in the view of space as the backdrop to human actions, activities and interactions, and in the automatic associations between spatial and material remains and social realms. Functionalism makes it very difficult to conceptualise either space or kinship as a locus of action and change. The underlying assumption that architecture fixes and stabilises social relations tends to treat settlements, buildings and their residential groups as static and prevents the recognition of variation across space and over time. Post-processual, alternative or social approaches to space, on the other hand, employ a large and eclectic array of other notions and analytical units in order to elicit social relations and practices from the archaeological data, from body to boundaries and social networks, from "House Societies" to landscapes, and from relational ontologies and the agency of things, often without considering the important relationships through which these social categories are constructed. Similarly, much of the growing socio-spatial analysis of houses, communities and everyday life, and particularly the archaeological applications of Lévi-Strauss's (1983, 1987) House Society model, is conducted independently of the issue of kinship.

Thus, despite innovative conceptual and analytical frameworks for space in archaeology and useful insights into social relations, particularly within social approaches, there is considerable compartmentalisation of theories and frames of research and methodologies. The common factor is that there has been little genuine interest in kinship. But with little such interest, much social discourse, particularly the one conducted in the name of "families", "lineages", "kin groups", "ancestors or ancestry", "origins or descent", "corporate groups" and so on, is likely to prove unconvincing in the long run.

The second theoretical problem concerns time and history. Archaeological approaches to time are also characterised by considerable fragmentation, varying theoretical traditions and research agendas, which impedes theorisation of any historical process. Like that of space, the subject of time has also always been topical in archaeology and has been discussed in great depth by many researchers and from various perspectives (see Souvatzi *et al.* 2019a for analytical review and references). However, in much of the archaeological literature, time is often conflated with history, when not treated traditionally as a device for the construction of chronological sequences against which things simply happen. This, in turn, brings us to the conceptualisation of history. The recent epistemological shift in archaeology, starting out with the postmodernist and postcolonial critiques, from the traditional focus on large scales of space and time towards the inclusion of smaller scales has led to the emergence of various bodies of historically oriented archaeological work (Souvatzi *et al.* 2019a, 2019b). However, history is approached obliquely through a variety of concepts and terms, most notably memory, temporality, relationality, transition and transformation, and archaeology is still struggling to find a coherent way to translate the archaeological record into historical evidence. One major theoretical obstacle is the influence of the grand models of social evolution and the Eurocentric view of history as a teleological, goal-directed, linear narrative of "progress". This influence can be seen, among other things, in the persistent archaeological focus on origins, transitions, dramatic take-offs and "revolutions" in order to analyse change. Clearly, there are severe limitations in using such top-down and ahistorical perspectives to develop a comprehensive archaeological theory of kinship as historical process. Apart from disregarding historical context, these perspectives effectively dismiss both non-Western societies and kinship as being outside history. In light of the above, then, it is not surprising that while archaeology is still struggling to find a coherent way to build analytical frames for a historical outlook on prehistoric societies, it has not turned to kinship.

The third theoretical obstacle is the ego-centred nature of many archaeological reconstructions of past social processes and the complete disregard of the issue of morality (Crumley 2017; Souvatzi 2021). Specifically, mainstream archaeological theory may place too much emphasis on individual action and on self-interest and pragmatism, focusing upon divisions and antagonism and privileging social asymmetry and domination, and the links between such relationships and the wider economy. Agency-based approaches have not always managed to extricate themselves from economistic thinking and have also tended to take individuals or individual social categories rather than groups as organisational or analytical units (e.g., see critical discussions in David 2001; Whittle 2003: 9-14, 51-52; Souvatzi 2008: 38-40), merely reorientating the focus from corporate groups to individual practices. This way of thinking presumes a fundamentally amoral human essence that leaves little space for altruism, solidarity, emotion and ethical notions limiting self-interest. It also fails to explain co-operative behaviour that does not conform to some straightforward economic logic.

However, what makes kinship different from other logics is that it is composed of both individual interests and group interests in a rich dialectic. Tensions, divisions, conflicts and inequalities are important themes of kinship relationships, but they are not the only ones. Kinship relationships are also about unities and the dialectical interweaving between the political economy and the moral economy. The discussion of morality and of the moral economy in anthropology draws attention to the fact that economic benefits, rewards, and ownership are only part of a much larger package that also includes emotion, shared value systems, and restrictions on self-interested or ego-centred behaviour (e.g., Cheal 1989; Wilk 1993; Hann 2001; Zigon 2007; Carrier 2018; Robbins 2021). In prehistoric archaeology, Whittle (2003: 68-69) argues that an idea of a moral community is necessary in order to understand more fully what might have been expected of smaller groupings and individuals who belonged to such a community. I have argued (Souvatzi 2017, 2021) that kinship, in both its political and moral dimension, must account for the social integration and cohesion evidenced in Neolithic communities, in addition to possibly reinforcing heterarchical ties in the long term.

A related, final note concerns the recent fascination with religion. While the role of kinship is overlooked, there is an increasing, cross-disciplinary interest in prehistoric religion, usually within neo-evolutionary frameworks, often aiming to explain co-operative behaviour (e.g., Hodder 2010, 2014, 2018; Laneri 2015; Sterelny 2020). This phenomenon causes further confusion, not least because the term "religion" is poorly and very variously defined, and the arguments are essentially speculative –e.g., see critical discussion in Bernbeck (2013) and Watkins (2019). We need to distinguish morality from religion. The view of religious ideology as a prerequisite for building reciprocity and the emphasis placed by some on religious ideals of transcendence essentially reflect modern Western notions. They may also constitute another version of individualism in archaeological theory. They assume an absence or a weakening of social institutions such that religion is called upon to act as some kind of regulator and to explain co-operative behaviour and collective identities.

I suggest that we need to take an explicit view of kinship as a social and historical construct linked to wider social structures, historical processes and transformations. One that is constituted through human acts, intentions and interactions, and that can be detected archaeologically through people's relations with space and material things and through their networks of interdependence. We need to reconceptualise kinship as an outcome of human agency and as a meaningful way for people to construct their relations to each other and their understandings of the world. As Brad Ensor (2019: 181) has written, looking for inferences for kin groupings "dwelling by dwelling, settlement by settlement, region by region and phase by phase is an empirical approach to construct a multi-scalar understanding of practices over time, thus avoiding normative generalisations that obscure variable group histories".

We should also focus more on the social production of space and the dialectical relationship between kinship and space. Habitation space and the spatial organisation of a community are important because they are deeply intertwined with dynamic social processes. In anthropology, Hamberger (2018) calls for renewed attention to kinship as a logic of structuring space, arguing that what we call "kinship systems" are, in fact, "different topologies of social space" (Hamberger 2018: 536). He also makes the important point that we should reject the idea of kinship as an ontologically prior classification that structures space and objectifies social relations (Hamberger 2018: 528-529). Indeed, if we move beyond the solely functional-typological –or purely symbolic– interpretation of given architectural forms, and their automatic association with pre-given social forms, we can approach space and kinship as something in flux and we can highlight their complex, mutual interplay. It is also important to realise that built space does not passively reflect social organisation, nor is it merely the representation of an unchanging kinship system. Social space is a vehicle for change, it is both the medium for and the outcome of action that fused social relations, and it is constructed by as much as it constructs and reconstructs any social environment.

This brings us to historical process and change, yet another field where the archaeological contribution is particularly crucial. There can be little theoretical or empirical help from anthropology or the other social sciences, because archaeology has the unique ability to witness the long-term sequence of events. It can also be argued that anthropological research has dealt with temporality almost exclusively in the short term. While anthropology still requires accounts that are historically informed, archaeology is in a privileged position to provide a historical perspective for the entire human past. To this end, kinship is one of the most appropriate analytical units. It is a locus where the short-term and the long-term interact and which can enable us to explore linkages between a whole range of spatial and temporal relationships.

3.3 The interplay of kinship, social space and history in the Neolithic

In this second part of the chapter, I offer some examples of how a focus on kinship may highlight new meanings for old patterns and may lead to new interpretations concerning the "big picture", using settlement data from Neolithic Greece and Turkey, spanning from the eighth to the fifth millennium BCE.

In the Neolithic of Greece and Turkey –and broadly of the whole of Europe and Southwest Asia– the most characteristic feature is the central importance of the household or the residential group and the village community. Thousands of settlements have been identified to date, containing substantial architectural remains, and there is little monumentality or funerary sites so far outside the settlement. This means that virtually all the remarkable abundance and variety of facilities, finds and food remains has been recovered from settlements, often *in situ*, together with a fairly considerable quantity of burial data. Another characteristic element is the complexity and diversity manifested in most types of the archaeological record, from settlement types and patterns to material culture, including the early development of craft-specialisation and long-distance exchange. A third significant feature is the presence of large-scale architectural works, including spatial demarcation, segmentation and boundaries, ranging from concentric stone enclosures to perimeter ditches, single or multiple.

Settlement types range from long-term restricted anthropogenic habitation mounds or *tells*, resulting from the vertical superimposition of closely spaced houses and compacted settlement layouts and the accumulation of habitation layers over hundreds or thousands of years, to comparatively shorter term, flat and horizontally shifting sites with widespread buildings and extensive open spaces, better known from Greece. There is a long discussion in prehistoric archaeology regarding the social formation of tells and non-tell settlements, revolving mostly around economic issues and the notions of continuity and social memory –see Souvatzi (2020) for details and references– while kinship remains the most heavily neglected factor. I have argued elsewhere (Souvatzi 2017) that these two different habitation patterns may well be due to different kinship systems. Here I would like to focus on tells, a settlement type common in the two regions under study, as the material expression of kinship relations and of history.

3.3.1 Habitation mounds (tells) and social relations

A key aspect of tells is that they combine two main means of tracing kinship –namely, co-residence and genealogy– in the same settlement form. The ancestral social space that covers a whole mound can also be seen as the material representation of stable lineages. Many tells with their ordered, and often segmented or otherwise partitioned, layouts, their abundance of agricultural surplus collectively and individually, and the frequent presence of central focal spaces point to unilineal descent groups. Long sequences of rectilinear and free-standing houses of a more or less uniform size and analogous contents seem to indicate conjugal –not to be equated with "nuclear"– family households, with variations regarding patrilocality and matrilocality. Kinship relationships, concepts of relatedness, understandings of kinship and identities were further constructed and realised through the day-to-day interaction in domestic and more public or communal spaces, the daily repetition of activities, the production, consumption and circulation of material products, the intra-site burials and rituals, and generally the shared experience of lived social space over hundreds of years. Through unilineal descent, households were linked to each other and to those preceding and those succeeding, were committed to certain forms of behaviour and enjoyed particular relationships, rights and duties. In addition, the recursive relationship of architecture with unilineal descent organisation may well be the main reason for the creation and maintenance of a specific and consistent settlement plan which formed over the years the social landmarks that are now known as *tells*.

However, there is no reason to assume that tells or their residential groups remained static and stable over time. The evidence shows that underneath the bigger picture of structure and consistency, shifts in intra-site organisation over time are standard for many tells. For example, in Greece, at Dikili Tash the average dwelling floor size (60 m^2) falls between the cross-cultural indicators for matrilocality (more than 80 m^2) and those for patrilocality (less than 43 m^2). The formal settlement layout of the later Neolithic phases, with large rectilinear post-framed dwellings arranged in regular rows, separated by narrow lanes and containing a rich and broadly analogous inventory of features and finds, points to a matrilineal descent group. One dwelling with three rooms of equal size, separate entrances and an almost identical internal organisation and range of material culture and facilities is compatible with matrilocal dwellings housing multiple inter-related households. Middle Neolithic Sesklo combines a tell 8.5 m high (Sesklo A) (Fig. 3.1) and a flat settlement spread below (Sesklo B). Sesklo A is distinguished by long and successive sequences of free-standing, small and rectangular buildings (up to 50 m^2) arranged around small courtyards or squares, a cross-cultural pattern of patrilocality. At Sesklo B, below the tell, the buildings were less long-lived and not free-standing but partly attached, giving the impression of complexes. The spatial organisation here seems rather random, implying bilocal households and bilateral descent. It seems likely that the division of Sesklo into two distinct spatial sectors (Sesklo A and B) might suggest affinal relationships between these two sectors rather

Figure 3.1. The tell settlement of Sesklo (Thessaly, Greece).

than a single lineage for the entire site. It also suggests the co-functioning of two levels of identity, one with patrilineal descent group and another with more emphasis on bilateral descent, respectively. Furthermore, both in the tell and in the flat component of Sesklo, the process of structuring and restructuring domestic space over the different building phases seems to have been incessant, suggesting that important re-organisation of internal areas and residential groups was played out against the long-term stability of the village (see Souvatzi 2008: 98-101 for details). Changes in village layouts reflect the development of new or the modification of existing social institutions. For example, at some tell or tell-like sites, such as Mandra in Thessaly, the small, scattered huts of the initial phases were later replaced by solid, above-ground rectangular buildings (Toufexis 2017: 42-124).

In Neolithic Turkey, Aşikli Höyük and Çatalhöyük East exemplify the merging of smaller separate villages into a single large, nucleated settlement, with each village's discrete social groups occupying different segments of the new settlement. These segments, and generally relationality at Çatalhöyük have been variously and vaguely identified as: "clustered neighbourhoods" (Düring 2007), "history houses" (Hodder and Pels 2010; Hodder 2019), "corporate kin-groups" (Carleton *et al.* 2013), "religious sodalities" and "flexible networks" (Mills 2014), "multi-family house clusters" (Kuijt 2018), "affiliations" (Mazzucato 2019), and so on. At a recent conference entitled *Contextualising the Neolithic in the Konya Plain*, held in Ankara in December 2023, Hodder (2023) also introduced the concepts of "neighbourhoodness", "nested structures" and "radial wedges". Overall, research at Çatalhöyük has focused on cross-cutting networks using a range of vague terms, systematically avoiding the identification of kin groups and looking to other factors as important structuring principles. This is partly due to bioarchaeological analyses based on dental morphology of the human remains buried at the site (e.g., Pilloud and Larsen 2011; Hillson *et al.* 2013) or using ancient genomes from Çatalhöyük among other Neolithic Anatolian sites (Yaka *et al.* 2021), which have found non-biological associations among burials in dwellings. This led to the conclusion that society was largely not kin-based but instead focused on the House, in the Lévi-Straussian sense of the term. However, these arguments clearly misconstrue kinship as solely biological relatedness, an equation that has been deconstructed in anthropology decades ago. At the same time, there has been a turn to religion in order to explain the marked site-wide connections. For example, it has been proposed that religious sodalities created affiliations between buildings and bridged different parts of the mound (Mills 2014).

Current kinship analysis conducted by Bradley Ensor and myself (Ensor and Souvatzi 2022) indicated that: a) the segments of the East Mound are merely spatial separations of residential groups; b) the rather haphazardly dispersed residential groups at any given time suggest bilocal families and bilateral descent rather than lineages; and c) the so-called "history houses" merely reflect the expectations for stem families reoccupying dwelling locations within the estates of bilocal groups. Affiliations and identities at Çatalhöyük were negotiated strategies using kindred and other networks rather than defined by exclusive corporate group memberships. Incidentally, this conclusion fits better with the overall picture of the mortuary practices, and particularly the frequent dismemberment of the human body and the movement of body parts around the site. Both practices imply that the burials constituted a process of constant shaping and re-shaping of identities. Interestingly, recent archaeogenomic analysis from 200 houses, announced at the aforementioned conference (Somel 2023), found no evidence of genetic continuity in succeeding buildings as well as very little genetic diversity across the site over time, which implies endogamy.

3.3.2 Kinship and concentricity

As another example of considering kinship as a major organisational principle, I would like to discuss another part of my current research: the social significance of a particular and very intriguing habitation pattern, the circular or concentric settlement, i.e., a settlement in which buildings are arranged in one or more rings (Souvatzi in press).

The circular village layout has been documented in a wide variety of societies all over the world and has attracted the attention of anthropologists since the nineteenthcentury (see Means 2007 for a detailed overview and analysis). Circular architecture occurs fairly widely in the prehistoric world, ranging from round houses and settlement enclosures to stone circles and passage tombs (e.g., Bradley 2012). However, the focus of research has always been on monumental, megalithic, ceremonial and more public constructions, whereas habitation sites have not gained as much attention. Let us see what happens if consider kinship as a dominant factor for circular habitation patterns.

In Greece, the well-known mound of Dimini (fifth millennium BCE) (Fig. 3.2) with its multiple stone enclosures constructed at different levels and configuring habitation terraces shows both circular organisation and consistent spatial segmentation, as well as a central space which remained unbuilt and unaltered throughout the lifetime of the settlement. My analysis of intra-site material distributions and the spatial patterning of activities indicated that the segments were made up of individual, probably conjugal family households (Souvatzi 2008: 146-149). Alternatively, Dimini exhibits the cross-culturally known circumferential patterning,

Figure 3.2. The concentric Neolithic settlement of Dimini (Thessaly, Greece), fifth millennium BCE.

where the habitation zone is divided into discrete segments likened to pie wedges consisting of different extended families (Dunnel 1983: 147-148; Means 2007: 61-65). In any case, despite its small size –*ca.* 10,000 m², with a population estimated at 200-300 people– the segmented and highly structured site's layout conforms with the cross-cultural indications for multi-lineage settlements (e.g., Chang 1958; Murdock 1967: 48). Palioskala (fifth millennium BCE), another extensively excavated mound 6 m high, is also surrounded and internally divided by multiple stone-built concentric enclosures, with the dwellings limited to the central parts of the mound and habitation extending also outside the mound (Toufexis 2016). Some large dwellings (60 m²) with three rooms or with two hearths (88 m²) might suggest matrilocal households. The central and uppermost part of the site was further surrounded by a pair of small enclosures and was occupied by a large building (72 m²), presumably communal. The layout of the site indicates a unilineal descent group, most likely one lineage. Furthermore, recent geophysical investigations, especially in eastern Thessaly, have provided complete plans of settlements, mostly tells, that were both surrounded and internally divided by a series of concentric enclosures, both ditched and built, often defining habitation sectors or rings (e.g., Kalayci *et al.* 2017; Sarris *et al.* 2017). In some cases, the dwellings were further separated by open zones. In most cases, the central and uppermost part of the mounds was left largely empty.

In Turkey, Aktopraklık B, Hacılar 2, Ilıpınar and some phases of Aşağipinar are all circular settlements, spanning together from the late seventh to the early fifth millennium BCE, encircled by a perimeter wall or a ditch and facing a focal space. At Aktopraklık a ring of small rectangular and single-roomed mudbrick dwellings adjacent with each other and showing a completely uniform size (35-40 m²), construction, interior organisation and contents with each other, surrounded a central courtyard which contained large communal ovens and human burials (Karul and Avcı 2013; Karul 2020). The whole arrangement points to conjugal families belonging to one lineage. Evidence for further ditches and rings of dwellings may suggest spatial segmentation, a concentric overall layout and multiple lineages. Aşağı Pınar (Layer 6) and Ilıpınar show a very similar organisation of adjacent and circularly arranged houses with the one at Aktopraklık, including systems of ditches and palisades surrounding and perhaps also dividing settlement space at Aşağı Pınar (Özdoğan and Schwarzberg 2020) and two-storied houses at Ilıpınar (Roodenberg and Alpaslan Roodenberg 2013).

Although the number of circular or concentric settlements is small –at least at present– compared to the thousands of other habitation sites in Neolithic Greece and Turkey, important observations and suggestions regarding the significance, history and organisation of this intriguing village type can still be made. First, despite differences in the material representation of circularity or concentricity –e.g., radial architectural segments or zones of rings– all the examples discussed here exhibit not just a circular, but indeed a concentric pattern, and they all share the distinction between communal central space and dwellings on the circles or the segments. Second, all these settlements meet the cross-cultural criteria for unilineal descent groups (see Souvatzi 2017 for details; cf., Ensor 2021: 126-130). Third, in all cases, individual households clearly had given up some of their autonomy in favour of some formal links, varying from smaller dwelling clusters that occupied entire rings or parts of them to larger dwelling segments which could have represented multi-lineage settlements (Ensor 2021: 126-131). Fourth, the central open spaces would have at once served for community gathering or rituals and materialised an ideology of social cohesion. Their generally large size ensured that there were few limits placed on participation. Finally, in most cases the concentric pattern was maintained for more than one building phase during the different communities' histories, indicating longevity as well as success over the long run.

I suggest that the circular or concentric habitation pattern acted as a flexible framework that accommodated social organisation at varying scales, minimally at three: the household; the larger social groups along the dwelling rings; and the community as a whole. These unique structures might have also been generated as new mechanisms and contexts for social integration in densely inhabited social landscapes, such as that of Thessaly – for instance, the region of hundreds of Neolithic settlements (e.g., Krahtopoulou 2019; Souvatzi 2022)– ensuring at the same time both the protection and the distribution of resources to their larger groups. In addition, the repeated process of large-scale construction of enclosures and ditches over long periods of time connected different households into a larger corporate group, through exchange of labour and resources at different times. Rather than habitually explained in the traditional terms of defence, fortification, territoriality and social hierarchy, enclosure construction can be seen as the material representation of lines of descent or the spatial mapping of group genealogies, thus, as social history. Significantly, there is no evidence that concentricity –or enclosures, for that matter– in the Neolithic emerged as a stepwise progression from egalitarianism to hierarchisation or that it was associated with inter-site hierarchies and the development of central places.

3.3.3 Kinship and social organisation
The Neolithic societies were definitely complex. They show many social and economic elements that are thought, according to the neo-evolutionary models, to characterise only later periods and to be concomitant with political centralisation and hierarchical

organisation, including: agricultural diversification and intensity; large, permanently co-resident and enduring communities; settlement agglomeration and large-scale architecture; and (part-time) craft specialisation and long-distance exchange. Yet, there is no consistent evidence of either intra- or inter-site hierarchy, centralisation or differentiation. Neolithic communities as a whole had a long and successful history of resistance to changes defined as a linear, cumulative processes towards hierarchisation.

So, how were communities held together? How did they succeed to solve tensions, achieve cohesion and remain in coexistence for such remarkably long periods? How was space allocated within such highly structured settlements, given also the clear manifestation of community wide standards in architecture and the spatial structuring of activities within each of them? Who was given the authority to exert a degree of communal rights to production, distribution, storage, and perhaps also, land ownership (Souvatzi 2013a)? On a larger scale, to what level of social organisation can the access to resources and exchange networks –often long-distance– be related? And is all this to be understood solely in terms of practical or economic reasons or is it better perceived primarily in terms of social interaction between people?

The most plausible answer to all these questions is kinship, in both its political and moral dimension. Among non-hierarchical societies, kinship provides the potential and the motives for craft-specialisation, for multiple modes of production and (re)distribution, for co-operation and interaction, and for networks of alliance, exchange and marriage strategies. Humans developed cooperative behaviour and ideas of indirect reciprocity as early as the Palaeolithic. For example, Gamble (2008, 2013) demonstrates how in Upper Palaeolithic societies, kinshipping and hospitality, realised through people's relations with material things and through their material transactions, created social life as well as networks of interdependence involving large-scale mobility and exchange. This enabled them to build larger moral communities later, in the Neolithic to create stronger forms of bonding within and among large, settled populations. Indeed, the formation and maintenance of social relationships and networks of intra- and inter-site interaction seems to have been a major aim in Neolithic social organisation.

3.4 Discussion and conclusions

The theoretical arguments and the archaeological evidence discussed here indicate that we need to identify the material, spatial and temporal dimensions of different kinship groups in archaeology in more specific, contextual and theoretically-informed ways instead of resorting to abstract references and vague interpretations. It seems that some of the long-held inherited models and straightforward associations concerning kinship, space and history are simplistic, and often ethnocentric, and require serious rethinking. A good place to start is the very notion of kinship. It has to be treated as an analytical issue rather than as a supra-contextual given, and such a treatment also entails a concern with its definition and understanding as a social and historical process operating at a variety of time and space scales. A shift from a focus on individualism towards the consideration of co-operation and collective strategies can also be critical. It allows us to see individual relationships reflecting larger social relationships and to consider the interplays between the macro- and micro-levels.

Other social categories and spatio-temporal analytical units such as houses, with all their various interpretative labels, from "history houses" (Hodder and Pels 2010) to "sociological houses" (Hendon 2010: 57), body, identity, personhood, networks, "historical ontologies" and "cycles of political and ritual development" (Robb and Pauketat 2013: 24-30) and so on, are important themes in social analysis. The problem is that some of them work either in the very long term or in the very short term, others are ethnographically or historically over-specific and therefore not of wide applicability –such as Lévi-Strauss' (1983, 1987) House Society model (see critiques in Ensor 2011 and Souvatzi 2017)– and yet others tend to abstraction. More importantly, they cannot be fully understood without considering the active role of kinship in the construction of social relations, agencies and histories. In my previous work I have focused on the household as a social process and as an analytical means through which we can interpret social organisation from the bottom up (Souvatzi 2008, 2012b, 2013b), but I have encountered many examples which suggest that individual households were interconnected into larger socio-economic groupings and depended on wider social institutions. I also realised that an archaeological approach to kinship is crucial to the wider kinship research. Anthropology and the other social sciences can provide important insights into the different uses and meanings of kinship in contemporary societies, but ethnographic knowledge cannot be simply projected into the past. Archaeology is in a privileged position to discern the "matter-space-time interdependence" (Adam and Kemp 2019: 226), including insights into the factors underlying the historical variations of kinship. It is the materiality, spatiality and temporality of kinship that connects it to key historical phenomena and that creates great conceptual and analytical potential for archaeology.

Acknowledgements

I thank Antonio Blanco González and Eva Alarcón García for inviting me to give a keynote speech at the *Social Archaeology of Kinship in Iberia and Beyond. Recent Multistranded Advances from Household Archaeology to aDNA* workshop at the University of Salamanca, Spain, and for their great hospitality. Thanks also to Brad Ensor for the exchange of ideas and his comments while preparing the talk, and to all the participants who made it such a fruitful and enjoyable meeting.

References

Bernbeck, R. (2013): Religious Revolutions in the Neolithic? 'Temples' in Present Discourse and Past Practice. In Kaniuth, K., Lohnert, A., Miller, J.L., Otto, A., Roaf, M. and Sallaberger, W. (eds.), *Tempel Im Alten Orient*. Wiesbaden: Harrassowitz, pp. 33-48.

Bickle, P., Van Vleet, K.E., Souvatzi, S., Cintas-Peña, M., Rebay-Salisbury, K., Ensor, B.E., Schauer, P., Khalil, U., Shaw, D. and Hofmann, D. (in press): Moving to stay in (a woman's) place: Was patrilocality the dominant mode of post-marital residence across later European Prehistory? *Current Anthropology*.

Bradley, R. (2012): *The Idea of Order: The Circular Archetype in Prehistoric Europe*. Oxford: Oxford University Press. DOI: 10.1093/oso/9780199608096.001.0001.

Carleton, W.C., Conolly, J. and Collard, M. (2013): Corporate kin-groups, social memory, and "history houses"? A quantitative test of recent reconstructions of social organization and building function at Çatalhöyük during the PPNB. *Journal of Archaeological Science* 40, pp. 1816-1822. DOI: 10.1016/j.jas.2012.11.011.

Carrier, J.G. (2018): Moral economy: What's in a name. *Anthropological Theory* 18(1), pp. 18-35. DOI: 10.1177/1463499617735.

Chang, K.C. (1958): Study of the Neolithic social grouping: Examples from the New World. *American Anthropologist* 60, pp. 298-334. DOI: 10.1525/aa.1958.60.2.02a00080.

Cheal, D. (1989): Strategies of resource management in household economies: Moral economy or political economy? In Wilk, R. (ed.), *The Household Economy: Reconsidering the Domestic Mode of Production*. Boulder, CO: Westview Press, pp. 11-22.

Crumley, C. (2017): Assembling conceptual tools to examine the moral and political structures of the past. *The SAA Archaeological Record* 17, pp. 22-24.

Cummings, V. and Fowler, C. (2023): Materialising descent: Lineage formation and transformation in Early Neolithic Southern Britain. *Proceedings of the Prehistoric Society* 89, pp. 1-21. DOI: 10.1017/ppr.2023.2.

David, B. (2000): An agency of choice? Review of Agency in Archaeology. *Cambridge Archaeological Journal* 11, pp. 270-271. DOI: 10.1017/S0959774301230169.

Dunnell, R. (1983): Aspects of the spatial structure of the Mayo Site (15-JO-14) Johnson County, Kentucky. In Dunnell, R.C. and Grayson, D.K. (eds.), *Lulu Linear Punctated: Essays in Honor of George Irving Quimby*. Michigan: University of Michigan, pp. 109-165. DOI: 10.3998/mpub.11395134.

Düring, B.S. (2007): Reconsidering the Çatalhöyük community: From households to settlement systems. *Journal of Mediterranean Archaeology* 20(2), pp. 155-182. DOI: 10.1558/jmea.v20i2.155.

Ensor, B.E. (2011): Kinship theory in archaeology: From critiques to the study of transformations. *American Antiquity* 76(2), pp. 203-227. DOI: 10.7183/0002-7316.76.2.203.

Ensor, B.E. (2019): Prehistoric histories of Hohokam kin groups. In Souvatzi, S., Baysal, A. and Baysal, E.L. (eds.), *Time and History in Prehistory*. London and New York: Routledge, pp. 172-191.

Ensor, B.E. (2021): *The not very patrilocal European Neolithic. Strontium, aDNA, and archaeological kinship analyses*. Oxford: Archaeopress.

Ensor, B.E. and Souvatzi, S. (2022): Kinship at Çatalhöyük. Paper presented at the *28th European Association of Archaeologists Annual Meeting* (Budapest, 31 August-3 September 2022).

Fowler, C. (2022): Social arrangements. Kinship, descent and affinity in the mortuary architecture of Early Neolithic Britain and Ireland. *Archaeological Dialogues* 29, pp. 67-88. DOI: 10.1017/S1380203821000210.

Gamble, C. (2008): Kinship and material culture: Archaeological implications of the global human diaspora. In Allen, N.J., Callan, H., Dunbar, R. and James, W. (eds.), *Early Human Kinship: From Sex to Social Reproduction*. Oxford: Blackwell, pp. 27-40. DOI: 10.1002/9781444302714.

Gamble, C. (2013): Deep time, history and the human-hominin imagination. In Robb, J. and Pauketat, T.R. (eds.), *Big Histories, Human Lives: Tackling Problems of Scale in Archaeology*. Santa Fe: School for Advanced Research Press, pp. 57-75.

Hamberger, K. (2018): Kinship as logic of space. *Current Anthropology* 59(5), pp. 525-548. DOI: 10.1086/699736.

Hann, C. (2001): From Volksgeist to radical humanism: Culture and value in economic anthropology. *Reviews in Anthropology* 30, pp. 1-30.

Hendon, J.A. (2010): *Houses in a Landscape: Memory and Everyday Life in Mesoamerica*. Durham and London: Duke University Press.

Hillson, S.W., Larsen, C.S., Boz, B., Pilloud, M.A., Sadvari, J.W., Aragwal, S.C., Glencross, B., Beauchesne, P., Pearson, J.A., Ruff, C.B., Garofalo, E.M., Hager, L.D. and Haddow, S.C. (2013): The human remains I: Interpreting community structure, health and diet in Neolithic Çatalhöyük. In Hodder, I. (ed.), *Humans and Landscapes in Çatalhöyük: Reports from the 2000-2008 Seasons*. London: British Institute at Ankara and Cotsen Institute of Archaeology Press, pp. 339-396.

Hodder, I. (ed.) (2010): *Religion in the Emergence of Civilization: Çatalhöyük as a Case Study*. Cambridge: Cambridge University Press.

Hodder, I. (ed.) (2014): *Religion at Work in a Neolithic Society: Vital Matters*. Cambridge: Cambridge University Press.

Hodder, I. (ed.) (2018): *Religion, History and Place in the Origin of Settled Life*. Boulder: University Press of Colorado.

Hodder, I. (2018): Contested history-making as part of the building of social networks at Neolithic Çatalhöyük, Turkey. In Souvatzi, S., Baysal, A. and Baysal, E.L. (eds.), *Time and History in Prehistory*. London and New York: Routledge, pp. 250-262. DOI: 10.4324/9781315531854-14.

Hodder, I. (2023): New results regarding social organization at Çatalhöyük. Paper presented at the conference *Contextualizing the Neolithic: Regional Approaches to Sedentism and Domestication in the Konya Plain (British Institute at Ankara and Bilkent University, Ankara, 8-10 December 2023)*.

Hodder, I. and Pels, P. (2010): History houses: A new interpretation of architectural elaboration at Çatalhöyük. In Hodder, I. (ed.), *Religion in the Emergence of Civilization: Çatalhöyük as a Case Study*. Cambridge: Cambridge University Press, pp. 163-186.

Hrnčíř, V., Vondrovský, V. and Květina, P. (2020): Post-marital residence patterns in LBK: Comparison of different models. *Journal of Anthropological Archaeology* 59, p. 101190. DOI: 10.1016/j.jaa.2020.101190.

Johnson, K. (in press): Disentangling kinship from biogenetic relatedness in bioarchaeology. In Souvatzi, S., Bickle, P. and Cveček, S. (eds.), *Prehistoric Kinship: Contemporary Perspectives in Archaeology and Bioarchaeology*. Cambridge: Cambridge University Press.

Kalayci T., Simon, F.X. and Sarris, A. (2017): A Manifold Approach for the Investigation of Early and Middle Neolithic Settlements in Thessaly, Greece. *Geosciences* 7(3), p. 79. DOI: 10.3390/geosciences7030079.

Karul, N. and Avcı, M. (2013): Aktopraklık. In Özdoğan, M., Basgelen, N. and Kuniholm, P. (eds.), *The Neolithic in Turkey: Northwestern Turkey and Istanbul*. Istanbul: Archaeology and Art Publications, pp. 45-68.

Karul, N. (2020): Living in an Enclosed Settlement. Settlement Pattern and Social Organization in Aktopraklık. In Tasič, N., Urem-Kotsou, D. and Burič, M. (eds.), *Making Spaces into Places: The North Aegean, the Balkans and Western Anatolia in the Neolithic*. Oxford: British Archaeological Reports, pp. 225-235.

Krahtopoulou, N. (2019): Archaeology in Greece 2018-2019: Current approaches to the Neolithic of Thessaly. *Archaeological Reports* 65, pp. 73-85. DOI: 10.1017/S0570608419000048.

Kuijt, I. (2018): Material Geographies of House Societies: Reconsidering Neolithic Çatalhöyük, Turkey. *Cambridge Archaeological Journal* 28(4), pp. 565-590. DOI: 10.1017/S0959774318000240.

Laneri, N. (ed.) (2015): *Defining the Sacred: Approaches to the Archaeology of Religion in the Near East*. Oxford: Oxbow. DOI: 10.2307/j.ctvh1dspq.

Lévi-Strauss, C. (1983): *The Way of the Masks*. London: Jonathan Cape.

Lévi-Strauss, C. (1987): *Anthropology and Myth: Lectures 1951-1982*. Oxford: Blackwell.

Mazzucato, C. (2019): Socio-Material Archaeological Networks at Çatalhöyük: A Community Detection Approach. *Frontiers in Digital Humanities* 6, paper 8. DOI: 10.3389/fdigh.2019.00008.

Means, B.K. (2007): *Circular Villages of the Monongahela Tradition*. Tuscaloosa: The University of Alabama Press.

Meller, H., Krause, J., Haak, W. and Risch, R. (eds.) (2023): *Kinship, sex, and biological relatedness. The contribution of archaeogenetics to the understanding of social and biological relations*. Heidelberg: Propylaeum. DOI: 10.11588/propylaeum.1280.c18044.

Mills, B.J. (2014): Relational networks and religious sodalities at Çatalhöyük. In Hodder, I. (ed.), *Religion at Work in a Neolithic Society: Vital Matters*. Cambridge: Cambridge University Press, 159-186.

Murdock, G.P. (1967): *Ethnographic Atlas*. Pittsburgh: University of Pittsburgh Press.

Özdoğan, E. and Schwarzberg, H. (2020): Contextualising the Neolithic house: A view from Aşağı Pınar in Eastern Thrace. In Tasič, N., Urem-Kotsou, D. and Burič, M. (eds.), *Making Spaces into Places: The North Aegean, the Balkans and Western Anatolia in the Neolithic*. Oxford: British Archaeological Reports, pp. 211-224.

Pilloud, M.A. and Larsen, C.S. (2011): "Official" and "practical" kin: Inferring social and community structure from dental phenotype at Neolithic Çatalhöyük, Turkey. *American Journal of Physical Anthropology* 145(4), pp. 519-530. DOI: 10.1002/ajpa.21520.

Robb, J. and Pauketat, T.R. (2013): From moments to millennia: Theorising scale and change in human history. In Robb, J. and Pauketat, T.R. (eds.), *Big Histories, Human Lives: Tackling Problems of Scale in Archaeology*. Santa Fe: School for Advanced Research Press, pp. 3-33.

Robbins, J. (2021): When did it become hard to be good? Axial dynamics and the problem of the moral self. In Robbins, J., Souvatzi, S. and Strathern, A. (eds.), *Society and Morality in Eurasia: from Prehistory to the Present Day*. Halle/Saale: Max Planck Institute for Social Anthropology, pp. 27-38.

Roodenberg, J. and Alpaslan Roodenberg, S. (2013): Ilipinar and Menteşe: Early farming communities in the Eastern Marmara. In Özdoğan, M., Basgelen, N. and Kunihom, P. (eds.), *The Neolithic in Turkey: Northwestern Turkey and Istanbul*. Istanbul: Archaeology and Art Publications, pp. 69-91.

Sarris, A., Kalayci, T., Donati, J., Garcia, C.C., Manataki, M., Cantoro, G., Karampatsou, G., Kalogiropoulou, E., Argyriou, N., Dederix, S., Manzetti, C., Nikas, N., Vouzaxakis, K., Rondiri, V., Arachoviti, P., Almatzi, K., Efstathiou, D. and Stamelou, E. (2017): Opening a New Frontier in the Neolithic Settlement Patterns of Eastern Thessaly, Greece. In Sarris, A., Kalogiropoulou, E., Kalayci, T. and Karimali, L. (eds.), *Communities, Landscapes, and Interaction in Neolithic Greece*. Ann Arbor: Berghahn Books, pp. 27-48. DOI: 10.2307/j.ctvw049k3.9.

Somel, M. (2023): Recent archaeogenomics results from Çatalhöyük: Biological ties and mobility. Paper presented at the conference *Contextualizing the Neolithic: Regional Approaches to Sedentism and Domestication in the Konya Plain. (British Institute at Ankara and Bilkent University, Ankara, 8-10 December 2023)*.

Souvatzi, S. (2008): *A Social Archaeology of Households in Neolithic Greece: An Anthropological Approach.* New York/Cambridge: Cambridge University Press.

Souvatzi, S. (2012a): Space, place, architecture: A major meeting point between social archaeology and anthropology? In Shankland, D. (ed.), *Archaeology and Anthropology: Past, Present and Future.* London: Association of Social Anthropologists, pp. 73-196.

Souvatzi, S. (2012b): Between the individual and the collective: Household as a social process in Neolithic Greece. In Parker, B.J. and Foster, C.P. (eds.), *Household Archaeology: New Perspectives from the Near East and Beyond.* Winona Lake: Eisenbrauns, pp. 15-43.

Souvatzi, S. (2013a): Land tenure, social relations and social landscapes. In Relaki, M. and Catapoti, D. (eds.), *An Archaeology of Land Ownership.* London: Routledge, pp. 21-45.

Souvatzi, S. (2013b): Diversity, homogeneity and the transformative properties of the house in Neolithic Greece. In Hofmann, D. and Smyth, J. (eds.), *Tracking the Neolithic House in Europe: Sedentism, Architecture and Practice.* New York: Springer, pp. 45-64. DOI: 10.1007/978-1-4614-5289-8_3.

Souvatzi, S. (2017): Kinship and Social Archaeology. *Cross-Cultural Research* 51(2), pp. 172-195. DOI: 10.1177/1069397117691028.

Souvatzi, S. (2020): Tells (and flat sites) as social agents: A view from Neolithic Greece. In Blanco-González, A. and Kienlin, T.L. (eds.), *Current Approaches to Tells in the Prehistoric Old World.* Oxford: Oxbow Books, pp. 125-138. DOI: 10.2307/j.ctv13pk5j9.11.

Souvatzi, S. (2021): Morality, egalitarianism and social complexity in the early farming societies. In Robbins, J., Souvatzi, S. and Strathern, A. (eds.), *Society and Morality in Eurasia: From Prehistory to the Present Day.* Halle/Saale: Max Planck Institute for Social Anthropology, pp. 15-26.

Souvatzi, S. (2022): The physical and social landscape of Neolithic Platia Magoula Zarkou. In Alram-Stern, E., Gallis, K. and Toufexis, G. (eds.), *Platia Magoula Zarkou. The Neolithic Period. Environment, Stratigraphy, Tools, Figurines and Ornaments.* Vienna: Austrian Academy of Sciences Press, pp. 590-608.

Souvatzi, S. (in press): Concentricity, circularity and kinship in the first farming communities: Examples from Neolithic Greece and Turkey. In Souvatzi, S., Bickle, P. and Cveček, S. (eds.), *Prehistoric Kinship: Contemporary Perspectives in Archaeology and Bioarchaeology.* Cambridge: Cambridge University Press.

Souvatzi, S., Baysal, A. and Baysal, E.L. (2019a): Is there prehistory? In Souvatzi, S., Baysal, A. and Baysal, E.L. (eds.), *Time and History in Prehistory.* London and New York: Routledge, pp. 1-27.

Souvatzi, S., Baysal, A. and Baysal, E.L. (eds.) (2019b): *Time and History in Prehistory.* London: Routledge. DOI: 10.4324/9781315531854.

Souvatzi, S., Bickle, P. and Cveček, S., (eds.) (in press): *Prehistoric Kinship: Contemporary Perspectives in Archaeology and Bioarchaeology.* Cambridge: Cambridge University Press.

Souvatzi, S. and Hadji, A. (eds.) (2014): *Space and Time in Mediterranean Prehistory.* New York: Routledge.

Sterelny, K. (2020): Religion: Costs, signals, and the Neolithic transition. *Religion, Brain and Behavior* 10(3), pp. 303-320. DOI: 10.1080/2153599X.2019.1678513.

Toufexis, G. (2016): Palioskala: A Late Neolithic, Final Neolithic and Early Bronze Age settlement in the eastern Thessalian plain (central Greece). In Tsirtsoni, Z. (ed.),

The Human Face of Radiocarbon: Reassessing Chronology in Prehistoric Greece and Bulgaria, 5000-3000 cal BC. Lyon: Maison de l'Orient et de la Méditerranée, pp. 361-380.

Toufexis, G. (2017): *Oikistiki Drastiriotita kai Organosi tou Horou stous Oikismous tis Neoteris Neolithiki sti Thessalia: Paradeigmata apo tous Oikismous ston Profiti Ilia Mandras, Macrychori, Galene and Rachmani.* PhD Dissertation, University of Thessaly.

Watkins, T. (2019): When do human representations become superhuman agents? In Becker, J., Beuger, C. and Müller-Neuhof, B. (eds.), *Iconography and Symbolic Meaning of the Human in Near Eastern Prehistory.* Vienna: Austrian Academy of Sciences Press, pp. 225-235.

Whittle, A. (2003): *An Archaeology of People. Dimensions of Neolithic Life.* London: Routledge.

Wilk, R.R. (1993): Altruism and Self-interest: Towards an Anthropological Theory of Decision Making. *Research in Economic Anthropology* 14, pp. 191-212.

Yaka, R., Mapelli, I., Kaptan, D., Doğu, A., Chyleński, M., Dilek Erdal, Ö., Koptekin, D. *et al.* (2021): Variable kinship patterns in Neolithic Anatolia revealed by ancient genomes. *Current Biology* 31, pp. 2455-2468. DOI:10.1016/j.cub.2021.03.050.

Zigon, J. (2007): Moral breakdown and the ethical demand: A theoretical framework for an anthropology of moralities. *Anthropological Theory* 7(2), pp. 131-150. DOI:10.1177/1463499607077295.

Part II

BIOMOLECULAR APPROACHES TO KINSHIP

Inferring Kinship in Ancient Populations: Advances Using Paleogenomic Techniques

Rosa Fregel

Department of Biochemistry, Microbiology, Cell Biology
and Genetics, University of La Laguna, Spain,
rfregel@ull.edu.es

Abstract

In the last decade, paleogenomics has been demonstrated to be a useful tool for studying the past, especially when applied within multidisciplinary research projects. One of the aspects of human populations that can be explored in detail using ancient DNA is "biological kinship" and social structure. Examples of the outcomes of this approach are the study of large extended family within a Neolithic mass grave in Koszyce (Poland) or the genetic characterization of entire necropolis, with relatives connected over several generations, such as the ones performed in the Avar (Carpathian Basin) and the Argar (Iberian Peninsula) societies. Because paleogenomic analyses should be performed in multidisciplinary teams, it has become necessary for researchers working in other disciplines to be familiar with concepts related to kinship inference. In this chapter, I present a brief review of how ancient DNA is recovered from human remains, what limitations palaeogenomic analyses have, and how DNA can be used for estimating kinship.

Keywords: *ancient DNA, paleogenomics, kinship inference, pedigree, identical by descent.*

4.1 Introduction

Since the first ancient genome was published in 2010 (Rasmussen *et al.* 2010), paleogenomic techniques have revolutionized the way past human populations are studied. Although initial publications were focused on exploring major episodes of our demographic history (Lazaridis *et al.* 2016), advances in the methods to obtain and analyze ancient DNA (henceforth aDNA) data have allowed researchers to start exploring human populations at a smaller scale. Within those approximations, it has become possible to identify "biological kinship" between individuals with a degree of reliability and sensibility never seen before (Schroeder *et al.* 2019; Villalba-Mouco *et al.* 2022; Gnecchi-Ruscone *et al.*

In Blanco-González, A. and Alarcón-García, E. (eds.) 2025, *A Social Archaeology of Kinship in Iberia and Beyond. Recent Multistranded Approaches from aDNA to Household Archaeology.* Leiden: Sidestone Press, pp. 73-92.

2024). However, to fully understand what those degrees of "biological kinship" can tell us about the social structure of ancient populations, it is necessary to take into consideration their archaeological and historical context and, therefore, paleogenomic studies should be conducted within the framework of highly multidisciplinary teams. This creates the need for researchers working on other disciplines, such as archaeology, history or anthropology, to be familiar with biological concepts related to aDNA analysis and kinship inference. In this chapter, I present a brief review of paleogenomic methodologies, with an emphasis on kinship analysis, in order to provide a reference for those interested in the use of aDNA to establish familial relationships.

Before we delve into the intricacies of aDNA analyses, it is important to clarify that, through this chapter, we are discussing how genetic information is used to infer kinship exclusively understood as biological relatedness. In this way, when aDNA studies conclude that two or more individuals are unrelated, it is only regarding their "biological kinship", whereas other types of cultural kinship relationships are still possible.

4.2 Obtaining ancient DNA

In this section, we are going to start by discussing what aDNA is, what characteristics it has, how it is obtained and the limitations regarding its study. The DNA molecule is composed of two chains that are wound into a spiral. Each chain is a polymer composed of four different subunits or nucleotides: adenine (A), guanine (G), cytosine (C) and thymine (T). These subunits are joined together to form a strand, and two strands are coiled into a spiral to form a DNA molecule. The two DNA strands are annealed in such a way that C and G, and A and T are always paired together (Fig. 4.1A).

4.2.1 Characteristics of aDNA

aDNA can be defined as the genetic information obtained from archaeological or paleontological remains. When an organism is alive, the DNA is protected inside the nucleus –except for mitochondrial DNA that is located within the cytoplasm– where enzymatic processes are in charge of maintaining its integrity. After the death of the organism, the DNA is exposed to the environment and starts to degrade. The degradation process leads to low DNA concentrations, as well as the presence of postmortem damage, which can render DNA unrecoverable.

The most common damage patterns observed in aDNA are fragmentation and chemical modifications. Briefly, fragmentation is caused by the loss of nuclotides, especially G, leading to the breakage of the DNA chain and the formation of single-stranded ends (Fig. 4.1A) (Dabney *et al.* 2013). In these damaged ends, Cs are exposed to being modified by a process that converts them into a different nucleotide (uracil or U) that is similar to thymine and that will cause C to T mutations when DNA is analyzed. Because of the expected postmortem damage, aDNA molecules are extremely short –less than 100 nucleotides– (Fig. 4.1C) and with an accumulation of C to T modifications at both ends of the DNA molecules (Fig. 4.1B).

4.2.2 DNA extraction and sequencing

Ancient DNA can be obtained from different sources, including skeletal remains, mummified tissue, hair and even sediments. However, DNA degradation can hinder the recovery of usable molecular information, so choosing a good source of well-conserved aDNA is key to

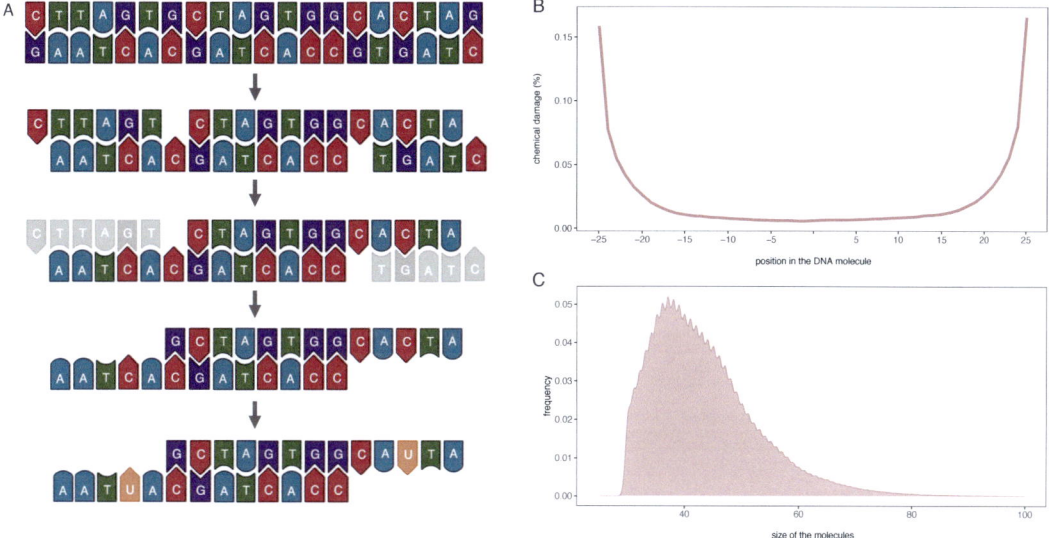

Figure 4.1. DNA degradation. A) Graphical representation of the DNA damage process. First, a section of double-stranded DNA molecule, showing the pairing of A and T, and C and G. Then, the molecule of DNA loses G nucleotides leading to the breakage of the DNA chain. Fragmented chains lead to the creation of single-stranded ends. Cs in the single-stranded ends are now susceptible of being chemically modified. Finally, chemical modification of the Cs leads to the formation of Us, which will eventually produce C to T mutations at the end of ancient molecules during the library preparation process (created with BioRender.com). B) Graphical representation of the chemical modification at the molecule ends expected from aDNA. C) Graph showing the expected size of aDNA molecules.

In this step, both ends of the ancient molecules are bound to synthetic DNA molecules, known as adapters. Finally, libraries are sequenced in a process in which all molecules bound to adapters are read to determine their sequence of nucleotides. The sequencing methodologies used today have the potential of characterizing all the DNA molecules contained in a sample, allowing for the sequencing of the complete genome of individuals, that is, all the DNA contained in their cells.

4.2.3 Data analysis

When sequencing the genomic DNA of an ancient individual, we are recovering tiny fragments of her or his DNA, just like pieces in a puzzle. When the sequencing finishes, we end up having hundreds of millions of DNA molecule sequences. Before we can determine what the aDNA can tell us, we need to reconstruct the ancient individual's genome. For that, we use bioinformatics tools to find the location of each sequence within the reference human genome in a process known as mapping (Fig. 4.2). Once the DNA is mapped to the reference genome, we can start determining the genetic composition of individuals and populations.

As we have seen before, one thing that characterizes aDNA studies is degradation. Because of degradation, when DNA is obtained from ancient human remains, a portion of the sequenced DNA usually comes from the environment, including that of soil bacteria or fungi, and only a small percentage is human endogenous DNA. If the conservation of the human remains is poor, the vast majority of the DNA would come from the environment (> 99.9%). Best-conserved remains can have endogenous DNA ranging between 10-40%. However, when analyzing aDNA from warm and humid areas, the expected percentage of endogenous DNA can be lower than 1%. For that reason, the percentage of endogenous DNA is one of the main limitations of aDNA analyses, as low endogenous contents hinder the recovery of whole genomes. In order to improve the obtention of endogenous DNA in degraded human remains, several enrichment technologies have been designed. These methodologies are directed at capturing human molecules to enrich libraries in endogenous DNA. In this way, we can reduce sequencing costs and improve results.

The quality of the genomic information retrieved from an ancient sample is measured using two values: coverage and depth. Although sometimes these two concepts are used as synonyms, they refer to different qualities of the data. Coverage represents the percentage of the reference genome that has been sequenced in the ancient sample. For example, if we only have been able to sequence half of the nucleotides of the reference genome, then the coverage would be 50% (or 0.5X). In Figure 4.3A, we have been able to read 38 positions out of the 56 nucleotides of the sequence. In that case, coverage would be 67.9% (38/56) which means that we do not know the nucleotide or nucleotides that are present in 32.1% of that sequence. On the other hand, as exemplified in Figure 4.3B, coverage will be 100% and we will have information for all the positions in the sequence. Depth is the average number of times a position in the genome has been read with independent sequences. A depth of 5X would mean that, on average each position of the genome has been read with five independent sequences. For example, in Figure 4.3B, the depth of the first position of the sequence would be 2 and the depth of the second

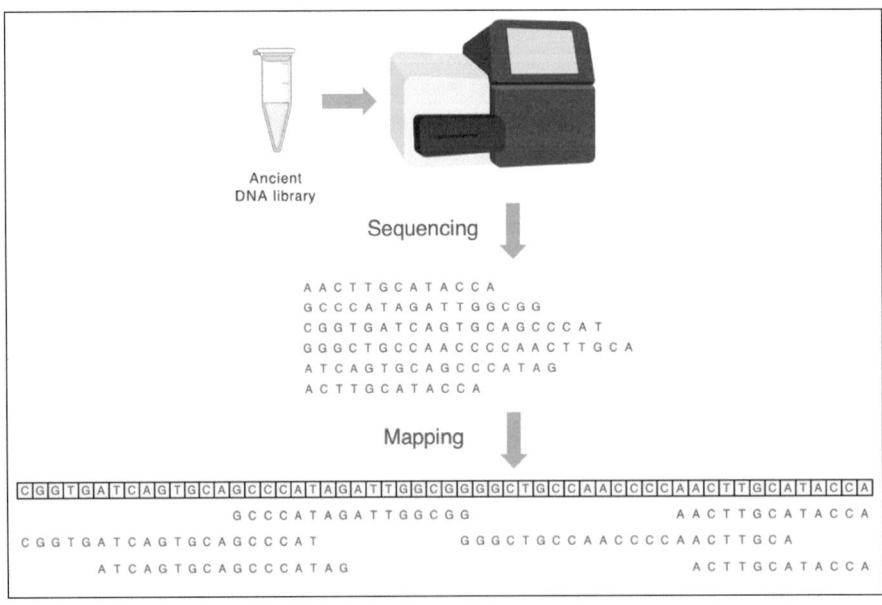

Figure 4.2. Graphical representation of the data generation phase. aDNA libraries are sequenced to generate sequencing data, that are later mapped to the human reference genome (created with BioRender.com).

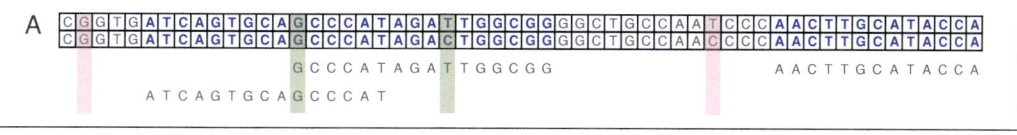

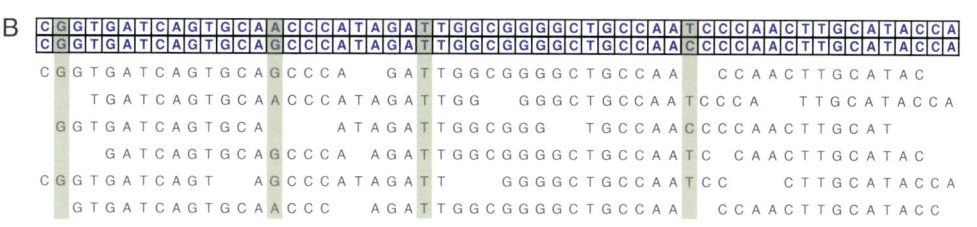

Figure 4.3. Representation of a low-coverage (A) and a medium-coverage (B) genome. In each section, the diploid genome of the individual is depicted in grey squares. Green rectangles correspond to variable sites that have been sequenced in the ancient samples, and red rectangles represent variable sites that have not been sequenced due to low coverage. Read positions in the genome are shown in blue.

one would be 3. Although the aim is obtaining high depth (> 15X) or, more commonly, medium-depth genomes (> 5X)[1], due to degradation, most of the ancient genomes have a coverage below 100% and a depth below 1X. In order to provide information on the quantity of DNA recovered from ancient individuals, coverage and depth should be provided for each sample.

4.2.4 Authentication

When analyzing ancient individuals, it is important to confirm that the DNA originates from the human remains and not from the manipulators or other sources of modern contaminant DNA. For that reason, the detection of postmortem damage in the aDNA is necessary to authenticate results. Usually, this is accomplished by showing that the DNA is fragmented (Fig. 4.1C) and that the DNA has C to T modifications at both ends of the molecules (Fig. 4.1B). As for coverage and depth, information on the level of fragmentation and chemical modification of ancient molecules must be provided to demonstrate the authenticity of the aDNA data.

4.3 Understanding pedigrees

Before we start looking at how to determine biological relatedness, it is important to explain how family relationships are presented in pedigree charts (Fig. 4.4A). A pedigree chart is a representation of genealogical relationships using standardized symbols. Each row of the pedigree chart corresponds to generations within a family. Individuals are represented using geometric shapes; with circles for females and squares for males[2]. Family relationships are indicated using lines: Parents are connected between them using horizontal lines, while parents and offspring are linked using vertical lines. Individuals

---

1   The limits for determining what a low-coverage, medium-coverage and high-coverage genome are not defined, but normally the minimum values are around those cited.

2   When the sex of the individuals is unknown it is represented using a triangle.

that are siblings –and, because of that share the same parents– are also linked using a horizontal line. In order to describe pedigrees, generations are usually identified using Roman numerals and each individual within each generation using Arabic numerals. To refer to a particular individual, you must indicate the Roman numeral of the generation followed by the Arabic number of the individual, separated by a point mark (e.g., I.1).

In the example (Fig. 4.4A), we show a pedigree representing individuals from three different generations. There, individuals I.1 and I.2 are the parents of individuals II.2 and II.3, who in turn have a sibling relationship between them. In that generation, individuals II.1 and II.2, and individuals II.3 and II.4 are the parents of III.1 and III.2, and III.3 and III.4, respectively. In this way, individuals III.1 to III.4 are all the grandchildren of individuals I.1 and I.2.

4.4 Types of DNA analyses

We can use different types of DNA to infer "biological kinship" from different perspectives. Within a human cell, DNA can be found in two different locations: the nuclear DNA (henceforth nDNA) in the nucleus and mitochondrial DNA (henceforth mtDNA) in the cytoplasm. nDNA is composed of linear DNA molecules organized in 23 chromosomes. For each chromosome, we have two copies: one inherited from our mother and one from our father. The chromosomes can be classified into autosomal chromosomes (from 1 to 22) and sexual chromosomes (X and Y). Female individuals inherit a copy of the chromosome X from both their parents, while males inherit a copy of the chromosome X from their mothers and a copy of the chromosome Y from their fathers. mtDNA is a small circular DNA molecule that is located inside the mitochondria found in the cytoplasm of cells. Because of that, mtDNA is inherited from mothers to daughters and sons. Now, we will look at how we can use different types of DNA to gather information on biological relatedness.

4.4.1 Mitochondrial DNA

Because mtDNA is passed on from mothers to daughters and sons, we can use it to determine familial relationships from the maternal side. In the example shown in Figure 4.4B, the female individual I.1 passes her mtDNA to her son (individual II.2) and daughter (individual II.3). In the next generation, only her daughter passes individual I.1's mtDNA to her children. In this way, mtDNA is inherited following the maternal line: from individual I.1 to individual II.3 and, finally, to individual III.4 who will be able to pass it on to her children.

mtDNA has been studied for decades and its distribution around the world has been well characterized (Kivisild 2015). The existing mtDNA variation in human populations is organized into haplogroups, which are groups of mtDNA sequences that share a common origin based on a phylogenetic analysis (van Oven 2015). Briefly, haplogroup names are given using a code of letters and numbers. Major haplogroups are represented with one letter, and more derived lineages are represented by alternating numbers and letters. For example, major haplogroup U is subdivided into U1, U2, U3, U4, U5, U6, U7 and U8. More derived lineages within U1 would then be U1a and U1b, and even more derived lineages would be identified as U1a1 or U1b1.

One thing that is important to consider when analyzing familial relationships using mtDNA is that sharing the same mtDNA does not prove that two individuals are related on the maternal side. Haplogroups can appear in populations at different frequencies, with some of them being very common. In that way, two individuals may have the same

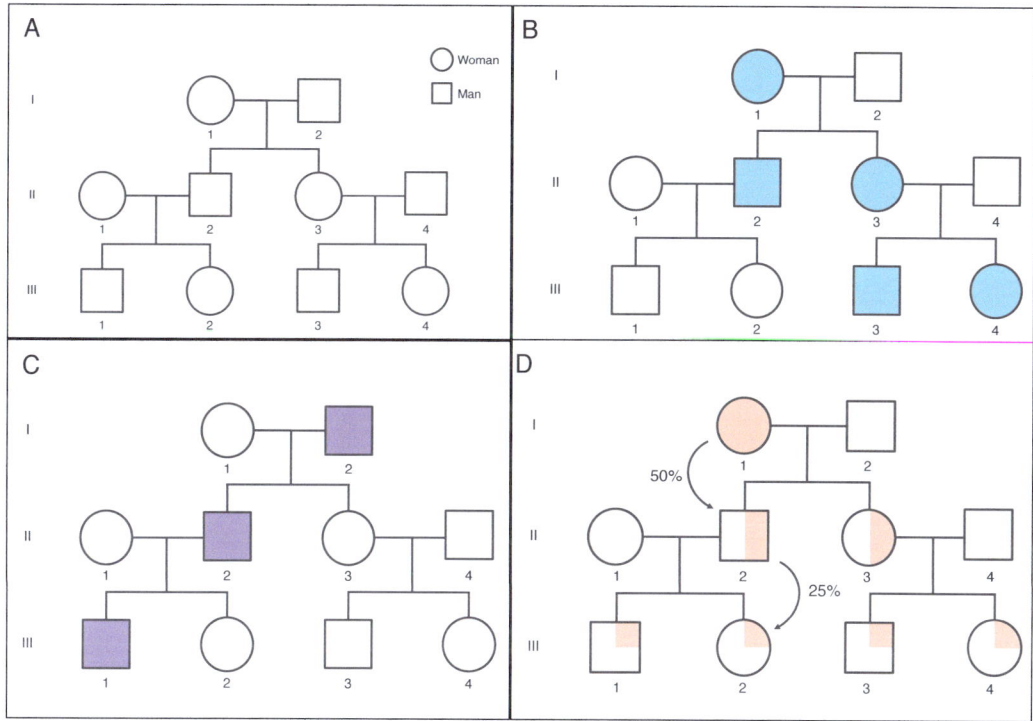

Figure 4.4. Pedigrees. A) Example of a pedigree chart showing family relationships for three generations. B) Pedigree showing the inheritance pattern of a mtDNA lineage (represented in blue). C) Pedigree showing the inheritance pattern of a Y chromosome lineage (represented in violet). D) Pedigree showing the inheritance pattern of the autosomal DNA of individual I.1 (represented in orange).

mtDNA just because it has a relatively high frequency in this population. In that situation, these two individuals would share a common ancestor many generations back, but they would not be directly related. Now, it is a good moment to present two concepts that are important for kinship analysis. When the mtDNA of two individuals is compared and is the same, it can be because they are related from the maternal side. For example, in Fig. 4.4B individuals II.2 and III.3 share the same mtDNA because it comes from the same recent ancestor. When this happens, we say these mtDNA lineages are identical by descent (IBD). On the other hand, as mentioned before, the mtDNA of two individuals can also be the same not because they share a family member, but because this lineage is relatively frequent in the population. In this case, to find a common ancestor we would have to go many generations back. In this case, we say that the mtDNAs of these individuals are identical by state (IBS).

For that reason, when two individuals have the same mtDNA, we can conclude that a maternal relationship for these individuals is possible, but we need additional evidence to confirm it. On the other hand, if two individuals have different mtDNA sequences, we can conclude that they are not related from the maternal side.

4.4.2 The Y chromosome

The Y chromosome is considered to be the paternal counterpart of mtDNA, as it is inherited from fathers to sons. In this way, we can use the Y chromosome to infer a direct paternal relationship between individuals. In Figure 4.4C, individuals I.2's Y chromosome is inherited by his son (II.2) and his grandchild (III.1), following the paternal line. Classification of Y chromosome lineages into haplogroups is done similarly to that already explained for the mtDNA. Also, as we already discussed for the mtDNA, when two individuals have the same Y chromosome, we can determine that a paternal relationship for these individuals is possible, although additional evidence is needed to confirm it. In the same way, we can conclude that two individuals with different Y chromosomes are not related on the paternal side.

4.4.3 Autosomal chromosomes

As explained before, we have two copies of each autosomal chromosome, one of them inherited from our mother and the other one from our father. In this way, autosomal DNA is inherited at 50% from each parent. In Figure 4.4D, we show how individual I.1 passes 50% of her autosomal DNA on to her daughter and son. In the following generation, her grandsons (individuals II.2 and II.3) inherit 25% of individual I.1 autosomal DNA. Finally, her great-grandsons (individuals III.1-III.4) inherit 12.5% of it. In this way, the inheritance of the genetic information of an individual by her or his descendants will diminish over time as "biological kinship" relationships become more and more distant.

4.5 Measuring relatedness for autosomal DNA

Based on the way autosomes are inherited, we can estimate the percentage of DNA that is expected to be shared between two individuals for all possible kinship relationships. For instance, an individual and her/his great-great-grandparents share an average of 6.25% of DNA, while anyone with their third cousins shares just 0.781% (Fig. 4.5).

A way to look at kinship is based on the probability of individuals having the same copy of the DNA at a particular position. For that, we must introduce the concept of alleles. We can define alleles as each of two or more alternative forms of a gene that arise by mutation and are found at the same place on a chromosome. In any position of any chromosome, two different alleles can exist, for example, one with a T and the other with a C. As everyone has two copies of each autosomal chromosome, we all have two alleles for each particular position in our genome, one inherited from our mother and one from our father. Alleles can be named in different ways, but for simplicity, we are going to call them by the initial of the nucleotides of the position that varies between alleles. In the previous example, we will have alleles T and alleles C. In that way, for any individual in the population, these two alleles can be the same (CC or TT) or different alleles (CT).

Now, for each "biological kinship" relationship we can estimate the probability of sharing one allele identical by descent (defined as k1), two alleles identical by descent (k2) or none (k0). In order to understand how alleles are inherited, we are going to use a simple pedigree (Fig. 4.6). As we are only interested in seeing how alleles are inherited, all of them have been identified as "A" and numbered 1 to 2 in the first generation, and with different colors depending on the recent ancestor they are coming from. As we share 50% of our DNA with our parents, we will always have one allele in IBD with them, in such a way all parent-child relationships have a k1 probability of 1, leaving k0 = 0 and k2 = 0. For example, individuals I.1 and II.1 have a mother-son relationship, and they share one allele in IBD (violet A1). For sibling relationships, we have all three possible outcomes: that they share 0, 1 or 2 alleles in IBD. When two siblings share zero alleles in IBD, it is because they

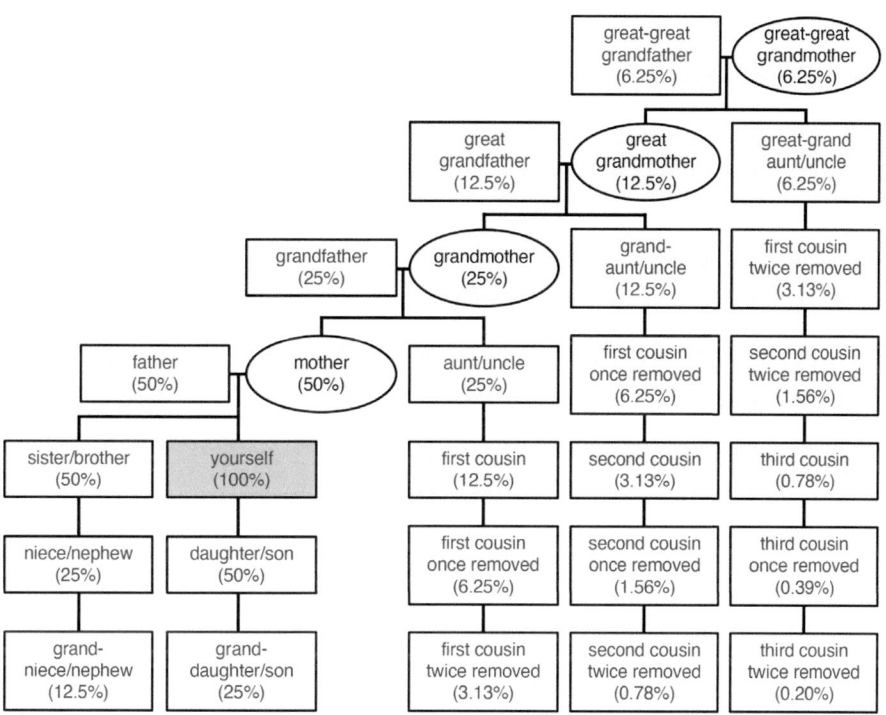

Figure 4.5. Pedigree showing the average percentage of DNA shared by any individual and her/his relatives.

each inherited a different allele from their mother and their father. We have an example of this outcome with individuals II.3 and II.4: individual II.3 inherited allele A1 from his mother and A2 from his father, while II.4 inherited allele A2 from his mother and A1 from his father, having 0 alleles in IBD. On the other hand, when two siblings share one allele in IBD, it is because they each inherited a different allele from one of their parents and the same from the other. This is what happens with siblings III.1 and III.2: they have the same allele A1 from their mother and different alleles from their father. Finally, there is the possibility of sharing both alleles in IBD when they inherit the same alleles from both their parents, as it happens with II.1 and II.2. When probabilities of each outcome are calculated for siblings, we obtain that k0 = 0.25, k1 = 0.50, and k2 = 0.25. We can extend this calculation for each biological relationship. For example, for grandparent-grandchild, we would have k0 = 0.50, k1 = 0.50, and k2 = 0. In the pedigree, we can see that grandparents I.1 and I.2 share either one or zero alleles with their grandchildren III.1 and III.2 (Fig. 4.6).

"Biological kinship" estimation analysis can be presented using the three relatedness coefficients k0, k1 and k2, and the kinship coefficient θ. The kinship coefficient represents the probability that two alleles, each selected at random from two different individuals are IBD. As we already discussed for the mtDNA, two alleles can be identical due to sharing a recent common ancestor, but also due to older common ancestry (Waples *et al.* 2019). For that reason, determining if a particular allele is IBD or not is something that cannot be known just by looking at DNA results. Instead, what can be done is estimate the values of the relatedness coefficients and the kinship coefficient based on the expectations we have of the statistics for different familial relationships as we

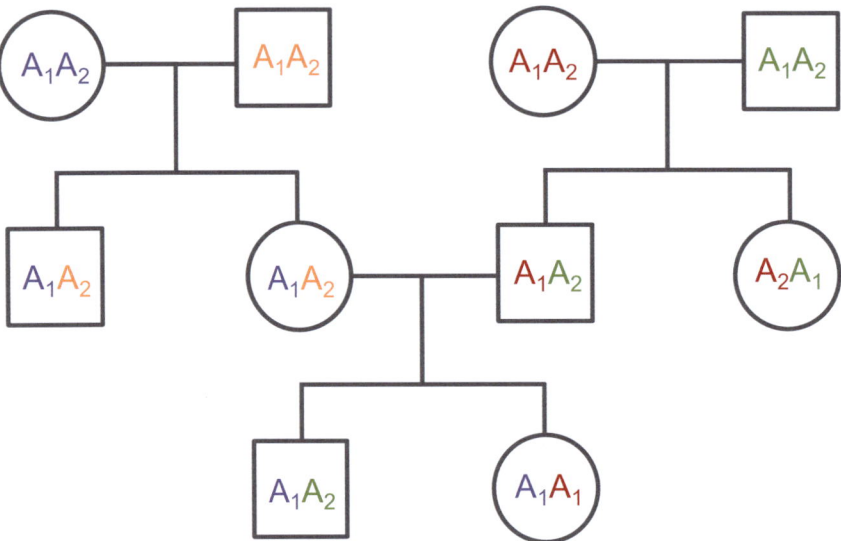

Figure 4.6. Pedigree showing how alleles from autosomal chromosomes are inherited. Alleles from the individuals from the first generation are numbered and shown in different colors to follow how they are passed to the following generations.

discussed before (Hill and Weir 2011). In this way, the kinship coefficient is estimated based on the proportions of the DNA where a pair of individuals share 0, 1 or 2 alleles identical by descent ($\theta = k_1/4 + k_2/2$). For example, parent-offspring and sibling-sibling pairs are expected to have a kinship coefficient of 0.25, while for first cousins the value is expected to be 0.0625.

4.6 Inferring relatedness for ancient individuals

There are several methods for inferring relatedness in ancient individuals (Lipatov *et al.* 2015; Monroy Kuhn *et al.* 2018; Fernandes *et al.* 2021; Žegarac *et al.* 2021, Popli *et al.* 2023; Rohrlach *et al.* 2023; Ringbauer *et al.* 2024). For this review, we are going to explain how one of the most-used methods for inferring "biological kinship" in aDNA studies works. The *Relationship Estimation from Ancient DNA* (*READ*) software was published by Monroy-Kuhn *et al.* (2018) and was aimed at determining "biological kinship" relationships with aDNA genomes with a coverage higher than 10%. As explained before, humans are diploid in such a way that each individual has two copies of all autosomes. For that reason, for every genetic variant, individuals can have the same allele (e.g., AA) or different alleles (e.g., AG). However, because the aDNA data tends to have a low coverage, most of the time it is impossible to determine if an individual has the same allele or different alleles in a particular position. If we have a look at the low-coverage genome in Figure 4.3A, we can see that the positions, if covered, are mostly read by only one molecule. In this scenario, it is not possible to discern if individuals have the same allele in both chromosomes or two different alleles in each one of them. For that reason, we usually work with "pseudo-haploid" genomes in which we just choose the only allele available if the coverage is 1, or one allele at random if the coverage is higher than 1.

Because genomes are low coverage and we are dealing with pseudo-haploid genomes, the *READ* method is not aimed at estimating k1, k2 and k0 as we previously discussed.

Instead, *READ* infers kinship relationships by determining the proportion of non-matching alleles (P0), an estimator also known as the pairwise mismatch rate (PMR). The P0 value is calculated for each pair of individuals and reflects the proportion of bases in which the two individuals have different alleles –for example, a P0 value of 50% would mean that individuals have different alleles in half the positions we are examining. The P0 value is calculated in sections or windows in which the information obtained for each of the individuals is compared. In Figure 4.7, we are comparing individual 1 (upper side) and individual 2 (lower side) for a particular DNA window, being the first one medium-coverage and the second one low-coverage. For calculating P0, we only pay attention to those positions we know are variable in the human genome (indicated with asterisks in Figure 4.7). In this way, we first get the pseudo-haploid calling for each position and each individual, choosing one allele at random when several reads are available. Then, when the individuals are compared, we can have three situations: a) we have information for both individuals and both of them have the same allele (shown in green in Fig. 4.7); b) we have information for both individuals and they have different alleles (shown in red); c) we lack information for one or two of the individuals (shown in grey). The lower the coverage of the individuals' genomes, the higher the proportion of positions lacking information to perform the comparison will be. In the example depicted in Figure 4.7, we have 9 variable positions, 3 do not have information for both individuals and 6 have information to compare them. Within them, 3 are matches –the two individuals have the same allele– and 3 are mismatches –they do not have the same allele. In this case, P0 would be 50%.

This analysis would be repeated for as many windows as necessary to cover the entire genome, calculating P0 for each window. From these values, *READ* estimates the average P0 and normalizes it to determine what threshold of the P0 corresponds to four biological descent levels: twins (or duplicated individuals in our dataset), first-degree, second-degree and unrelated. For normalizing, we need to know the average P0 value expected for unrelated individuals, which would be different for different populations. For example, in a highly diverse population, the average P0 value expected for unrelated individuals would be higher than that of an isolated population that has low values of diversity and a certain degree of consanguinity. As this value is unknown, the average P0 value expected for unrelated individuals is estimated based on the median of all average P0 across all the individuals of the population.

The output of the *READ* analysis is shown in Figure 4.8. In this plot, the non-normalized average P0 values are shown for each pair of individuals in the population, with bars indicating the two standard errors of the mean. The solid horizontal line represents the P0 median value used for normalization, while the dashed lines show the cutoffs used to classify the individuals into different kinship groups. Finally, gray areas around dashed lines indicate 95% confidence intervals for the kinship cutoffs. In this example, we see pairs of individuals with the P0 values we would expect for identical twins, first-degree relatives, second-degree relatives and unrelated individuals. It is important to note that the higher the number of positions lacking information for comparison, the larger the standard errors, leading to uncertainty in the kinship level inferred for these pairs of individuals. We have an example of this output in the last comparison in Figure 4.8 (pair 6). In this case, error bars indicate that it is impossible to discern if the individuals of pair 6 are related by a first or second-degree relationship.

Although the aim of the present review is not to offer a detailed view of all the methods available for inferring "biological kinship" in ancient individuals, we will

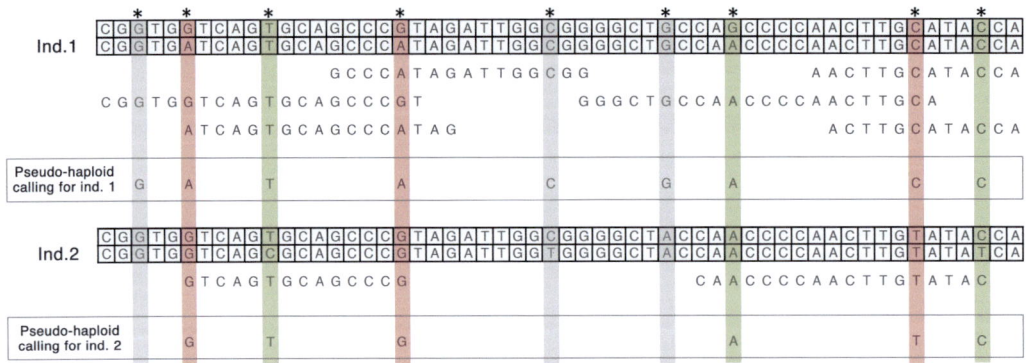

Figure 4.7. Example of the P0 estimation in a particular window when comparing aDNA data from two individuals.

briefly present some of them to provide some aid in interpreting aDNA results. Another method that uses pseudo-haploid genomes and is also based on the P0 or PMR value is BREADR (Rohrlach *et al.* 2023). This method provides a probability for each degree of relatedness for each pair of individuals, including identical twins (or the same individual), first-degree, second-degree or unrelated. In this sense, by providing probabilities for kinship levels, it allows researchers to consider different scenarios within the archaeological site's context. Because BREADR uses pseudo-haploid genomes for calculating PRM, individuals with coverage as low as 4% (0.04X) can be analyzed. Other methods are aimed at estimating k0, k1 and k2, and the coefficient of relatedness as explained before. In this case, instead of relying on pseudo-haploid genomes, we need to infer the genotypes of the individuals –for example, if individuals are AA, AG or GG– from the aDNA sequencing data for each one of the positions considered. The advantage of these methods is that they allow the detection of kinship relationships that surpass the second degree. For example, lcMLkin (Lipatov *et al.* 2015; Žegarac *et al.* 2021) can infer relatedness up to the fifth degree –e.g., a great-great-great-grandparent and great-great-great-grandson relationships. However, the main disadvantage is that we need high-coverage genomes to achieve this level of sensitivity, precluding the application of this method in sites with highly degraded human remains. In the case of lcMLkin, to identify kinship at the fifth degree, the two individuals that are compared should have a minimum depth of 2X. In this way, when the sequencing depth is lower, the ability of lcMLkin to identify more distant relationships is also reduced.

Finally, more recent methods perform kinship inference based on the identification of identical-by-descent segments (IBDs). An IBDs is a fragment of DNA that two individuals inherit from the same ancestor. To explain this phenomenon, we need to introduce the concept of recombination. Recombination is the process that produces the interchange of genetic information between the two chromosome copies –the one coming from our mother and the one from our father– during the formation of eggs and sperm. Because of recombination, the chromosomes a daughter or son receives from their parents are not the same as their parents received from the grandparents. We can see the effects of recombination in Figure 4.9A, where a segment of a particular chromosome is depicted for each individual, with a color code indicating the ancestor from whom

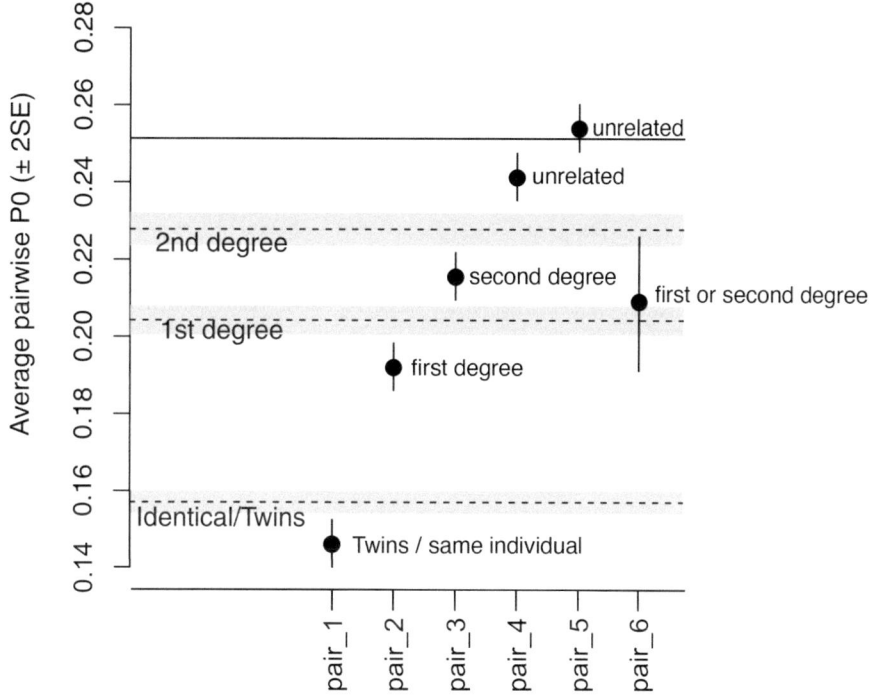

Figure 4.8. Example of an output of the *READ* analysis.

they are coming. In the first generation, we see that one of the parents has dark and light blue chromosomes, while de other one has dark and light green chromosomes. When we follow how this DNA fragment is inherited by their children, we see that they passed on a chromosome fragment that is the product of the recombination of both of the grandparents' chromosomes on both sides. For example, instead of inheriting a completely dark or light blue chromosome, they receive a chromosome with both dark and light blue sections. This process is repeated generation after generation, in such a way our genome is something like a puzzle composed of fragments of the genomes of our recent ancestors. When two individuals are related is because they share the same ancestor at some point in time, and in this way, they may share IBDs. In Figure 4.9B, we see that the individuals in the fourth generation share the same great-grandparents. If we have a look at the chromosome fragments they inherited, we see that they share two IBDs: one dark green from one of their great-grandparents and one dark blue from the other great-grandparent. One interesting aspect of IBDs is how their size is correlated with time. In Figure 4.9A, you can see how the chromosome segments of the first-generation couple are inherited, leading to smaller fragments over time. What this means is that the closer the biological relationship, the longer the IBDs. On the other hand, individuals with a more distant familial relationship are expected to have shorter IBDs. This allows IBD methods to infer more distant relationships than previous methods. For example, ancIBD uses IBDs to infer biological relatedness up to the sixth degree (Ringbauer *et al.* 2024). One instance of a sixth-degree relationship would be third cousins (see Fig. 4.5 for reference). As happens with lcMLkin, the ancIBD method can explore more distant relationships, but

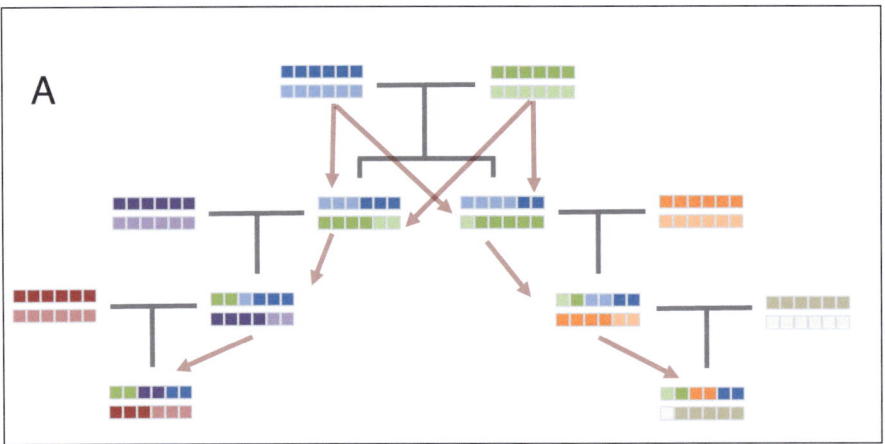

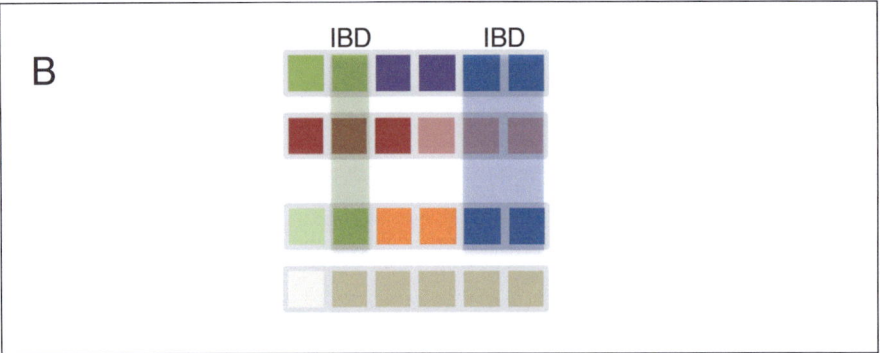

Figure 4.9. Inheritance of segments identical-by-descent (IBD) by related individuals.

with the disadvantage that it requires high-coverage data. However, it can trace distant kinship relationships, and can provide valuable information regarding kinship within sites, as well as migration patterns within geographical areas.

4.7 "Reconstructing" kinship using paleogenomic data

To show how "biological kinship" analyses are performed in aDNA research, we are going to present a hypothetical project in which we are determining familial relationships in a small burial with four graves and five individuals: four adults and one infant (Fig. 4.10A). To start, we identify each individual with the grave name and for grave 4, we will name the adult 4a and the infant 4b. After the excavation and the bioanthropological work have been performed, samples are taken from the individuals and identified with the names given to each grave. In the lab, the DNA is extracted, libraries are built and then sequenced to determine the percentage of endogenous DNA. Let's assume all the individuals are relatively well conserved and provide enough DNA data for determining the mtDNA and Y-chromosome haplogroups, as well as generating a low-coverage nuclear genome.

First, we compare the sexual chromosomes to determine the molecular sex of the individuals. Let's assume that, in our dataset, individuals 1 and 3 are male, while individuals 2, 4a and 4b are females (Fig. 4.10A). A possible following step is comparing the mtDNA and Y-chromosome haplogroups between the burial individuals to determine

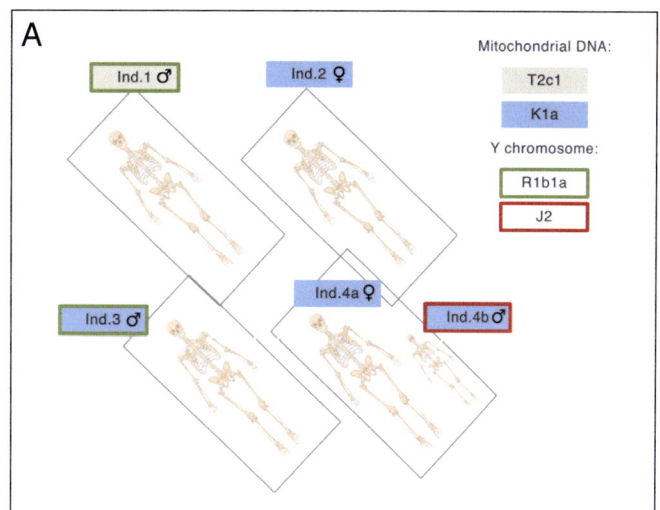

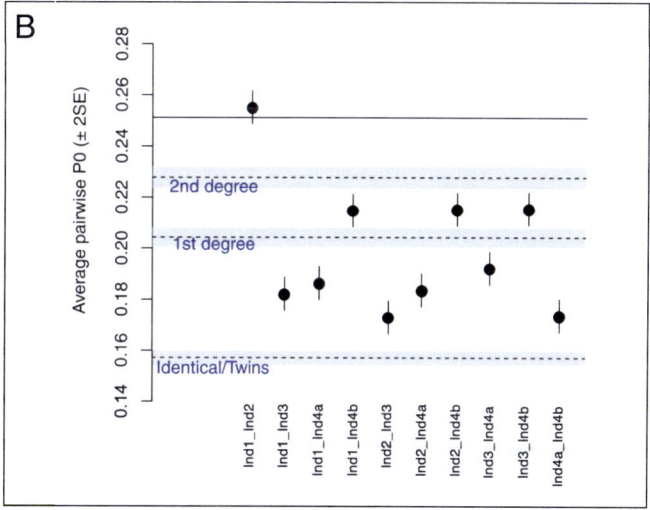

Figure 4.10. Example of a "biological kinship" analysis. A) Representation of the burial site with the information obtained from each individual, including their molecular sex and mtDNA and Y-chromosome haplogroups (created with BioRender.com). B) Results of the READ analysis for all pairs of individuals. C) Inferred pedigree from the DNA data obtained from individuals.

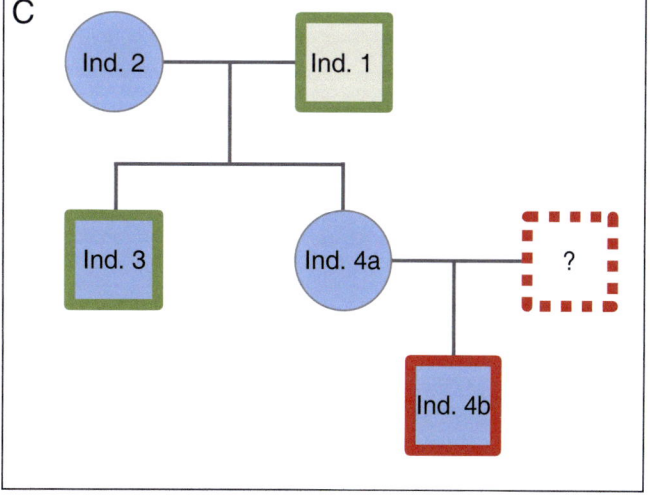

possible relationships from the maternal and paternal lines, respectively. We detect two different mtDNA haplogroups in the burial: T2c1 for individual 1 and K1a for the remaining. This result suggests a possible maternal relationship for four out of five individuals is possible. For the Y chromosome, we also detect two different haplogroups in the male individuals: R1b1a for individuals 1 and 3, and J2 for individual 4b. Again, this result is compatible with individuals 1 and 3 having a family relationship from the paternal side.

To confirm if the individuals are related up to the second degree, we can use *READ* software as described before. When the individuals are analyzed using *READ* (Fig. 4.10B), we observe that only a pair of individuals are biologically unrelated: individuals 1 and 2. The remaining pair are all related in the first or the second degree. As individuals 1 and 2 are unrelated between them but related in the first degree to individuals 3 and 4a, one possibility is that individuals 1 and 2 are the parents of 3 and 4a. For that to be true, we need three things to also be true: a) if individual 2 is the mother of individuals 3 and 4a, the three of them should have the same mtDNA; b) if individual 1 is the father of the male individual 3, the two of them should have the same Y chromosome; c) if individuals 3 and 4a are siblings, they should be related in the first degree. We confirm that individuals 2, 3 and 4a have the same mtDNA, individuals 1 and 3 have the same Y-chromosome and that individuals 3 and 4a are related in the first degree. In that way, all the information analyzed so far points to individuals 1 and 2 being the parents of 3 and 4.

Finally, we need to determine what relationship individual 4b has with the remaining individuals. As the infant is buried with individual 4a, one possibility is that 4a is his mother. For that to be true: a) individuals 4a and 4b should have the same mtDNA and a first-degree relationship; b) in our pedigree individual 3 is the brother of individual 4a, so individual 3 has to be the maternal uncle of 4b and should have the same mtDNA and a second-degree relationship with him; c) individuals 1 and 2 are the parents of individual 4a, so they have to be the grandparents of individual 4b and thus should have a second-degree relationship with him. We confirm that both individuals 4a and 4b have the same mtDNA and a first-degree relationship, so they can be a mother and a son. Also, individual 3 and 4b have the same mtDNA and a second-degree relationship, so individual 3 may be the maternal uncle of individual 4b. Finally, we confirm that individual 4b has a second-degree relationship with both individuals 1 and 2 and the same mtDNA as individual 2, making it possible for them to be his grandparents from the maternal side. In this way, we can draw the pedigree with individuals 1 and 2 in the first generation, their son (individual 3) and daughter (4a) in the second generation, and their grandson in the third generation (4b) (Fig. 4.10C). Once the pedigree is drawn, we see that the father of individual 4b is not present in the burial. Anyway, we know that this individual should belong to the J2 haplogroup of the Y chromosome to have passed it on to his son. This information can be recorded in the pedigree indicating that the father of the infant individual was not present at the burial, as shown in Figure 4.10C.

At this point, if the genome coverage of the individuals allows it, we could also explore more distant family relationships using previously published data from the same region and the same period, and methods such as lcMLkin or ancIBD. For example, if the burial we are studying belongs to the Neolithic period, we could gather all publicly available data on this period –e.g., Mallick and coauthors (2024)– and determine if individuals related up to the sixth degree are detected.

4.8 Future prospects

The use of paleogenomics, fueled by the advantages of next-generation sequencing methodologies, has already proved its worth for shedding light on kinship and social structure of past populations. One landmark publication of aDNA kinship inference was the 2019 analysis of the 15 individuals buried in the Neolithic mass grave at Koszyce (Poland), revealing they all were part of a large "extended family" (Schroeder *et al.* 2019). From that point on, the field has moved to explore larger burials, discovering extended pedigrees that connect family members through up to seven (Rivollat *et al.* 2023) and nine generations (Gnecchi-Ruscone *et al.* 2024), and establishing distant genetic relationships up to the sixth-seventh degree between individuals from different archaeological sites (Villalba-Mouco *et al.* 2022).

Apart from the development of next-generation sequencing approaches, the paleogenomics revolution has also been driven by the design of both improved laboratory methods for recovering degraded DNA and more powerful bioinformatic tools to analyze low-coverage data, providing a resolution on biological progeny and social structure never seen before. If advances in new methods follow a similar pattern to that seen in recent years, our capacity to understand the past will definitely increase in the future, with better methods to recover DNA from even more degraded remains and more sensitive tools for inferring more distant relationships.

Acknowledgements

This research was funded by the Spanish Ministry of Science, Innovation, and Universities grant PID2021-123080NB-I00 (MCIN/AEI/10.13039/501100011033 and "ERDF A way of making Europe").

References

Dabney, J., Meyer, M. and Pääbo, S. (2013): Ancient DNA damage. *Cold Spring Harbor Perspectives in Biology* 5(7), p. a012567. DOI:10.1101/cshperspect.a012567.

Fernandes, D.M., Cheronet, O., Gelabert, P. and Pinhasi, R. (2021): TKGWV2: An ancient DNA relatedness pipeline for ultra-low coverage whole genome shotgun data. *Scientific Reports* 11(1), p. 21262. DOI:10.1101/10.1038/s41598-021-00581-3.

Gnecchi-Ruscone, G.A., Rácz, Z., Samu, L., Szeniczey, T., Faragó, N., Knipper, C., Friedrich, R., Zlámalová, D., Traverso, L., Liccardo, S., Wabnitz, S., Popli, D., Wang, K., Radzeviciute, R., Gulyás, B., Koncz, I., Balogh, C., Lezsák, G. M., Mácsai, V., Bunbury, M.M.E., Spekker, O., le Roux, P., Szécsényi-Nagy, A., Mende, B.G., Colleran, H., Hajdu, T., Geary, P., Pohl, W., Vida, T., Krause, J. and Hofmanová, Z. (2024): Network of large pedigrees reveals social practices of Avar communities. *Nature* 629, pp. 376-383. DOI:10.1038/s41586-024-07312-4.

Hill, W.G., and Weir, B.S. (2011): Variation in actual relationship as a consequence of Mendelian sampling and linkage. *Genetics Research* 93(1), pp. 47-64.DOI:10.1017/S0016672310000480.

Kivisild, T. (2015): Maternal ancestry and population history from whole mitochondrial genomes. *Investigative Genetics* 6, paper 3. DOI:10.1186/s13323-015-0022-2.

Lazaridis, I., Nadel, D., Rollefson, G., Merrett, D.C., Rohland, N., Mallick, S., Fernandes, D., Novak, M., Gamarra, B., Sirak, K., Connell, S., Stewardson, K., Harney, E., Fu, Q., Gonzalez-Fortes, G., Jones, E.R., Roodenberg, S.A., Lengyel, G., Bocquentin, F., Gasparian, B.,

Monge, J.M., Gregg, M., Eshed, V., Mizrahi, A.S., Meiklejohn, C., Gerritsen, F., Bejenaru, L., Bluher, M., Campbell, A., Cavalleri, G., Comas, D., Froguel, P., Gilbert, E., Kerr, S.M., Kovacs, P., Krause, J., McGettigan, D., Merrigan, M., Merriwether, D.A., O'Reilly, S., Richards, M.B., Semino, O., Shamoon-Pour, M., Stefanescu, G., Stumvoll, M., Tonjes, A., Torroni, A., Wilson, J.F., Yengo, L., Hovhannisyan, N.A., Patterson, N., Pinhasi, R. and Reich, D. (2016): Genomic insights into the origin of farming in the ancient Near East. *Nature* 536, pp. 419-424. DOI: 10.1038/nature19310.

Lipatov, M., Sanjeev, K., Patro, R. and Veeramah, K.R. (2015): Maximum Likelihood Estimation of Biological Relatedness from Low Coverage Sequencing Data. *bioRxiv*, p. 023374. DOI: 10.1101/023374.

Mallick, S., Micco, A., Mah, M., Ringbauer, H., Lazaridis, I., Olalde, I., Patterson, N. and Reich, D. (2024): The Allen Ancient DNA Resource (AADR): A curated compendium of ancient human genomes. *Scientific Data* 11(1), p. 182. DOI: 10.1038/s41597-024-03031-7.

Monroy Kuhn, J.M., Jakobsson, M. and Gunther, T. (2018): Estimating genetic kin relationships in prehistoric populations. *PLoS One* 13, p. e0195491. DOI: 10.1371/journal.pone.0195491.

Parker, C., Rohrlach, A.B., Friederich, S., Nagel, S., Meyer, M., Krause, J., Bos, K.I. and Haak, W. (2020): A systematic investigation of human DNA preservation in medieval skeletons. *Scientific Reports* 10, p. 18225. DOI: 10.1038/s41598-020-75163-w.

Popli, D., Peyrégne, S. and Peter, B.M. (2023): KIN: a method to infer relatedness from low-coverage ancient DNA. *Genome Biology* 24(1), p. 10. DOI: 10.1186/s13059-023-02847-7.

Rasmussen, M., Li, Y., Lindgreen, S., Pedersen, J.S., Albrechtsen, A., Moltke, I., Metspalu, M., Metspalu, E., Kivisild, T., Gupta, R., Bertalan, M., Nielsen, K., Gilbert, M. T., Wang, Y., Raghavan, M., Campos, P.F., Kamp, H.M., Wilson, A.S., Gledhill, A., Tridico, S., Bunce, M., Lorenzen, E.D., Binladen, J., Guo, X., Zhao, J., Zhang, X., Zhang, H., Li, Z., Chen, M., Orlando, L., Kristiansen, K., Bak, M., Tommerup, N., Bendixen, C., Pierre, T.L., Gronnow, B., Meldgaard, M., Andreasen, C., Fedorova, S.A., Osipova, L.P., Higham, T. F., Ramsey, C.B., Hansen, T.V., Nielsen, F.C., Crawford, M.H., Brunak, S., Sicheritz-Ponten, T., Villems, R., Nielsen, R., Krogh, A., Wang, J. and Willerslev, E. (2010): Ancient human genome sequence of an extinct Palaeo-Eskimo. *Nature* 463, pp. 757-762. DOI: 10.1038/nature08835.

Ringbauer, H., Huang, Y., Akbari, A., Mallick, S., Olalde, I., Patterson, N. and Reich, D. (2024): Accurate detection of identity-by-descent segments in human ancient DNA. *Nature Genetics* 56(1), pp. 143-151. DOI: 10.1038/s41588-023-01582-w.

Rivollat, M., Rohrlach, A.B., Ringbauer, H., Childebayeva, A., Mendisco, F., Barquera, R., Szolek, A., Le Roy, M., Colleran, H., Tuke, J., Aron, F., Pemonge, M.H., Späth, E., Télouk, P., Rey, L., Goude, G., Balter, V., Krause, J., Rottier, S., Deguilloux, M.F. and Haak, W. (2023): Extensive pedigrees reveal the social organization of a Neolithic community. *Nature*, 620, pp. 600-606. DOI: 10.1038/s41586-023-06350-8.

Rohrlach, A.B., Tuke, J., Popli, D. and Haak, W. (2023): BREADR: An R Package for the Bayesian Estimation of Genetic Relatedness from Low-coverage Genotype Data. *bioRxiv*, 2023.2004.2017.537144. DOI: 10.1101/2023.04.17.537144.

Schroeder, H., Margaryan, A., Szmyt, M., Theulot, B., Włodarczak, P., Rasmussen, S., Gopalakrishnan, S., Szczepanek, A., Konopka, T., Jensen, T.Z.T., Witkowska, B., Wilk, S., Przybyła, M.M., Pospieszny, Ł., Sjögren, K.G., Belka, Z., Olsen, J., Kristiansen, K.,

Willerslev, E., Frei, K.M., Sikora, M., Johannsen, N.N. and Allentoft, M.E. (2019): Unraveling ancestry, kinship, and violence in a Late Neolithic mass grave. *Proceedings of the National Academy of Sciences USA* 116(22), pp. 10705-10710. DOI:10.1073/pnas.1820210116.

Van Oven, M. (2015): PhyloTree Build 17: Growing the human mitochondrial DNA tree. *Forensic Science International: Genetics Supplement Series* 5, pp. e392-e394. DOI:10.1016/j.fsigss.2015.09.155.

Villalba-Mouco, V., Oliart, C., Rihuete-Herrada, C., Rohrlach, A.B., Fregeiro, M.I., Childebayeva, A., Ringbauer, H., Olalde, I., Celdrán Beltrán, E., Puello-Mora, C., Valério, M., Krause, J., Lull, V., Micó, R., Risch, R. and Haak, W. (2022): Kinship practices in the early state El Argar society from Bronze Age Iberia. *Scientific Reports* 12(1), p. 22415. DOI:10.1038/s41598-022-25975-9.

Waples, R.K., Albrechtsen, A. and Moltke, I. (2019). Allele frequency-free inference of close familial relationships from genotypes or low-depth sequencing data. *Molecular Ecology* 28(1), pp. 35-48. DOI:10.1111/mec.14954.

Žegarac, A., Winkelbach, L., Blöcher, J., Diekmann, Y., Krečković Gavrilović, M., Porčić, M., Stojković, B., Milašinović, L., Schreiber, M., Wegmann, D., Veeramah, K.R., Stefanović, S. and Burger, J. (2021): Ancient genomes provide insights into family structure and the heredity of social status in the early Bronze Age of southeastern Europe. *Scientific Reports* 11(1), p. 10072. DOI:10.1038/s41598-021-89090-x.

Kinship and Residence in Copper Age Iberia: Insights from aDNA and Strontium Isotope Analyses

Marta Cintas-Peña

Department of Prehistory and Archaeology,
University of Seville, Spain,
marcinpen@us.es

Ana M. Herrero-Corral

Department of Prehistory, Archaeology, Social
Anthropology, and Historical Sciences and
Techniques, University of Valladolid, Spain,
anamercedes.herrero@uva.es

Abstract

Societies that inhabited Iberia in the third millennium BCE transitioned through profound changes related to economy, settlement patterns, religion or social organization. These transformations would lead to the conformation of a completely different system in the Bronze Age. Understanding both the Copper Age itself, as well as its collapse, implies having a certain knowledge on how its social structure operated, for which addressing issues such as kinship and residence are crucial. This is currently possible thanks to the application of aDNA and $^{87}Sr/^{86}Sr$ analyses, techniques that at different speeds are being employed in the research of Chalcolithic contexts. In this paper we aim to summarize the current state of knowledge in relation to kinship and residence in Chalcolithic Iberia by providing a complete record of aDNA and $^{87}Sr/^{86}Sr$ information, evaluating the data critically, and contextualizing it within the European framework. Our analysis suggests, first, the existence of a quantitatively important difference between data available for $^{87}Sr/^{86}Sr$ and that for aDNA; secondly, a limitation of usefulness of the data to address residence or kinship; thirdly, a more complex Iberian than European scenario. Despite the current difficulties and limitations, prospects seem to be promising, and it is reasonable to expect a big leap forward in our comprehension of Copper Age kinship and residence in the near future.

Keywords: *kinship, residence, Copper Age, aDNA, strontium isotopes*.

5.1 Introduction

5.1.1 Kinship and residence

Kinship has been, since the foundations of anthropology as a discipline, one of its most privileged fields of study. Considered the fabric of any society, it was conceptualized by

In Blanco-González, A. and Alarcón-García, E. (eds.) 2025, *A Social Archaeology of Kinship in Iberia and Beyond. Recent Multistranded Approaches from aDNA to Household Archaeology.* Leiden: Sidestone Press, pp. 93-110.

George P. Murdock (1949) as "the general subject matter of relationships, be they social and/or biological" (Ensor *et al.* 2017: 752). Put in other words, kinship refers to the relationships between persons based on descent (by blood or *consanguineal*) or marriage (*affinal*), forming the basis of social, economic, and political structure (Stone 2010: 5). Thus, the proper understanding of any present or past society implies having a certain knowledge on kinship, as well as being aware of the fact that it may change through time –contemporary times are a good example of how kinship can undergo profound changes and transformations (cf., Edwards *et al.* 1999).

Addressing kinship means studying the relationships established through descent and residence. All human groups organize descent and marriage. In the case of descent, it is achieved in one of three possible modes: 1) cognatic, through both women and men; 2) patrilineal or agnatic, through males; and 3) matrilineal or uterine, through females. In the case of marriage, it is arranged through patterns of exogamy or endogamy, and the so-called *marital residence rules*, which determine where married couples should live after marriage. Among different options, patrilocal and matrilocal are the more common types, depending on whether the new couple live with or near the husband's parents or the wife's parents, respectively (Ember *et al.* 2022). Specific modes of descent are more likely to be associated to particular patterns of postmarital residence. This is the case, for example, of patrilineality and patrilocality, which use to be related. However, residence patterns may vary among matrilineal groups (Divale 1975). As a consequence, descent and residence are commonly analyzed together, since it is not possible to address them separately when researching kinship, neither in the present nor in the past, though no fixed correlation exists.

Unaffordable until very recently, in the last years and thanks to the application of new techniques, kinship has also become a key issue for prehistoric archaeologists (Ensor *et al.* 2017). Looking at biological relatedness –one of the ways in which kinship is expressed– in the past is now possible through the application of genomics (Risch *et al.* 2023). Together with ancient DNA (henceforth aDNA), an indirect proxy to address kinship are strontium isotopes, since they inform us about mobility, allowing the identification of endogamy or exogamy, as well as marital residence patterns.

5.1.2 Addressing kinship in prehistory

Since the pioneering work of Haak and others (2008) on the biological relationships among members of the famous Eulau burials, it has taken almost a decade for paleogenetic studies to become fully established. However, these studies were primarily concerned with major migrations and ancestral origins, rather than attempting to reconstruct the internal social organisation of groups inhabiting Europe in later prehistory. In the last five years, the field of archaeogenetics has undergone a significant shift in focus towards kinship relationships between individuals buried in the same grave (Schroeder *et al.* 2019), cemetery (e.g., Žegarac *et al.* 2021), or even region (e.g., Mittnik *et al.* 2019). These works generally cover a chronological framework from the Neolithic (e.g., Fowler *et al.* 2022; Rivollat *et al.* 2023) to the Early Bronze Age (Zedda *et al.* 2023). This was a crucial period for the emergence of social hierarchies in which the nature of social organisation, residential patterns, and other social dynamics played a central role in establishing the foundations of social stratification.

Despite their different geographical distribution, the broad chronological framework and the inherent diversity of the funerary contexts analysed in these studies, many of the results and interpretations derived from these studies are repeated in most of the publications –but see Ensor (2021) for a critique. Typically, these groups consist of one or more extended families, with the paternal line being followed when it comes to burial practices. This is evidenced archaeologically by the fact that all or most of the men share the same Y-chromosome haplogroup, while there is a great diversity of mitochondrial DNA (henceforth mtDNA) lineages. Women who do not descend from those men of the main lineage not only have diverse haplogroups but also lack ancestors buried in the same site, pointing to what these publications refer to as "female exogamy" (e.g., Mittnik *et al.* 2019; Sjögren *et al.* 2020; Zegarac *et al.* 2021; Penske *et al.* 2024) but should be properly named "female postmarital mobility" (see Ensor this volume). The aforementioned data, when considered in conjunction with those derived from strontium analyses, typically indicate the existence of a virilocal residential pattern. This system would entail women migrating from their birthplaces to the birthplace of their male reproductive partner (Rivollat *et al.* 2023). However, these data must be interpreted with caution, as there are still few studies that have systematically analyzed the same individuals for both DNA and strontium. Indeed, there are already several instances where –despite the absence of statistically significant differences in strontium results between men and women– they still point to a patrilocal residence system (Schroeder *et al.* 2019: 10708; Penske *et al.* 2024: 4-7; see Ensor 2021 for a critique). Similarly, the presence of reproductive-age women from the male lineage in the family tomb does not align with the theory of female postmarital mobility and patrilocality (Rivollat *et al.* 2023). These unresolved issues, along with the considerable number of individuals detected in all the studies who appear to be integrated into the group despite having no genetic relationship with the main lineage members, suggests the existence of a less rigid and more fluid kinship system, with a broader range of possibilities based on social criteria.

Focusing on Iberia, it is regrettable that research is not yet equipped to respond to many of the aforementioned questions. To date, apart from a few studies focusing on reconstructing kinship ties in specific graves of the Northern Meseta region, no studies on kinship in the third millennium BCE in the Iberian Peninsula have been undertaken. A greater number of studies have been conducted regarding the residential and mobility patterns of the Chalcolithic populations of Iberia through strontium isotopes analyses. Nevertheless, these data may inform us on kinship only to the extent that residence and descent may be related –e.g., patrilineality and patrilocality or matrilineality and matrilocality– but each hypothesis in each specific case would need to be verified or refuted from other archaeological or genetic information.

In the next section, the genetic and strontium data available to date will be presented and discussed. Our aim is triple: 1) to provide a complete record of the information for the third millennium BCE in Iberia; 2) to critically evaluate the data, in terms of methodology and interpretation; and 3) to frame them within the European context.

5.2 Available data on kinship and residential patterns for Copper Age Iberia

5.2.1 Mobility and residential patterns though strontium analyses

The first paper dealing with the application of strontium isotopes analysis to an Iberian archaeological context appeared in 2009 (Díaz-Zorita Bonilla *et al.* 2009). Fifteen years later

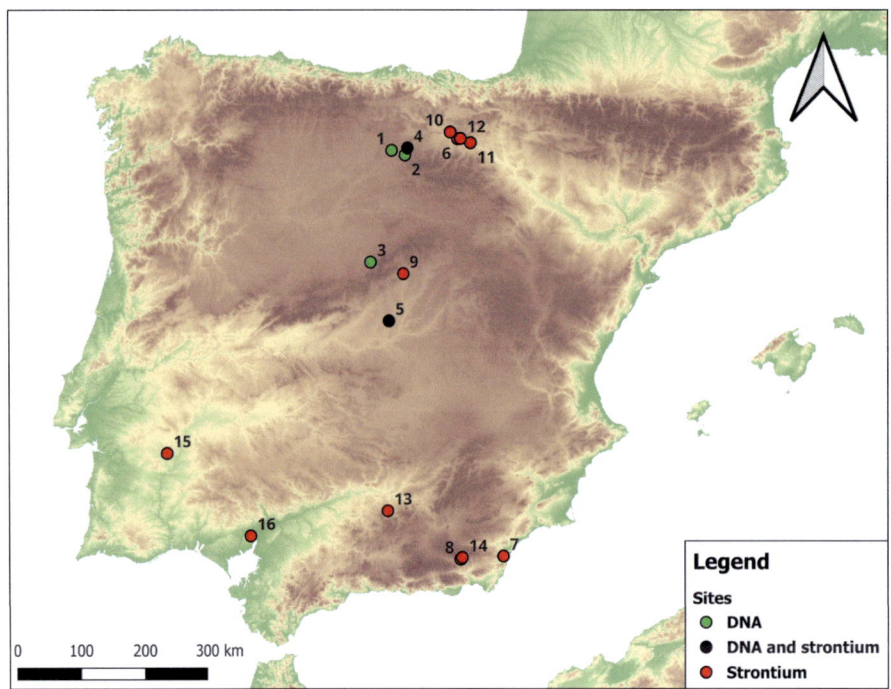

Figure 5.1. Location of Copper Age sites for which there are strontium isotopes data, DNA data or both. 1. Arroyal I; 2. El Mirador; 3. Los Areneros; 4. El Hundido; 5. Humanejos; 6. Alto de la Huesera; 7. Camino del Molino; 8. Chabola de la Hechicera; 9. El Rebollosillo; 10. Las Yurdinas; 11. Longar; 12. Los Husos I; 13. Marroquíes Bajos; 14. Peña Larga; 15. Perdigões; 16. Valencina. Author: Manuel E. Costa Caramé.

more than 100 sites –most of them of a late prehistoric chronology– have been studied employing such technique (see Díaz-del-Río *et al.* 2022 for a detailed compilation). In the case of the Copper Age, up to now, there is strontium data available from 555 individuals from 28 different Chalcolithic sites in Iberia (Fig. 5.1 and Tab. 5.1). However, the main interest in most of these papers is not the study of residence with regards to prehistoric kinship, rather centering on an approach regarding general mobility, rarely discussing results in terms of residence. This aim explains the fact that in many cases sampled individuals are sexually undetermined, which makes these results useless for the research on postmarital residence patterns and kinship.

Specifically, the sex of 255 of the 555 individuals (*ca.* 46%) has been identified. Although it is still a low number for the whole of Iberia in a period of *ca.* 900 years, it is remarkable that it almost doubles the figures of the last compilation made three years ago (Cintas-Peña and García Sanjuán 2022), which included 131 sexed individuals. Among the 255 individuals, 118 (46.3%) are females or likely females and 137 (53.7%) are males or likely males, with a sex ratio of 116.1. According to the local baselines provided in each of the publications, 179 (70.2%) individuals would be local and 76 non-locals (29.5%), though figures vary notably between different sites. If we distinguish by sex, the percentages are quite similar: 80 (*ca.* 68%) of the females or likely females are local *versus* 38 (32.2%) whose strontium ratio would exceed that of the local baseline considered; in the case of males, 99 (*ca.* 74%) would be locals and 38 (27.7%) non-locals or foreigners. Thus, between a third and a fourth of the

Site	NMI total	NMI sex	F/F?	F/F? local	F/F? non-local	M/M?	M/M? local	M/M? non-local	Reference
Alto de la Huesera	7	7 (100%)	3	2 (66.6%)	1 (33.3%)	4	3 (75%)	1 (25%)	Fernández-Crespo *et al.* 2020
Bolores	6	0 (0%)	-	-	-	-	-	-	Waterman *et al.* 2014
Camino del Molino	93	45 (48.4%)	19	16 (84.2%)	3 (15.8%)	26	23 (88.5%)	3 (11.5%)	Merner 2017
Cebolinhos 1	2	0 (0%)	-	-	-	-	-	-	Valera *et al.* 2020
Chabola de la Hechicera	2	2 (100%)	1	0 (0%)	1 (100%)	1	1 (100%)	0 (0%)	Fernández-Crespo *et al.* 2020
Comenda 1	1	0 (0%)	-	-	-	-	-	-	Valera *et al.* 2020
Cova da Moura	13	0 (0%)	-	-	-	-	-	-	Waterman *et al.* 2014
Cueva de los Cristales	4	0 (0%)	-	-	-	-	-	-	Villalba-Mouco *et al.* 2020
El Hundido	2	2 (100%)	0	0 (0%)	0 (0%)	2	1 (50%)	1 (50%)	Ortega *et al.* 2021
El Rebollosillo	16	3 (18.8%)	1	1 (100%)	0 (0%)	2	2 (100%)	0 (0%)	Díaz del Río *et al.* 2017
Feteira II	10	0 (0%)	-	-	-	-	-		Waterman *et al.* 2014
Gózquez 047	3	0 (0%)	-	-	-	-	-	-	Díaz del Río *et al.* 2017
Humanejos	44	44 (100%)	16	14 (87.5%)	2 (12.5%)	28	23 (82.1%)	5 (17.9%)	Cintas-Peña *et al.* 2023
La Pijotilla	17	0 (0%)	-	-	-	-	-	-	Díaz-Zorita Bonilla 2017
Lapa da Rainha 2	2	0 (0%)	-	-	-	-	-	-	Waterman *et al.* 2014
Las Yurdinas II	11	11 (100%)	4	3 (75%)	1 (25%)	7	6 (85.7%)	1 (14.3%)	Fernández-Crespo *et al.* 2020
Longar	7	6 (85.7%)	2	2 (100%)	0 (0%)	4	4 (100%)	0 (0%)	Fernández-Crespo *et al.* 2020
Los Husos I	3	3 (100%)	2	2 (100%)	0 (0%)	1	1 (100%)	0 (0%)	Fernández-Crespo *et al.* 2020
Marroquíes Bajos	100	36 (36%)	20	17 (85%)	3 (15%)	16	16 (100%)	0 (0%)	Fernández-Crespo *et al.* 2020
Paimogo I	12	0 (0%)	-	-	-	-	-		Waterman *et al.* 2014
Palacio III	8	0 (0%)	-	-	-	-	-	-	Díaz-Zorita Bonilla *et al.* 2009
Peña Larga	2	1 (50%)	1	1 (100%)	0 (0%)	0	0 (0%)	0 (0%)	Fernández-Crespo *et al.* 2020
Perdigões	72	35 (48.6%)	15	2 (13.3%)	13 (86.7%)	20	6 (30%)	14 (70%)	Zalaite 2016; Valera *et al.* 2020; Cintas-Peña *et al.* 2025
Pico Ramos	24	0 (0%)	-	-	-	-	-	-	Sarasketa-Gartzia *et al.* 2018
San Juan	21	0 (0%)	-	-	-	-	-	-	Villalba-Mouco *et al.* 2018
Valencina	67	60 (89.6%)	34	20 (58.8%)	14 (41.2%)	26	13 (50%)	13 (50%)	Díaz-Zorita Bonilla 2017; García Sanjuán *et al.* in prep.
Vidigueiras 2	2	0 (0%)	-	-	-	-	-	-	Valera *et al.* 2020
Zambujal	3	0 (0%)	-	-	-	-	-	-	Waterman *et al.* 2014
TOTAL	555	255 (45.9%)	118	80 (67.8%)	38 (32.2%)	137	99 (72.3%)	38 (27.7%)	

Table 5.1. Strontium isotopes data for Iberian Copper Age.

population sampled would be more mobile than the rest, and this proportion is repeated both in the case of males and females. The distribution is not statistically significant (chi squared: 0.60443, p=0.43689; Fisher's exact: 0.49296). Thus, at Iberian level it is not possible to identify any difference between males and females.

A more detailed analysis reveals a slightly different scenario (Fig. 5.2). In most sites (n=10) local females or males far outnumber non-locals. In fact, in 4 of them (El Rebollosillo, Longar, Los Husos I and Peña Larga) there are no outsiders or migrant individuals, which suggests a limited mobility of the people linked to these places. As a consequence, postmarital residence patterns associated to mobility cannot be inferred. In the remaining 7, the strontium values of some individuals indicate they were more mobile. Females of Alto de la Huesera, Camino del Molino, Chabola de la Hechicera, Las Yurdinas II or Marroquíes Bajos were more likely non-locals than their male counterparts, but in no case the difference is big enough to support the hypothesis of female postmarital mobility

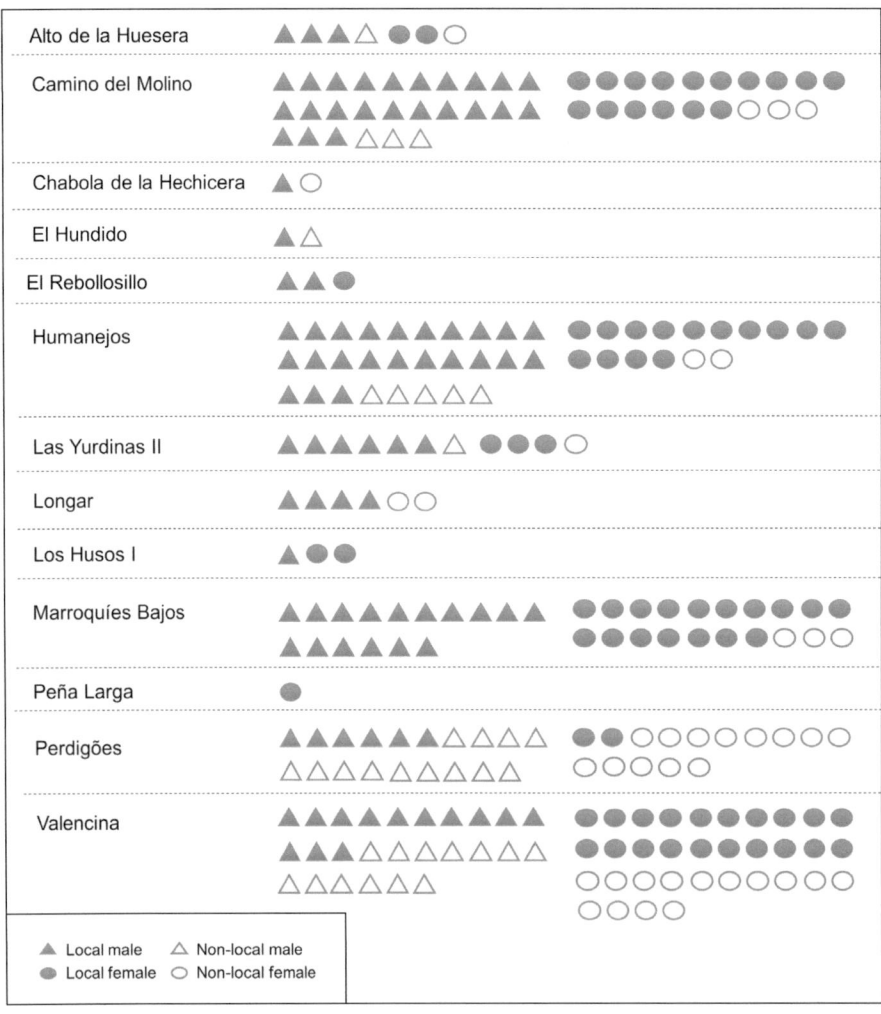

Figure 5.2. Local and non-local individuals of Copper Age Iberia according to available strontium isotopes data.

A SOCIAL ARCHAEOLOGY OF KINSHIP IN IBERIA AND BEYOND

and patrilocality. Something similar but in the opposite way can be said concerning the data from Humanejos. At this site non-local males outnumber non-local females, but the distance among them is not high enough to propose male postmarital mobility or matrilocality. In El Hundido there are 2 individuals, one with a foreign origin, and one local.

Although in most places local individuals overpass non-locals there are two exceptions: Perdigões and Valencina. In the former, 13 (*ca.* 87%) of 15 females and 14 (70%) of 20 males are non-locals; in the latter, 14 (41%) of 34 females and 13 (50%) of 26 males are outsiders. The highest degree of mobility in both sites in comparison to the rest –together with the presence of exogenous materials – would lend support to the hypothesis that they are aggregation centers involved in large scale interaction networks (Valera *et al.* 2020: 11). Both Perdigões and Valencina would have had the capacity to attract people and exotic materials. Although the *mobility of dead* is not equivalent to the *mobility of living*, the high proportion of non-local individuals in funerary contexts suggests that in such places there was a much higher degree of mobility than in other areas, which connects better with the interpretation of these sites as aggregation places – the dead cannot walk. Furthermore, the fact that in both sites there is a high percentage of both non-local males and females allow to reject previous proposals for female postmarital mobility and patrilocality, or bilocality biased to patrilocality (Cintas-Peña and García Sanjuán 2022).

5.2.2 Reconstruction of "genetic kinship" relationships

The limited number of archaeogenetic studies conducted for the Chalcolithic period in the Iberian Peninsula have not focused on reconstructing kinship relationships, but rather on issues of ancestry and migratory processes. These processes, however, are also closely related to the different types of social organization or marital strategies maintained by various groups. These data can therefore assist in elucidating certain issues such as the rise of the Bell Beaker elite and their subsequent decline in the interior of Iberia with the transition to the Bronze Age (Villalba-Mouco *et al.* 2022). As previously stated, the current data on kinship systems during the third millennium BCE in Iberia is limited. The available information must be extracted from the genetic data obtained in the major archaeogenetic studies published in the last five years (Haak *et al.* 2015; Szécsényi-Nagy *et al.* 2017; Olalde *et al.* 2018, 2019). While these data enable the identification of some "genetic kinship" relationships between two or more individuals from the same site, they do not permit the exclusion of relatives among the remaining individuals – affinal or non-consanguineal relationships should also be considered, as should social kinship. This is due to the fact that in numerous instances, sample preservation prevents more in-depth analyses of kinship, particularly when these studies did not prioritize this aspect.

Therefore, it is evident that the work undertaken by Sara Palomo's team between 2015 and the present-day merits particular emphasis. Despite the limitations of their findings –due to sample preservation concerns and methodological issues– it is notable that they represent the only studies to date that have focused specifically on kinship relationships in Chalcolithic populations of the Iberian Peninsula. Of the 24 individuals sampled for the pre-Bell Beaker and Bell Beaker Chalcolithic periods, only 10 yielded results for mtDNA. These individuals belong to the pre-Bell Beaker sites of Los Cercados (Valladolid) and Los Areneros (Segovia), with 3 and 7 individuals in secondary context, respectively. While only one possible kinship relationship was detected between two of the individuals from Los Areneros (Tab. 5.2), these analyses also allowed for the exclusion of maternal

Site	Archaeological ID	Sample ID	Sex	Degree of relationship	Chronology	Reference
Los Areneros	–	1ARE and 6ARE	males	second degree cousins?	2916-2696 cal BCE/ 2826-2580 cal BCE	Palomo Díez *et al.* 2024
El Mirador	MIR24 and MIR26	I1302 and I1314	males	first degree father-son	2568-2346 cal BCE	Olalde *et al.* 2019
El Mirador	MIR1 and MIR 11 and MIR14	I1271 and I1274 and I1277	males	second-third degree relatives	2900-2300 cal BCE	Olalde *et al.* 2019
El Mirador	MIR14 and MIR11	I1277 and I1274	males	first degree relatives	2900-2300 cal BCE	Olalde *et al.* 2019
Humanejos	Hume 10 and Hume 5	I6588 and I6539	males	second-third degree relatives	2341-2042 cal BCE	Olalde 2019
Humanejos	HUME 2 and HUME 4	I6628 and I6630	fe-males	second-third degree relatives	2480-2344 cal BCE	Olalde 2019
Humanejos	HUME 4 and HUME 8	I6630 and I6596	fe-male and male	second-third degree relatives	2480-2344 cal BCE	Olalde 2019
Arroyal I	Roy1 and Roy 3	I0458 and I0460	males	first degree father-son	2456-2204 cal BCE/ 2460-2206 cal BCE	Olalde *et al.* 2018
El Hundido	EHU001 and EHU002	UE750 and UE 450	males	second third degree relatives	2287-2041 cal BCE / 2564-2299 cal BCE	Olalde *et al.* 2019

Table 5.2. "Genetic kinship" data for the Iberian Copper Age.

kinship among all the analysed individuals, as they did not share mtDNA (Palomo *et al.* 2024). In addition to the aforementioned sites, one of the pre-Bell Beaker multiple tombs at El Tomillar (Ávila) was analysed. This tomb contained the remains of eight individuals: three adults (two women and one man) and five infants aged between one and six months. Unfortunately, the poor preservation of the DNA meant that no data could be obtained regarding kinship relationships, not even in terms of mtDNA. A similar situation occurred with the famous Bell Beaker burial at Aldeagordillo (Ávila), where the poor quality of the aDNA preserved in the remains did not allow for any data to be recovered, either from the infant associated with Bell Beaker materials or from the other individuals found at different levels of the tumulus (Palomo *et al.* 2024).

Although the study did not focus on kinship relationships, the genetic study conducted on the pre-Bell Beaker and Bell Beaker Chalcolithic populations of the Cueva de El Mirador (Atapuerca, Burgos) also detected a high degree of variability in mtDNA haplogroups (Gómez Sánchez *et al.* 2014), indicating that many of them would not be maternally related. El Mirador, however, offers some of the few first-degree biological relationships found for Chalcolithic chronologies in the Iberian Peninsula. These include individuals MIR24 and MIR26, two men who could be father and son, and MIR14 and MIR11, also men, with a first-degree relationship. These two individuals would in turn be related in the second or third degree to a third man, MIR1 (Olalde *et al.* 2019).

A genetic study conducted on the population buried at Humanejos (Madrid) during the Bell Beaker period revealed the presence of steppe-related ancestry in most of the male individuals with Bell Beaker grave goods. Furthermore, the study identified second-degree genealogical relationships between two individuals associated with Bell Beaker burial goods in a multiple tomb (Tab. 5.2). These relationships could potentially be maternal or paternal –grandchild-grandparent, paternal uncle-nephew, or half-siblings with the same father. Two other second- or third-degree genetic relationships were detected among three individuals in a multi-grave burial contemporary to the Bell Beaker period, though lacking the characteristic grave goods (Tab. 5.2). But in this case, as in the previous ones, what is significant is not so much the presence of biological relationships, but rather their absence. For example, at a lower level in that same tomb where the two related men were found, an additional individual was subjected to genetic analysis, in this instance a male juvenile. This analysis revealed he was not genetically related to the other two men. Moreover, as observed in the preceding case studies, the 12 sampled individuals from Humanejos exhibited distinct mitochondrial genomes, thereby refuting the hypothesis of genetic relatedness solely through the maternal line (Olalde 2019: 278-280).

To this data set we must also add the second or third-degree relationship of two adult males associated with the Bell Beaker material analyzed at El Hundido (Burgos), as well as the possible father-son relationship of two individuals from the Chalcolithic Bell Beaker period at the site of Arroyal I (Burgos), although they cannot be directly associated with materials from this period (Tab. 5.2) (Olalde *et al.* 2018).

5.3 Discussion and conclusions

What do strontium isotopes and aDNA data tell us about kinship and residence in Copper Age Iberia? In the first place, there is a quantitatively important difference between the amount of data available for strontium isotopes (relatively high) and that of aDNA (relatively low); in the second place and due to the research objectives of most publications, the usefulness of the data when trying to address residence or kinship is limited; thirdly, the current scenario seems to be more complex than expected in comparison to what is happening in Europe.

Currently, there is no consensus on how to name those individuals whose strontium values exceed those of the local range or show greater variability than the rest: outliers, migrants, non-locals, outsiders, foreigners or more mobile ones. Regardless of how we call these individuals, the data grouped here for the whole of Iberia in the third millennium BCE – excluding undetermined individuals– suggests that they would be around a third (29.8%) of the total, both males and females. Locals are the majority in most sites (n=11), the minority in one site (Perdigões) and the half in another (Valencina). In none of the sites are the values of men and women so far apart as to suggest patrilocal or matrilocal residential patterns. It contrasts deeply to what has been proposed for different regions of Europe during the third millennium BCE, where a higher female mobility has been interpreted in terms of "female exogamy" – or more accurately "female postmarital mobility" – and patrilocality (cf., Knipper *et al.* 2017; Sjögren *et al.* 2020). Was Iberia an exception in the European context? In these cases, where there is presence of outliers, the more consistent hypothesis to suggest in relation to residence would be bilocality. However, it should be born in mind that mobility may be explained by other reasons such as shepherding, nomadism, fostering or gathering, as it was in fact proposed for Perdigões (Valera *et al.* 2020). All these possibilities assume that those who moved were the living individuals, but it cannot be ruled out that the dead were moved by the living.

Compared to 255 individuals from 13 sites – or 555 individuals from 28 sites if we take all of them into account, including those for which there is no sex data – for which there is strontium data, in the case of aDNA the information regarding specifically genetical relationships is limited to 16 individuals from 5 sites. In two cases, Humanejos and El Hundido, we have both strontium and aDNA. As it happens for strontium isotopes, the main research objective followed in aDNA publications is not the reconstruction of kinship relationships, but issues of migration or ancestry. Regardless of research aims conducted, biological relationships could be identified in Los Areneros, El Mirador, Humanejos, Arroyal I and El Hundido. In most cases, such relationships are between male individuals, whether they are first, second, or third-degree relationships. Also, the female haplogroups analyzed show greater variability than the male ones, which suggests a closer biological relatedness among male individuals and longer distance among female individuals. Such results could be explained by male-centered kinship (patrilineality), but the limited data invites caution.

As important as the presence are the absences. Although the genetic studies available regarding Chalcolithic populations in Iberia are limited, there is already a considerable number of studies carried out for this period in the rest of Europe. These publications tend to focus on describing the general patterns detected in these populations, as outlined in the introduction to this work. However, they also include exceptions that escape the "norm", demonstrating a complex and changing kinship system. In addition to reproducing impressive genealogical trees comprising up to five generations (Penske *et al.* 2024), which all descend from the same male ancestor, these studies have also identified individuals who, despite sharing a burial space and the same funerary treatment, lack any genetic link with the rest of the family. Although most of these cases could be attributed to young women who have become part of the lineage and have not yet had the opportunity to reproduce, a significant number of cases involving men or children do not align with this hypothesis. Although it is commonly assumed that kinship is much more than genetic ties between people, and that social kinship –which is more complex and mutable– is a better fit for the heterogeneous panorama observed in recent European prehistory, there are still very few works that attempt to investigate these questions in depth (Johnson 2019). Kinship as social relatedness is characterized by fluidity, enabling individuals to join and exit groups throughout the life cycle, involving different practices, such as commensality, co-residence, or the sharing of status (Johnson and Paul 2016).

Anthropological and ethnographic research has demonstrated the significance of this social kinship, whereby social relationships play a pivotal role in the formation of familial bonds. In a multitude of cultures, practices such as ritual kinship, adoption, godparenthood or fostering demonstrate the central position that non-biological connections play in the social structuring. A case worthy of note –and displaying numerous similarities with certain exceptions observed within the European Chalcolithic archaeological record– which do not align with the strictly genetically formed patrilineal family unit, is that of the Nuer people (Evans-Pritchard 1990[1951]). The pastoralists and herdsmen of the swamps and savannahs of Southern Sudan employ a range of strategies such as the "blind-eye adultery" or the "woman-woman marriage" to ensure the continuation of the patrilineal lineage, with the primary objective being the production of male offspring. These strategies encompass not only the practice of polygyny –a man marrying multiple women– to produce offspring but also the acceptance of a non-biological son as one's own in order to make him an heir. The desire to obtain this heir persists beyond the individual's death. In the event of a Nuer's man death prior to the birth of children, his wife is married to a close relative of the same paternal line and the child is then regarded as the son and heir of the deceased. Childlessness is also a significant issue for Nuer women. In case of female infertility, the woman has the

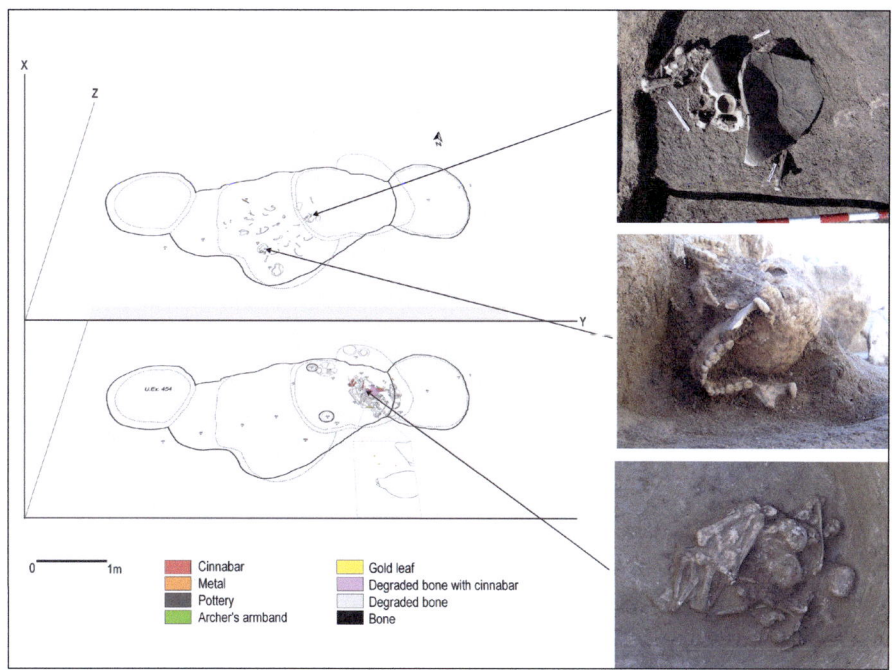

Figure 5.3. The two burial phases of the multiple grave 7 from the Bell Beaker cemetery necropolis of Humanejos (Madrid, Spain). Photo: Sara Genicio. Plans: Raúl Flores. Design: Miriam Luciañez Triviño.

right to terminate the marriage and return to her father's family. In the absence of offspring, the woman is regarded as a man among her natal patrilineal kin. Consequently, she may consent to marry another woman and look for a man to engage in sexual relations with her, thus facilitating the desired offspring and perpetuating the paternal line (Stone 2010: 89-92). This is just an example that demonstrates the intricate nature of kinship systems, which extend far beyond the boundaries of the biological family unit.

Back to the archaeological record, we can also find strong evidence that point to a greater complexity in kinship systems than previously thought, in which different factors beyond blood ties were undoubtedly involved. This seems particularly revealing in the case of Bell-Beaker tombs. People with Bell Beaker grave goods exhibited a notable degree of social complexity, with evidence suggesting the existence of social hierarchies and emerging elites (Garrido-Pena 2014). However, the power held by these elites appears to have been relatively unstable and not strongly institutionalized (Garrido-Pena 2006). This is particularly evident in certain regions, such as the interior of Iberia, where the end of the Bell Beaker period coincides with a significant societal collapse at the onset of the Bronze Age. This collapse indicates a period of social instability, possibly driven by the lack of robust power structures and the challenges in maintaining social cohesion. Familial strategies and alliances beyond blood relations were likely key elements in efforts to consolidate power. The presence of richly furnished child burials with Bell Beaker artifacts (Herrero and Garrido 2019), contrary to what happens in the pre-Beaker Copper Age (Cintas-Peña *et al.* 2018), suggests some form of status inheritance, at least for certain descendants, indicating an attempt to perpetuate elite status across generations. Although the initial hypothesis would be that these are genetically linked lineages or families –as has already been detected

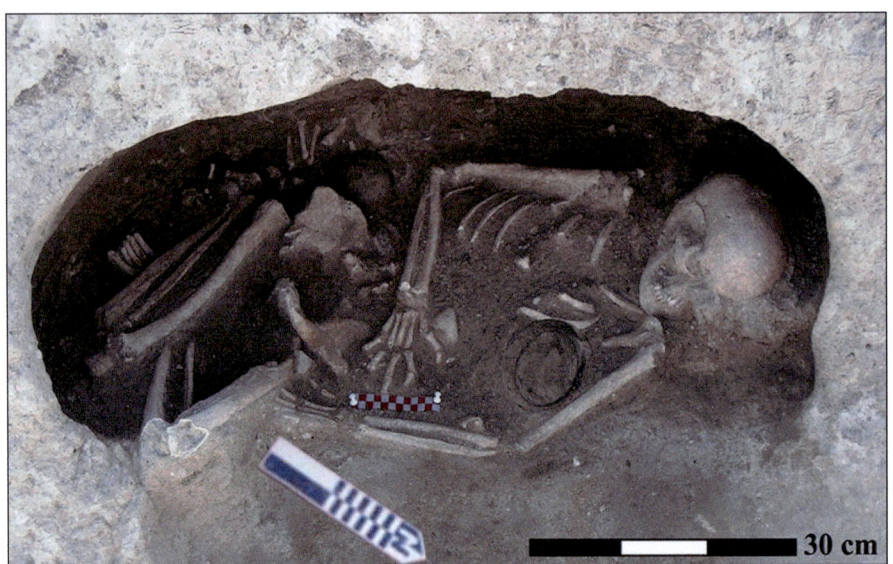

Figure 5.4. Double Beaker tomb of a woman and a child from the "Área funeraria 2", Camino de las Yeseras (Madrid) and their grave goods (Liesau *et al.* 2015: 110-112).

in some other European studies (Zedda *et al.* 2023)– the limited genetic analyses that have been conducted in Iberia indicate otherwise. As previously shown, no first-degree consanguineal relationships have been identified between individuals associated with Bell-Beaker items buried in the same tomb or cemetery. Only a limited number of second- or third-degree relationships between men from the sites of El Hundido and Humanejos have been identified (Tab. 5.2). At the latter site, further genetic analyses have been carried out on additional individuals with Bell-Beaker items from the same necropolis. In these cases, it is notable, for example, that there is no genetic link between a male juvenile buried in a multiple tomb and the other two adult males from the same structure for whom aDNA samples have also been obtained (Olalde 2019) (Fig. 5.3).

A further striking case is that of the double burial in the "Area Funeraria 2" of Camino de las Yeseras (Madrid). The bodies of an adult woman (20-30 years) and a girl of 1-5 years

A SOCIAL ARCHAEOLOGY OF KINSHIP IN IBERIA AND BEYOND

of age were interred with great care, seemingly at the same time, with their respective Bell-Beaker ceramic offerings (Liesau *et al.* 2015) (Fig. 5.4). Once again, genetic analysis of the two individuals showed that they were not genetically linked to each other (Olalde *et al.* 2018: supplementary data). The nature of the relationship between the two individuals is unknown. However, it is evident that they were connected in some manner, leading to their careful burial together. Furthermore, in the case of the little girl, the act of burial was a clear indication of her status as a member of the Bell-Beaker elite, as well as a means of establishing a relationship with the woman. In any case, there are still very few kinship analyses based on aDNA available in Iberia for this period. In order to gain a more complete understanding of kinship and residence patterns in these groups, further aDNA and strontium analyses of the same individuals would be necessary. This would enable the construction of a more comprehensive picture of this undoubtedly complex system which, to date, still presents more unknowns than certainties.

It is encouraging, however, to note that several projects are currently underway with the objective of further investigating these issues in the forthcoming years. For instance, the project SKIN aims to investigate the social organisation of groups from the third and second millennia BCE in Iberia through the analysis of kinship relationships between adults and children buried in the same funerary structures. In contrast to previous research, SKIN is designed to concentrate on those forms of kinship that extend beyond the biological, with a particular focus on infant individuals and related issues such as alloparental care, foster care and adoption (Herrero-Corral in press). Similarly, but centred on the application of strontium isotopes, the project WOMAM analyzed the relationship between social complexity, gender inequality and residential patterns –and, by extension, kinship– increasing the available data on strontium isotopes for sex determined individuals (cf., Cintas-Peña and García Sanjuán 2020; Cintas-Peña *et al.* 2023). Finally, the project DEMOS, whose co-IP is one of the authors of this paper seeks to make a significant contribution to the understanding of the role of monumentalised central places –such as Antequera (Málaga) and Valencina (fourth-third millennia BCE)– in early complex societies by providing new data on aDNA, strontium isotopes, radiocarbon-dating, etc. These three examples suggest that the available data on aDNA and strontium isotopes for third millennium BCE Iberia will experience a significant increase in the short term, allowing for a more complete and complex analysis of the issues discussed throughout this paper. The future is promising for the archaeology of kinship in Copper Age Iberia.

Acknowledgements

This contribution is part of the research projects SKIN (Social Kinship and Cooperative Care in Recent Prehistory Iberia), funded by the European Commission (HORIZON-MSCA-2021-PF-01-101062307) and led by one of the authors (AHC) and WOMAM (Women, Men and Mobility. Understanding Gender Inequality in Prehistory), funded by the European Commission (HORIZON-MSCA-GA891776) and also led by one of the authors (MCP).

References

Cintas-Peña, M. and García Sanjuán, L. (2022): Women, residential patterns and early social complexity. From Theory to Practice in Copper Age Iberia. *Journal of Anthropological Archaeology* 67, p. 101422. DOI: 10.1016/j.jaa.2022.101422.

Cintas-Peña, M., García Sanjuan, L., Díaz-Zorita Bonilla, M., Herrero Corral, A.M., and Robles Carrasco, S. (2018): The non-adult population at the Copper Age settlement of Valencina de la Concepción (Seville, Spain): A demographic, contextual and sociological approach. *Trabajos de Prehistoria* 75(1), pp. 85-108. DOI: 10.3989/tp.2018.12205.

Cintas-Peña, M., Garrido Pena, R., Herrero-Corral, A.M., Flores Fernández, R., Waterman, A.J., Díaz-Zorita Bonilla, M., Díaz-del-Río, P. and Peate, D.W. (2023): Isotopic Evidence for Mobility in the Copper and Bronze Age Cemetery of Humanejos (Parla, Madrid): A Diachronic Approach Using Biological and Archaeological Variables. *Journal of Archaeological Method and Theory* 31, pp. 1152-1184. DOI: 10.1007/s10816-023-09633-6.

Cintas-Peña, M., Shaw Evangelista, L., Valera, A.C., García Sanjuán, L., Rebay-Salisbury, K., Koenig, C., Olsen, J.V. and Kanz, F. (2025): Fresh insights into sex-specific mobility in Copper Age Perdigões (Reguengos de Monsaraz, Portugal) provided by amelogenin peptide analysis. *Archaeological and Anthropological Sciences* 17, p. 92. https://doi.org/10.1007/s12520-025-02205-7

Díaz-del-Río, P, Consuegra, S., Audije, J., Zapata, S., Cambra, Ó., González, A., Waterman, A., Thomas, J., Peate, D., Odriozola, C., Villalobos, R., Bueno, P. and Tykot, R.H. (2017): Un enterramiento colectivo en cueva del III milenio AC en el centro de la Península Ibérica: el Rebollosillo (Torrelaguna, Madrid). *Trabajos de Prehistoria* 74(1), pp. 68-85. DOI: 10.3989/tp.2017.12184.

Díaz-Zorita Bonilla, M., Waterman, A.J. and Knudson, K.J. (2009): Explorando la movilidad y los patrones dietarios durante la Edad del Cobre en el suroeste de la Península Ibérica: estudio preliminar bioarqueológico del Tholos de Palacio III (Almadén de la Plata, Sevilla). In Plo Cerdá, M. and García-Prósper, E. (eds.), *Investigaciones histórico-médicas sobre salud y enfermedad en el pasado*. Madrid: Sociedad Española de Paleopatología, pp. 671-676.

Díaz-Zorita Bonilla, M. (2017): *The Copper Age in South-West Spain: A bioarchaeological approach to prehistoric social organisation*. Oxford: British Archaeological Reports.

Divale, W.T. (1975): An Explanation for Matrilocal Residence. In Dana, R. (ed.), *Being Female: Reproduction, Power and Change*. The Hague: Mouton Publishers. pp. 99-108.

Edwards, J., Franklin, S., Hirsch, E., Price, F. and Strathern, M. (1999): *Technologies of Procreation: Kinship in the Age of Assisted Conception*. London: Routledge.

Ember, C.R., Droe, A. and Russell, D. (2022): Residence and Kinship. In Ember, C.R. (ed.), *Explaining Human Culture*. Human Relations Area Files. https://hraf.yale.edu/ehc/summaries/residence-and-kinship (accessed 09/07/2024).

Ensor, B.E., Irish, J.D. and Keegan, W.F. (2017): The Bioarchaeology of Kinship: Proposed Revisions to Assumptions Guiding Interpretation. *Current Anthropology* 58(6), pp. 739-761. DOI: 10.1086/694584.

Ensor, B.E. (2021): *The Not Very Patrilocal European Neolithic. Strontium, aDNA, and Archaeological Kinship Analyses*. Oxford: Archaeopress.

Evans-Pritchard, E.E. (1990 [1951]): *Kinship and Marriage Among the Nuer*. New York: Oxford University Press.

Fernández-Crespo, T., Snoeck, C., Ordoño, J., de Winter, N.J., Czermak, A., Mattielli, N., Lee-Thorp, J.A. and Schulting, R.J. (2020): Multi-isotope evidence for the emergence of cultural alterity in Late Neolithic Europe. *Science Advances* 6(4), p. eaay216. DOI: 10.1126/sciadv.aay2169.

Fowler, C., Olalde, I., Cummings, V., Armit, I., Büster, L., Cuthbert, S., Rohland, N., Cheronet, O., Pinhasi, R. and Reich, D. (2021): A high-resolution picture of kinship practices in an Early Neolithic tomb. *Nature* 601, pp. 584-587. DOI: 10.1038/s41586-021-04241-4.

García Sanjuán, L., Montero Artús, R., Shaw Evangelista, L., Watermann, A., Rebay-Salisbury, K., Kanz, F., Emslie, S., Villalba-Mouco, V., Lozano Rodríguez, J.A., Haak, W. and Cintas-Peña, M. From Bone Chemistry to Human Demography: Uncovering Copper Age Society at Valencina (c. 2900-2650 BC). Under evaluation.

Garrido-Pena, R. (2006): Transegalitarian societies: An ethnoarchaeological model for the analysis of Copper Age Bell Beaker using groups in Central Iberia. In Díaz del Río, P. and García Sanjuán, L. (eds.), *Social Inequality in Iberian Late Prehistory*. Oxford: British Archaeological Reports, pp. 81-96.

Garrido-Pena, R. (2014): Bell Beakers in Iberia. In Almagro, M. (ed.), *Iberia. Protohistory of the far West of Europe: From Neolithic to Roman Conquest*. Burgos: Universidad de Burgos, pp. 113-124.

Gómez-Sánchez, D., Olalde, I., Pierini, F., Matas-Lalueza, L., Gigli, E., Lari, M., Civit, S., Lozano, M., Vergès, J.M., Caramelli, D., Ramírez, O. and Lalueza-Fox, C. (2014): Mitochondrial DNA from El Mirador Cave (Atapuerca, Spain) Reveals the Heterogeneity of Chalcolithic Populations. *PLoS One* 9(8), p. e105105. DOI: 10.1371/journal.pone.0105105.

Haak, W., Brandt, G., de Jong, H.N., Meyer, C., Ganslmeier, R., Heyd, V., Hawkesworth, C., Pike, A.W., Meller, H., and Alt, K.W. (2008): Ancient DNA, strontium isotopes, and osteological analyses shed light on social and kinship organization of the Later Stone Age. *Proceedings of the National Academy of Sciences USA* 105(47), pp. 18226-18231. DOI: 10.1073/pnas.0807592105.

Haak, W., Lazaridis, I., Patterson, N., Rohland, N., Mallick, S. *et al.* (2015): Massive migration from the steppe was a source for Indo-European languages in Europe. *Nature* 522, pp. 207-211. DOI: 10.1038/nature14317.

Herrero-Corral, A.M. and Garrido-Pena, R. (2019): The Inheritors: Bell Beaker Children's Tombs and Their Social Context. *Journal of Mediterranean Archaeology* 32(1), pp. 63-87. DOI: 10.1558/jma.39328.

Herrero-Corral, A.M. (2025): El estudio de las relaciones de parentesco no biológico y el cuidado cooperativo de los individuos infantiles durante la Prehistoria Reciente peninsular. El proyecto SKIN como propuesta metodológica. In Sánchez Romero, M., Alarcón García, E. and Rivera-Hernández, A. (eds.), *Pequeños cuerpos con grandes biografías. Una mirada a la infancia desde la arqueología y la antropología*. Granada: Comares, pp. 137-153.

Johnson, K.M. (2019) Opening up the family tree: Promoting more diverse and inclusive studies of family, kinship, and relatedness in bioarchaeology. In Buikstra, J. (ed.), *Bioarchaeologists speak out. Deep time perspectives on contemporary issues*. Cham: Springer, pp. 201-230. DOI: 10.1007/978-3-319-93012-1_9.

Johnson, K.M. and Paul, K.S. (2016): Bioarchaeology and Kinship: Integrating Theory, Social Relatedness, and Biology in Ancient Family Research. *Journal of Archaeological Research* 24, pp. 75-123. DOI: 10.1007/s10814-015-9086-z.

Knipper, C., Mittnik, A., Massy, K., Kociumaka, C., Kucukkalipci, I., Maus, M., Wittenborn, F., Metz, S.E., Staskiewicz, A., Krause, J., and Stockhammer, P.W. (2017): Female exogamy and gene pool diversification at the transition from the Final Neolithic to the Early Bronze Age in Central Europe. *Proceedings of the National Academy of Sciences USA* 114, pp. 10083-10088. DOI: 10.1073/pnas.1706355114.

Liesau, C., Blasco, C., Ríos, P. and Flores, R. (2015): La mujer en el registro funerario campaniforme y su reconocimiento social. *Trabajos de Prehistoria* 72, pp. 105-25. DOI: 10.3989/tp.2015.12146.

Merner, C. (2017): *Reconstructing Southeast Iberian Copper Age Mobility: A Strontium Isotope Analysis of the Camino del Molino Mass Burial*. MA dissertation. Memorial University of Newfoundland.

Mittnik, A., Massy, K., Knipper, C., Wittenborn, F., Pfrengle, S. *et al.* (2019): Kinship-based social inequality in Bronze Age Europe. *Science* 366, pp. 731-734. DOI: 10.1126/science.aax6219.

Murdock, G.P. (1949): *Social structure*. New York: Macmillan.

Olalde, I. (2019): Anexo III. Estudio genómico de los individuos de Humanejos. In Garrido, G., Flores, R. and Herrero, A.M. (eds.), *Las sepulturas campaniformes de Humanejos (Parla, Madrid)*. Madrid: Comunidad de Madrid, pp. 277-283.

Olalde, I., Mallick, S., Patterson, N., Rohland, N., Villalba Mouco, V. *et al.* (2019): The genomic history of the Iberian Peninsula over the past 8000 years. *Science* 363, pp. 1230-1234. DOI: 10.1126/science.aav4040.

Olalde, I., Brace, S., Allentoft, M.E., Armit, I., Kristiansen, K. *et al.* (2018): The Beaker phenomenon and the genomic transformation of northwest Europe. *Nature* 555, pp. 190-196.

Ortega, L.A., Guede, I., Zuluaga, M.C., Alonso-Olazabal, A. and Jiménez-Echevarría, J. (2021): Strontium and oxygen isotopes to trace mobility routes during the Bell Beaker period in the north of Spain. *Scientific Reports* 11, p. 19553. DOI: 10.1038/s41598-021-99002-8.

Palomo-Díez, S., Esparza-Arroyo, A., Gomes, C., Rickards, O., Labajo-González, E., Perea-Pérez, B., Martínez-Labarga, C. and Arroyo-Pardo, E. (2024): Family History in the Iberian Peninsula during Chalcolithic and Bronze Age: An Interpretation through the Genetic Analysis of Plural Burials. *Genealogy* 8(25). DOI: 10.3390/genealogy8010025.

Penske, S., Küßner, M., Rohrlach, A.B. *et al.* (2024): Kinship practices at the early Bronze Age site of Leubingen in Central Germany. *Scientific Reports* 14, p. 3871. DOI: 10.1038/s41598-024-54462-6.

Risch, R., Haak, W., Krause, J. and Meller, H. (2023): Kinship, sex, and biological related-ness. The contribution of archaeogenetics to the understanding of social and biolog-ical relations. In Meller, H., Krause, J., Haak, W. and Risch, R. (eds.), *Kinship, sex, and biological relatedness. The contribution of archaeogenetics to the understanding of social and biological relations*. Heidelberg: Propylaeum, pp. 9-25. DOI: 10.11588/pro-pylaeum.1280.c18044.

Rivollat, M., Rohrlach, A.B., Ringbauer, H., Childebayeva, A., Mendisco, F. *et al.* (2023): Extensive pedigrees reveal the social organization of a Neolithic community. *Nature* 620, pp. 600-606. DOI: 10.1038/s41586-023-06350-8.

Sarasketa-Gartzia, I., Villalba-Mouco, V., Le Roux, P., Arrizabalaga, Á. and Salazar-García, D.C. (2018): Late Neolithic-Chalcolithic socio-economical dynamics in northern Iberia. A multi-isotope study on diet and provenance from Santimamiñe and Pico Ramos archaeological sites (Basque Country, Spain). *Quaternary International* 481, pp. 14-27. DOI: 10.1016/j.quaint.2017.05.049.

Schroeder, H., Margaryan, A., Szmyt, M., Theulot, B., Włodarczak, P. *et al.* (2019): Unraveling ancestry, kinship, and violence in a Late Neolithic mass grave. *Proceedings of the National Academy of Sciences USA* 116(22), pp. 10705-10710. DOI: 10.1073/pnas.1820210116.

Sjögren, K.G., Olalde, I., Carver, S., Allentoft, M.E., Knowles, T. *et al.* (2020): Kinship and social organization in Copper Age Europe. A cross-disciplinary analysis of archaeology, DNA, isotopes, and anthropology from two Bell Beaker cemeteries. *PLoS One* 15(11), p. e0241278. DOI: 10.1371/journal.pone.0241278.

Szécsényi-Nagy, A., Roth, C., Brandt, G., Rihuete-Herrada, C., Tejedor-Rodríguez, C. *et al.* (2017): The maternal genetic make-up of the Iberian Peninsula between the Neolithic and the Early Bronze Age. *Scientific Reports* 7, p. 15644. DOI: 10.1038/s41598-017-15480-9.

Stone, L. (2010): *Kinship and Gender. An introduction.* Fourth edition. Boulder: Westview Press.

Valera A.C., Žalaitė, I., Maurer, A.F., Grimes, V., Silva, A.M., Ribeiro, S. Santos, J.F. and Barrocas Dias, C. (2020): Addressing human mobility in Iberian Neolithic and Chalcolithic ditched enclosures: The case of Perdigões (South Portugal). *Journal of Archaeological Science: Reports* 30, p. 102264. DOI: 10.1016/j.jasrep.2020.102264.

Villalba-Mouco, V., Sauqué, V., Sarasketa-Gartzia, I., Pastor, M.V., Le Roux, P.J., Vicente, D., Utrilla, P. and Salazar-García, D.C. (2018): Territorial mobility and subsistence strategies during the Ebro Basin Late Neolithic-Chalcolithic: A multi-isotope approach from San Juan cave (Loarre, Spain). *Quaternary International* 481, pp. 28-41. DOI: 10.1016/j.quaint.2017.05.051.

Villalba-Mouco, V., Bea, M., Montes, L., and Salazar-García, D.C. (2020): Mobility across the pre-Pyrenean mountain ranges during the Chalcolithic through strontium isotopes in human enamel: La Cueva de los Cristales (Sarsa de Surta, Huesca, Spain). *Journal of Archaeological Science: Reports* 31, pp. 102343. DOI: 10.1016/j.jasrep.2020.102343.

Villalba-Mouco, V., Oliart, C., Rihuete-Herrada, C., Rohrlach, A.B., Fregeiro, M.I., Childebayeva, A., Ringbauer, H., Olalde, I., Celdrán Beltrán, E., Puello-Mora, C., Valério, M., Krause, J., Lull, V., Micó, R., Risch, R. and Haak, W. (2022): Kinship practices in the early state El Argar society from Bronze Age Iberia. *Scientific Reports* 12(1), p. 22415. DOI: 10.1038/s41598-022-25975-9.

Waterman, A.J., Peate, D.W., Silva, A.M. and Thomas, J.T. (2014): In search of homelands: Using strontium isotopes to identify biological markers of mobility in late prehistoric Portugal. *Journal of Archaeological Science* 42(1), pp. 119-127. DOI: 10.1016/j.jas.2013.11.004.

Žalaitė, I. (2016): *Exploring Chalcolithic diet and mobility of humans and animals from Perdigões site.* MA dissertation. Evora: Universidade de Evora.

Zedda, N., Meheux, K., Blöcher, J., Diekmann, Y., Gorelik, A.V. *et al.* (2023): Biological and substitute parents in Beaker period adult–child graves. *Scientific Reports* 13, p. 18765. DOI: 10.1038/s41598-023-45612-3.

Žegarac, A., Winkelbach, L., Blöcher, J., Diekmann, Y., Krečković Gavrilović, M. *et al.* (2021): Ancient genomes provide insights into family structure and the heredity of social status in the Early Bronze Age of southeastern Europe. *Scientific Reports* 11(1), p. 10072. DOI: 10.1038/s41598-021-89090-x.

6

Kinship and Gender in Bronze Age Argaric Society

Eva Celdrán Beltrán

Department of Prehistory, Archaeology, Ancient History, Medieval History, and Historical Sciences and Techniques, University of Murcia, Spain, evaceldranbeltran@um.es

Vicente Lull

Department of Prehistory, Autonomous University of Barcelona, Spain, vicenc.lull@uab.cat

Rafael Micó

Department of Prehistory, Autonomous University of Barcelona, Spain, rafael.mico@uab.cat

Camila Oliart

Department of Prehistory, Autonomous University of Barcelona, Spain, camila.oliart@uab.cat

Cristina Rihuete-Herrada

Department of Prehistory, Autonomous University of Barcelona, Spain, cristina.rihuete@uab.cat

Miguel Valério

Department of Prehistory, Autonomous University of Barcelona, Spain, miguel.valerio@uab.cat

Abstract

This chapter explores kinship and gender in the Early Bronze Age society of El Argar (2200-1550 BCE) in southeast Iberia, with a focus on the site of La Almoloya (Murcia, Spain). Our interdisciplinary approach combines archaeological and archaeogenetic analyses, using recent genetic statistical tools. Before analyzing the data, we review the concepts of "kinship" and "family", challenging the Modern-era view of heteronormative patriarchal structures as "natural", and provide a historiographical overview of past hypotheses on Argaric kinship. Results suggest that Argaric communities were open and shaped by high mobility –mainly female– and social alliances often expressed in double burials. While patrilocality is attested at La Almoloya –now further supported by the detection of new third-degree genetic relationships– other social structures likely coexisted. Expanding the DNA sample both geographically and chronologically is crucial for verifying whether these conclusions, drawn primarily from a single site, hold more broadly. This would deepen our understanding of the social transformations in southeast Iberia during the El Argar period.

Keywords: *kinship, biological relatedness, archaeogenetics, Bronze Age, La Almoloya.*

In Blanco-González, A. and Alarcón-García, E. (eds.) 2025, *A Social Archaeology of Kinship in Iberia and Beyond. Recent Multistranded Approaches from aDNA to Household Archaeology.* Leiden: Sidestone Press, pp. 111-138.

6.1 Introduction

The concepts of "kinship" and "family" refer to basic principles of social organization involving sexuality, reproduction, and offspring. They have implications for various aspects, such as residence and the socialization of children. They are also involved in moral and ethical prescriptions and in scientific research projects. Their ubiquity and relevance explain why "kinship" and "family" have been used in so many ways and taken on a variety of meanings, and also why they are at the core of intense political and scientific debates. This vast multiplicity is beyond our scope; we will assess here how certain assumptions about both terms affect archaeological research. Although the semantic trajectories of "kinship" and "family" have converged, they are not synonymous and had different points of departure. The etymological root of *kinship* is *kin*, a term whose Indo-European source emphasized relatedness by birth (Klein 1971: 845-846). The semantic range of "kinship", however, has expanded beyond strictly blood ties to include relationships of various kinds and scales (political, residential, etc.). Thus, in current usage, the term is no longer restricted to consanguinity and includes both biological relatives and others.[1] "Family" followed the opposite path. Its Latin source, *familia*, referred to the Roman household, specifically the group under the authority of the male head (*paterfamilias*). *Familia* often denoted servants or slaves (*famulus* "servant" is etymologically "the one from the house") while excluding free individuals. Only later did it come to include the *agnati*, individuals from the same household related by blood through males. By the sixteenth century CE, the English word *family* also encompassed persons related by blood or marriage (Saller 1984). In this case, the relationships initially implied political and economic dependence, but over time they came to include all kinds of bonds –both by marriage and blood– beyond those defined by a master's ownership.

Thus, the meanings of "kinship" and "family" have converged from different starting points, leading to a few considerations that we will elaborate on in this text:

a. Consanguinity ties are involved in kinship and family, but they are neither sufficient nor necessary for setting up the social relationships these terms represent.
b. There is much historical variability in these two concepts and the realities they name, showing that there is no biological imperative behind them. In other words, no form of family or kinship is "natural". There is no single and mandatory "natural" foundation for kinship or familial relationships.
c. Both kinship and familial ties refer to other social relationships that are broader and more diverse. As the configuration of familial units depends on shifting social factors, it cannot be claimed that such units have a founding role in any society. Quite the opposite.
d. Once univocal determinations are abandoned ("God", "Nature"), the expression of kinship/familial ties reveals itself as subject to historical variation, which research must try to uncover in each case.

The main goal of this chapter is to explore kinship in a prehistoric society, focusing on El Argar Early Bronze Age (2200-1550 BCE) in southeast Iberia. Our approach combines

---

1   The same trajectory is seen in our native Neo-Latin languages: Spanish and Portuguese *parentesco* ("kinship") and Catalan *parentiu* come from Latin *parens*, a term used mostly in plural to denote the progenitors or immediate biological ancestors (and ultimately from *parere* "to give birth; to conceive"; see Lewis and Short 1879). It was only later that these terms came to denote non-consanguineous relations as well.

archaeological and archaeogenetic data, with much protagonism given to the site of La Almoloya (Pliego, Murcia, Spain). However, as in any archaeological research –perhaps even more so with research on kinship– the concepts and categories we use determine the results as much as the actual data. Thus, before addressing the empirical work, it is important to clarify and position ourselves regarding several key notions founding the research.

6.2 Human "nature" and kinship

Since Hume's critique of causality, philosophy and science have explored how certain conditions can generate others without necessarily determining them. Nonetheless, Western thought remains largely causalist. As a result, certain "first causes", "God" and later "Nature", usually explain in last instance past and present major human affairs. But how can one conceive of natural causality in the explanation of human affairs without resorting to divine intentions? Modernity has sought to achieve this through a dual operation: (a) separating the "human" from the "social", and (b) considering the human dimension as a previous and founder instance for social bonds, and keeping always the role of ultimate cause. Thus, we should imagine the earliest human existence in a pre-political and pre-social "state of nature". Hobbes saw humans as selfish, individualistic machines in constant struggle for survival. The only lasting interaction amidst this violent "state of nature" ("the war of all against all") would be the relationship between a mother and her offspring—a strange, disruptive element in the Hobbesian ontology. In contrast, for Locke, the "state of nature" had the heterosexual nuclear family as the only effective relationship for the basic functions of procreation and child-rearing, as enacted by natural law. This idea has been fundamental in establishing a universal concept of an alleged human essence. For Hobbes, Locke, and other modern thinkers like Rousseau, the departure from the "state of nature" occurred when human entities –family and the individual– overcame their natural inertia for convenience or interest. Political society emerged from a covenant between these entities, marking a turning point. Since then, humanity –defined by its individual and/or familial "natural" essence according to the heterosexual monogamous model– has been subordinated to the political laws that structure social life. From the nineteenth century onward, social and human sciences have sought to explain and illustrate humanity's transition from its natural state to its social and political condition. Anthropology and archaeology have provided empirical support for the mythical "state of nature", while still adhering to core philosophical assumptions, such as a humanity's simple and primitive infancy driven by biological needs (subsistence, reproduction), and a radical event that founded Law and, hence, society. This event, central to this chapter's topic, is the prohibition of incest.

We must critically reassess this belief: neither individuals nor families were the starting point for society; society is not the result of a contract between pre-existing parties. On the contrary, social relations –social production– precede any party or pact. Thus, Hobbes' *Leviathan* assumed a previous male domination over women, as Pateman (1988) duly noted. A prior "contract" among men that denied the importance and rights of the only true founding relationship, that of mother and offspring, without which men –first violent, then reasonable– could not even exist. Yet before any reasonable agreement was reached, reason itself had to be achieved through language. Language, in turn, could not emerge without close and enduring social relations, an entirely opposite situation to the "war of all against all". In short, the *Leviathan* is not the result of a primeval contract, but the last in a sequence of socially and historically formed relations.

If Hobbes' *Leviathan* legalized a prior social imposition –androcentric, but social nonetheless– Locke's "government" or "civil society" did the same for the male head of the family, who became political representative, judge, and ruler. Locke's approach is especially relevant because it includes two arguments that have implicitly or explicitly shaped mainstream research in the social and human sciences:

a. First, it naturalizes the heterosexual nuclear family, seeing it as a union driven by the biological imperatives of reproduction and child-rearing. It is easy to see that this argument is based on an unfounded preconception (prejudice): it assigns to the specific form of cooperation in nuclear families the role of a universal necessity (the social cooperation needed to sustain the lives of infants). The nuclear family and any other form of cooperation entail a prior social relationship, whose historicity is denied by invoking nature.

b. Second, it makes private property individual by considering it as the result of personal work on an object (land, raw materials). In doing so, it also naturalizes property as a consequence of the physical transformation of nature by a creative effort, both physical and "natural". Locke's argument relies on omission, abstraction, and appropriation: it denies that any worker is in turn the product of social relations, which provides the necessary capacity, skills, and means to work. This argument fictionalizes laborers as autonomous, self-sufficient entities –like a creative deity– and claims that the fruit of their individual labor sets up a Right to Property... Except if its fruit is another individual. Here, "Nature" is again called upon to deny that gestation is a form of labor, and consequently, to deny that the product of this labor belongs to the mother. Thus, Locke ideologically subverted the labor-property principle he proposed: as head of the family, the father claims ownership of new individuals without having contributed labor to their production. As with Hobbes, these supposedly foundational proposals conceal a prior, violent foundation that the appeal to "Nature" attempts, in vain, to silence.

Thus, Hobbes and Locke presented as the primary and founding political relationship what is actually a secondary relationship designed to establish and protect other prior asymmetric political relations: the nuclear family and private property. In both cases, the will to define the origin of society as an event following a natural human state did not erase the traces of this self-serving operation. Social and political relations already existed in the state of nature, even before individuals were born. There is no pre-social starting point for society. Society *produces* many things, even the ideologies that deny it. Lévi-Strauss's account of the founding of Law echoes modern-era accounts: the prohibition of incest led to the exchange of women and, therefore, to a contract among men. Yet, those who agreed to this exchange and made the pact were already owners of women. That is, they had previously allied and agreed to appropriate females' bodies. Notice the structural similarity between Lévi-Strauss's argument and the foundational myths of modern thinkers.

The claim that family exists naturally stems from the need of the currently dominant form of family as the basis for society. To justify its imposition, the hegemonic family needs the label "natural" as much as individualism needs "competition", which is also considered natural. Once "individual" and "family" were established as essential principles "by Nature", "society" can be seen as the final outcome of an agreement between individuals or families. The heteronormative patriarchal ruling systems that impose exclusive forms of family insist that these forms are "natural" and that, therefore, questioning them would be "anti-natural". Yet, it might be worth considering whether it is not more absurd to regard as "natural" practices like dowry or bride price, the symbolic handover of the woman to her husband, or even surrogate motherhood. If these relationships are fixed as natural, practices that often fit the definition of human trafficking become normalized. Through repetition, the body is filled with language rather than reality, creating subjugated subjects, as Judith Butler translates from Michel Foucault. "Everything which is usual appears natural", as the famous quote by Harriet Taylor and John Stuart Mill suggests.

But then, is there any original, natural bond between people? The only things we could consider natural are that we are born and die, we copulate like many living beings, and only some bodies can become pregnant and initially nourish other bodies. *If there is any bond that can be called a natural family, it is the one formed by mother-child* (cf., Fox 1983 [1967]: 27, 40). The role of men here is incidental, fortuitous, or random. Beyond sexual intercourse –which assisted reproductive technologies no longer require– their role is dispensable or potentially irrelevant. Therefore, the man/woman binary is unnecessary for the formation of society, even though it is regarded as its basic component. The defense of a "natural" binary clashes with current knowledge. The man/woman binary is a construct, as we cannot even assert from a strictly biological perspective, whether it is based on chromosomes, hormones, or gonads, that only two sexes occupy and illustrate the world (Fausto Sterling 2000; Ainsworth 2015).

We should agree that society as a whole embraces forms of cohabitation, whether sexual or not, and also determines who represses or allows interpersonal relationships. The most critical moment against open relationships among people occurs when the law enters the scene –through custom, interest, or decision– and stipulates how people should be and be seen, relate to one another, with whom, under what conditions, and even how bonds of belonging should be established. This law or norm is what defines gender relationships (Rubin 1975). Thus, kinship systems are crossed by the sex/gender system.

Kinship and family relationships are based on alliances supported, among other factors, by common residence. Of course, families perpetuate themselves through offspring –though that is not the only mechanism. However, before reaching that point, they undergo a long journey that begins by considering which pairings are suitable and which are not. A family formed in this constrained way hinders the relational freedom of individuals, restricts sexual freedom, and may become a reducing and repressive institution. We cannot speak of an essential and natural society, nor can we say that some societies are more natural than others, because nature is dynamic and shifting.[2] It is our relationships that make us natural. In fact, going against nature means fixing, determining, or deciding which of our ways of doing or thinking are natural and, therefore, correct. Regardless of what the laws and certain assumptions made in the context of human and social sciences say, *no human is more "natural" than another human; rather, we have become the humans we are today and will become the humans+ we will be.*[3]

---

[2] Some authors argue that there is vibrating life in inorganic matter (Bennett 2010) and others defend a flat ontology of mutual influence between organic and inorganic matter (Latour 1991; Harman 2018), thereby abandoning the animate/inanimate distinction (Barad 2007).

[3] Used as an adjective, the word for "humans" is gendered in Spanish (*humanos* and *humanas*). We use the neologism *humanes* to convey a gender-neutral form. As English "humans" lacks grammatical gender, with the mark "+" we seek to maintain this political gesture in the translation.

6.3 Kinship and archaeogenetics

Because we live in a heteropatriarchal normative system, it is crucial to recognize that scientific proposals may be constrained by a self-serving association of kinship, social identities, and ideological interests that favors the *statu quo*. Archaeology is no stranger to this. However, it benefits from being positioned halfway between two forms of understanding: one that entails reflection and another that involves knowledge. Genetic studies relevant to archaeology (archaeogenetics) are nothing more than a new research tool. And although it may appear to be a revolutionary field devoted to scientific neutrality, it risks being used in support of old identity politics or, quite simply, racist agendas (TallBear 2013; Saini 2020).

The virtue of archaeogenetics lies in its ability to determine the biological distance between individuals by means of direct evidence. To suggest that political and social relationships, in general, stem from this biological distance is an overreach that one must remain vigilant against. The temptation to link family and consanguinity is no less dangerous than that of connecting "race" and "people" or "ethnicity" and "culture". Genetic research on kinship, when it aims to be socio-culturally defining, sometimes reveals an essentialism that aligns with many of today's policies of exclusion. The purported support of genetics is as scientifically fallacious as it is politically dangerous and reactionary.

The genetic makeup of human bodies is not the signature of our ways of doing, gathering, or thinking. Making society has little to do with the DNA of individuals or groups. The material ties that produce any form of social life do not necessarily correspond with biological affinity. Similarities or distances in genetic makeup have little or nothing to do with social behaviors. Genetic material cannot generate categories of people or populations. Even so, genetics is in vogue. Yet, archaeology should not be concerned with following trends, but rather with questioning them and examining why today's society accepts the answers it gets. We believe that good research involves challenging what is considered certain and asking what society does not usually ask. If we limit ourselves to doing retrospectives of what suits us, there is no reason to pursue archaeology.

In this chapter, we will attempt to engage with scientific narratives whose primary focus ought to be the pursuit of precision. To achieve this in research, there is nothing better than the ethics of care. Care is a political behavior that pays attention to small details, a key attitude when we seek to understand where do we come from, what those who preceded us did, and how we arrived here.

6.4 The archaeology of El Argar and research on kinship

El Argar is an archaeological group that is highly exposed to the dangerous equation of blood, ethnicity, and culture. The reason is that it is one of the few entities from European prehistory that can be identified because it developed a solid material and relational homogeneity across a vast territory for centuries. As we will see, this close "material relatedness" teaches us that even in unique cases like El Argar, we cannot equate such relatedness with a "people" in the ethnobiological sense.

6.4.1 Social and archaeological theory

Prehistoric archaeology has relied on kinship concepts and relationships used in anthropology. Terms such as family, lineage, or clan have sporadically populated archaeological discourses, thanks to extrapolations between archaeological features and anthropological definitions. There have been few attempts in our discipline to disconnect from this theoretical and methodological dependency. More than two decades ago, some of us proposed a materialist approach according to which the empirical record of archaeology contains the expression of three types of social practices: economic, political, and kin-based (Castro *et al.* 1996). Kin-based or "socio-kinship" practices –*prácticas socio-parentales* in the Spanish original– referred to activities related to pregnancy, breastfeeding, maintenance of labor force, and education/socialization, as long as they were carried out by individuals with blood ties and/or other types of affinity. Conversely, if these ties were not involved, we referred to the activities as "sociopolitical practices". While the material and spatial expression of "socio-kinship" practices are the domestic units, sociopolitical practices take place in "supra-domestic" structures, such as hospitals and schools. In short, we discarded adopting anthropological concepts (lineage, clan, etc.) as references and aimed at discovering the hitherto unknown combinations involved in the production and socialization of subjects.

Indeed, unlike socioeconomic practices, which are oriented toward the production of objects (food and all kinds of implements), "socio-kinship" practices produce social subjects. In theoretical terms, these practices are linked to two areas of general production: basic production and the production for the maintenance of subjects (Castro *et al.* 1998, 2001). The first, also known as "reproduction", refers to the gestation of new individuals and, in the biological sense, is led by females. Conversely, the maintenance production involves subjects regardless of their sex or gender.[4]

Basic production, production for the maintenance of subjects, and their material expression in social practices highlighted aspects of general production in the Marxist sense that had been hidden or naturalized by research. They brought to light what a patriarchal ideological bias insisted on overlooking, in the pursuit of a holistic sociological knowledge that, paradoxically, achieved the opposite: knowledge that was always partial, exclusive, and exclusionary. To sum up:

a. Our interest in investigating the field called "kinship" was to understand the social mechanisms for the production of subjects. While the role of males is incidental or random, depending on how one views it, female bodies are basic to biological reproduction. Moreover, there are multiple possibilities when it comes to establishing the groups responsible for the maintenance and upbringing of individuals. In any case, the man/woman binary is not necessary for the formation of society.

b. We wanted to investigate the social relationships involved in child-rearing (the production for the maintenance of subjects) as they need not involve the mothers, or at least not exclusively. In the basic production and the production for the maintenance of subjects, the measure of biological relatedness/consanguinity among individuals is an element which empirical (in our case archaeological) research could objectify.

c. Once the degree of biological relatedness among subjects is known, archaeological research should continue, by opening research lines aimed at determining the political, economic, and symbolic components that shape social relationships.

---

4   We must stress that the theoretical content of the "production of maintenance" is quite different from the later proposal of "maintenance activities" (González Marcén *et al.* 2008). For a critique of the latter, see Lull *et al.* (2016: 44).

6.4.2 Kinship in the research of the late prehistory of southeastern Iberia

Research on kinship has appeared unevenly in the study of late prehistoric southeastern Iberia. For the Chalcolithic society of Los Millares, it was inferred from their modest size that its typical circular huts housed "families" or "domestic units", while the collective burials –in orthostatic and false-dome tombs, natural caves, and hypogea– would accommodate the members of a "lineage", that is, a corporate group who recognized a common ancestor (e.g., Chapman 1990). Overarching all this, the notions of motherhood and of a "mother goddess", symbolized in different media, would connote the social value of biological filiation and thus the centrality of women in social relations and their imaginary.[5] In short, anthropological extrapolation and interpretive approaches have driven the timid attempts to explore kinship during the Copper Age.

Paradoxically, research on kinship in El Argar has been limited by two of its most numerous and relevant archaeological traits. First, the abundance of individual tombs seems to symbolically highlight the individuality of subjects above any other social bond, making it difficult to identify such social ties in the archaeological record. Second, the identification of residential units –that is, structural containers that could point to bonds between individuals buried separately underneath them– often proves controversial. This is because documented living spaces are fragmentary –due to later building activity or erosion in hilltop sites– difficult to interpret as a whole when several individual spaces aggregate in building complexes, or contain a surprisingly high number of tools, suggesting "workshops" rather than "houses". Still, Argaric archaeology found a way to explore kinship: the analysis of double tombs.

In El Argar society, a double tomb is not merely a grave with two corpses. Its relative frequency and internal structure make it a material expression of the social bond between two individuals through a ritually sanctioned arrangement. About 10% of the Argaric burials are double, with some sites exceeding 20%. Nearly two-thirds of the documented double tombs contained adult individuals; a quarter included one adult and one child; and only about 8% contained two children (Lull *et al.* 2016). With few exceptions, tombs with two adults were meant for a male and a female,[6] buried at different times, as the remains of one of them is often found in a secondary position. The contiguity of the bodies, the combination of sexes, and sometimes the care taken in their deposition have been interpreted as evidence of a heterosexual monogamous marital bond and, by extension, the nuclear family. In short, adult double tombs would literally represent monogamous heterosexual couples. This chain of inferences established a view of Argaric kinship that has remained dominant since the late nineteenth century.[7]

---

5   A recent biochemical study (DNA and proteomics) on pre-Argaric populations buried in collective tombs at the megalithic cemetery of Panoría (Darro, Granada), which was published as this chapter was completed, suggests that women may have played a prominent role in social organization and funerary rituals, possibly indicating a matrilineal descent system (Díaz-Zorita Bonilla *et al.* 2024).

6   Exceptionally, double tombs containing two adult males have been found at Cerro de la Encantada 37 (Monsalve *et al.* 2014), Eras del Alcázar 9 (Jiménez-Brobeil *et al.* 2009), Cerro del Alcázar 12 (Robledo and Trancho 2003), Castellón Alto 110 (Rubio 2021), and La Bastida 18 (Lull *et al.* 2016). To the best of our knowledge, no tomb containing two adult females has been found so far.

7   See, e.g., Inchaurrandieta (1870: 809), Siret and Siret (1890: 206), Childe (1958: 284), Contreras *et al.* (1997: 134), and Schubart (2012: 42).

Interpreting a double tomb as a sex-affective relationship between two adults of different sex is reasonable from the perspective of the common-sense of our time. By the same token, the combination of an adult and a child in a burial would evoke a mother-child or father-child relationship. The arrangement of the skeletons could support this interpretation, as it has been seen that the two corpses were buried at the same time in the case of some tombs containing a female adult and a newborn. Finally, in the case of two children, one might assume they were siblings.

To test these widespread "common-sense" inferences, we began producing ^{14}C dates for male and female adult skeletons from double tombs in the late 1980s (Castro *et al.* 1993, 1994). At that time, we benefited from a novelty that today is routinary: the first AMS radiocarbon results, which allowed for dating without compromising the preservation of a large portion of the skeleton. The initial results from a small sample of five graves (Gatas 33 and 37, Fuente Álamo 75, Lorca-Los Tintes 2, and Lorca-Madres Mercedarias 11) suggested that each individual's time of death likely differed by at least several decades, making their coexistence unlikely (Castro *et al.* 1993-1994: 88-89; Lull 2000). Later, Bayesian statistical analysis of a larger sample (23 double tombs) suggested an intergenerational gap, although in some cases the temporal distance appeared short or nearly absent (Lull *et al.* 2013). This drew a complex and varied scenario.

The hypothesis suggesting a substantial temporal gap between adults challenged the monogamous model. Double tombs of adults could symbolize a relationship distinct from marriage, and it was ruled out that the two children or the adult and child in other types of double burials were as supposed. The chronological distance between adults opened the possibility that alternative genealogical relationships were represented. In the early stages of research, when the evidence was limited to a few graves from Gatas, Fuente Álamo, and Lorca with anthropological determinations, finding that the first individual buried was consistently female led to suggestions of a matrilocal and matrilineal structure, likely within "extended families" (Castro *et al.* 1993-1994; Lull 2000; Lull *et al.* 2016). Based on this, we proposed intergenerational biological ties, such as grandmother-grandson or grandfather-granddaughter. This proposal left unexplained the nature of the bond in cases where a significant chronological distance between the two individuals in a tomb could not be established.

The new hypothesis was supported by osteological analyses and the interpretation of certain funerary elements. First, an anthropometric study of a sample of skulls from the site of El Argar showed that the cranial variability among males was five times greater than among females, suggesting greater postmarital mobility among men and, therefore, matrilocality (Buikstra and Hoshower 1994). Only some males would have remained in their birthplaces for their entire lives. Second, the presence of copper awls –exclusively female items– in child burials (girls?) outnumbered the presence of axes –exclusively male items– in child burials (boys?), which suited better a hypothetical matrilineal transmission of rights (Lull *et al.* 2005).

From the start of our investigation, we were aware that the proposal of double tombs as expressions of biological and genealogical bonds had to be tested directly by means of genetic analyses. A range of alternatives with clearly defined contrasting implications then emerged:

a. If the two individuals were separated by an intergenerational time gap and had consanguineous ties, then the hypothesis would be confirmed.

b. If a significant temporal distance between the two individuals were impossible to observe, but genetic proximity existed, then the heterosexual monogamous model could be challenged, also allowing for alternative hypotheses of kinship relationships.

c. If the two individuals were not separated by a significant temporal distance and did not share blood ties, they could have had a sex-affective relationship in life, as suggested by the "common sense" hypothesis.

Since the late 1990s, we have carried out archaeogenetic analyses aimed at testing these hypotheses on pairs of adult skeletons from double tombs from the sites of Gatas (Almería), Fuente Álamo (Almería), and Lorca (Murcia).[8] Unfortunately, the poor preservation of bone material, the limitations of techniques for recovering aDNA at the time, or both, prevented us from obtaining positive results. Years later, prospects improved significantly due to new techniques for recovering and sequencing aDNA, along with the implementation of standards for the preservation of bone samples from the moment of their discovery. Thus, in 2015, we started a new program, this time in collaboration with the Institut für Anthropologie at the University of Mainz (K. Alt, C. Roth) which shortly afterward expanded to the Max Planck Institut für Menschheitsgeschichte (W. Haak, J. Krause, V. Villalba-Mouco), and finally yielded positive results (Villalba-Mouco *et al.* 2021, 2022).

The interdisciplinary combination of extensive archaeological, anthropological, and archaeogenetic research has enabled us to assess decades-old competing hypotheses, outline a new state of affairs, and propose future lines of research. We will address these three issues next.

6.5 Testing Kinship Hypotheses

The genetic sampling of the Argaric population has been uneven in terms of its representativeness and success rate. Up to date, 87 individuals have been successfully analyzed,[9] mainly from the area around the Espuña Mountain Range and the nearby Guadalentín Valley (La Almoloya, n=68; La Bastida, n=10; Lorca, n=3), but also from inner highland areas such as Cerro del Morrón (n=3) (Celdrán *et al.* 2023), Cerro de la Virgen (n=2), and Cerro de la Encina (n=1). The adult (n=64) and subadult (n=23) groups are evenly represented by sex (34 XX and 30 XY adults[10]; 13 XX and 10 XY children)[11] (Fig. 6.1 and Tab. 6.1). These individuals were buried in single (n=49) and double (n=37) graves, spanning the two main Argaric phases (*ca.* 2000-1750 cal BCE, 55% of individuals; *ca.* 1750-1550 cal BCE, 45%).[12]

---

8   These initial analyses were conducted in collaboration with J.E. Buikstra (New Mexico University and Arizona State University) in 2001, 2003, and 2005; with A. Malgosa (Unitat d'Antropologia of the Universitat Autònoma de Barcelona), in 2006; and with D. Reich (Harvard Medical School) and C. Lalueza-Fox (Instituto de Biología Evolutiva, CSIC-Universitat Pompeu Fabra), in 2014.

9   The genetic data is curated and publicly available at the Harvard University dataverse (Mallick and Reich 2023).

10   We use the terms "male" and "female" following the sex determinations made in the context of the osteological study of skeletons from Argaric sites. Similarly, the distinction between XY ("male") and XX ("female") chromosomes is used when archaeogenetic data are available. Thus, these terms strictly refer to a biological consideration produced by scientific methods involved in archaeological research. Research on gender categories in Argaric times remains one of the goals of our group, though it is still ongoing.

11   This is particularly significant considering that children typically represent 40% of the Argaric burials. In this case, the child sample accounts for 26% of the total.

12   An older sample is also included, originating from a secondary context within the fortification complex of La Bastida (BAH77-112.1).

Site (municipality, province)	Samples analysed	Valid results
La Almoloya (Pliego, Murcia)	86	68
La Bastida (Totana, Murcia)	27	10
Lorca (Lorca, Murcia)	13	3
Cerro del Morrón (Moratalla, Murcia)	3	3
Fuente Álamo (Cuevas de Almanzora, Almería)	8	0
Monteagudo (Murcia, Murcia)	2	0
Cerro de la Virgen (Orce, Granada)	3	2
Cerro de la Encina (Monachil, Granada)	Unknown	1
Total	**>142**	**87**

Table 6.1. Genetic samples from Argaric settlements.

We have explored Argaric biological relatedness along two dimensions within the framework of our collaboration with the Max Planck Institute. First, we assessed the biological distance of the Argaric population as a whole in relation to coeval populations (Villalba-Mouco *et al.* 2021). Second, we estimated biological relatedness among Argaric individuals, both at the intra-site and the inter-site levels (Villalba-Mouco *et al.* 2022).

At the intra-site level, La Almoloya stands as one of the largest genetic series from a single prehistoric community studied so far. Remarkably, positive results account for more than half of the skeletal collection (53%) that had been excavated at the time the sampling was conducted. Our inferences on kinship derive from this collection. However, they should be considered preliminary. Verifying and refining our understanding of "socio-kinship" practices in El Argar requires additional samples from La Almoloya and other Argaric sites, as well as other independent indicators.

One novelty of this chapter concerns the identification of third-degree relationships –i.e., cousins, great-grandparents and great-grandchildren, or great-uncle/aunt and great-niece/nephew– through the application of two new statistical tools: BREADR (Rohrlach *et al.* 2023; version 1.0.2), and READv2 (Alaçamlı *et al.* 2024). This will help broaden our previous proposal of biological relatedness at La Almoloya. Both tools are based on PMR (Pairwise Mismatch Rate) and proactively identify third-degree relationships, although their presentation of results differs slightly. While BREADR specifies whether the degree of kinship is hypothetically maintained or rejected, sometimes showing that a relationship between two individuals may fall between different degrees, READv2 simply indicates whether the relationship is first, second, or third degree, without specifying the probability. For third-degree relationships, the two tools agree on 15 cases, yield compatible results in 5 cases,[13] and contradict each other in 20 cases (Tab. 6.2 and Fig. 6.4). In 15 cases, READv2 suggests a third-degree relationship, while BREADR determines the relationship to be fourth-degree or more distant. In 5 cases, READv2 indicates that the two individuals are unrelated, while BREADR maintains the hypothesis of a third-degree relationship. Four out of the 16 third-degree relationships identified by both BREADR and READv2 were also validated through IBD (identical-by-descent segments).[14] In this study,

---

13   Third- to fourth-degree relatives: AY75-AY50, AY80/0-AY22/2, AY14-AY60/2. Third- to fifth-degree relatives: BA23/2-MN2/1. third- to ninth-degree relatives: AY13-BA6/2.

14   AY80/0-AY80/1, AY32-AY102, AY80/2-AY28, and AY46-AY90/1.

only the cases where the two software tools (BREADR and READv2) show agreement have been selected.

Only 36 out of 87 individuals are involved in first- or second-degree relationships,[15] and several of these individuals show third-degree relationships. Just 7 additional individuals take part in them (AY5, AY25, AY32, AY46, AY60/2, AY85/2, AY102). In the previous study (Villalba-Mouco *et al.* 2022), some second-degree relationships had been identified (e.g., AY90/1-AY94/2 and AY60/1-AY94/2), but the absence of additional connections hampered the reconstruction of a pedigree. Now, thanks to READv2 and BREADR, we have detected third-degree relationships among some of the individuals involved (AY90/1-AY46, AY60/1-AY90/1, AY94/2-AY46), making the reconstruction possible (Figs. 6.5 and 6.6, proposals A and A'). Another set of individuals related in the third degree has also emerged: AY85/2-AY32, AY32-AY102, AY32-AY25, AY5-AY32, AY85/2-AY25, AY5-AY25, and AY60/2-AY25.[16] We can reconstruct alternative pedigrees for both sets, and they all necessarily involve the presence of adult siblings at La Almoloya. More specifically, the identification of infant first cousins, namely the 2-3 years-old XY AY25 and the XX neonate AY85/2, suggests that their parents, who were siblings, could have cohabited at the site (Figs. 6.5 and 6.6, proposals B and B'). A third adult cousin of AY25 and AY85/2 is identified (AY32 in proposal B or AY5 in proposal B') whose mitochondrial DNA was different from AY25, so necessarily they were cousins on their father's side.

In a permanent, long-lasting settlement, one might expect an exponential increase in genetic relatedness over time, including relationships in more distant degrees (i.e., the number of more removed relatives should increase). However, this is not the case at La Almoloya, where almost the same number of first-, second- and third-degree relationships are identified (Tab. 6.2). We have determined that 41 out of 68 individuals are genetically related up to the third degree. Hence, biological relatedness appears to remain relatively low, mainly reflecting the closest bonds (first degree) and being restricted to a particular group within the community. The small number of second- and (especially) third-degree relationships should also be viewed in light of the spatial proximity of tombs of individuals with close biological ties (Villalba-Mouco *et al.* 2022: Figs. 2a and 3). The latter suggests significant ritual focus on the closest blood relatives, often involving children, or children and their parents.

However, the correlation between spatial closeness and biological relatedness does not apply when considering adult individuals of different sexes buried in close proximity, namely in double burials. In these cases,[17] the individuals are not genetically related until the sixth or seventh degrees (Villalba-Mouco *et al.* 2022); that is, they did not share close biological ancestry. Instead, thanatological, taphonomical and/or radiocarbon analyses have revealed that the two adults in a double grave may have been contemporaneous. Moreover, in three cases (AY22, AY80, AY38) it has been shown that they had offspring also buried at La Almoloya.

---

15 The paired relationships BA12/2-BA23/1 (first degree) and AY30/1-CM1/2 (second degree) have been excluded due to genetic and chronological discrepancies (Villalba-Mouco *et al.* 2022).

16 AY25-AY102 may also be related in the fourth or fifth degree (BREADR).

17 first-, second-, and third-degree relationships have been ruled out for 11 double tombs from La Almoloya (AY22, AY24, AY38, AY42, AY60, AY68, AY80, AY90, AY94, and AY94) and in one from Cerro del Morrón (Tomb 1), as well as up to sixth/seventh-degree relationships in two burials from La Almoloya (AY80 and AY90).

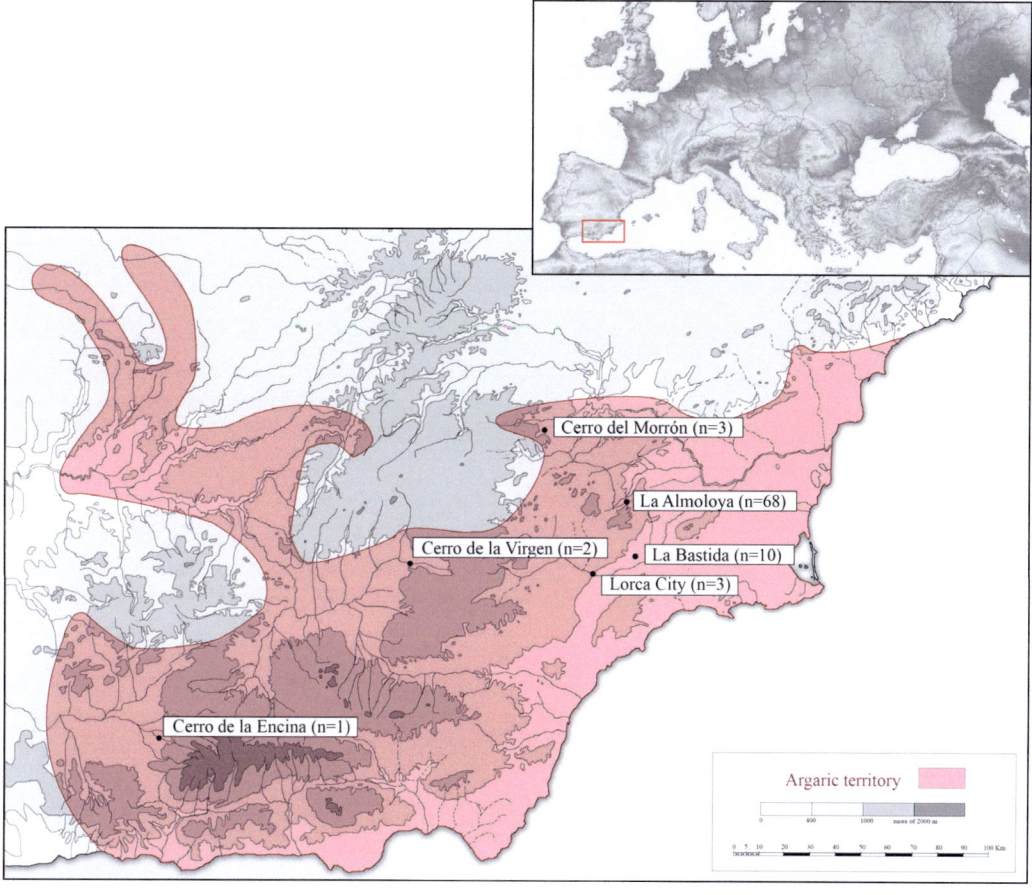

Figure 6.1. Map of the El Argar territory showing the sites with available genetic data (source: ASOME-UAB).

Therefore, our previous hypothesis that double tombs expressed intergenerational biological ties is not supported by the new data. Evidence beyond genetic analyses had already weakened the scenario of intergenerational matrilineality on a matrilocal basis. Thus, it was shown that the time elapsed between two interments in double graves was likely short in most cases. The application of more restricted statistical tests, such as the "Combine" analysis (OxCal 4.4) and the "test sample significance" (Calib 8.2), indicated that only 15-20% of the pairs of radiocarbon dates show significant differences.[18] Similarly, the trend of burying the female first in double graves, which was observed in the initial study at Gatas, Fuente Álamo, and Lorca (Lull *et al.* 2016) and supported the matrilocal hypothesis, is now more nuanced in light of recent excavations. Currently, the tendency attested at La Almoloya is just the opposite, as only in 8 out of 26 cases was the female buried first.

---

18   Interestingly, these significant differences have been observed in some of the pairs of radiocarbon dates produced in the late twentieth century with broader standard deviations. This raises the possibility of substantial measurement errors among the early AMS dates.

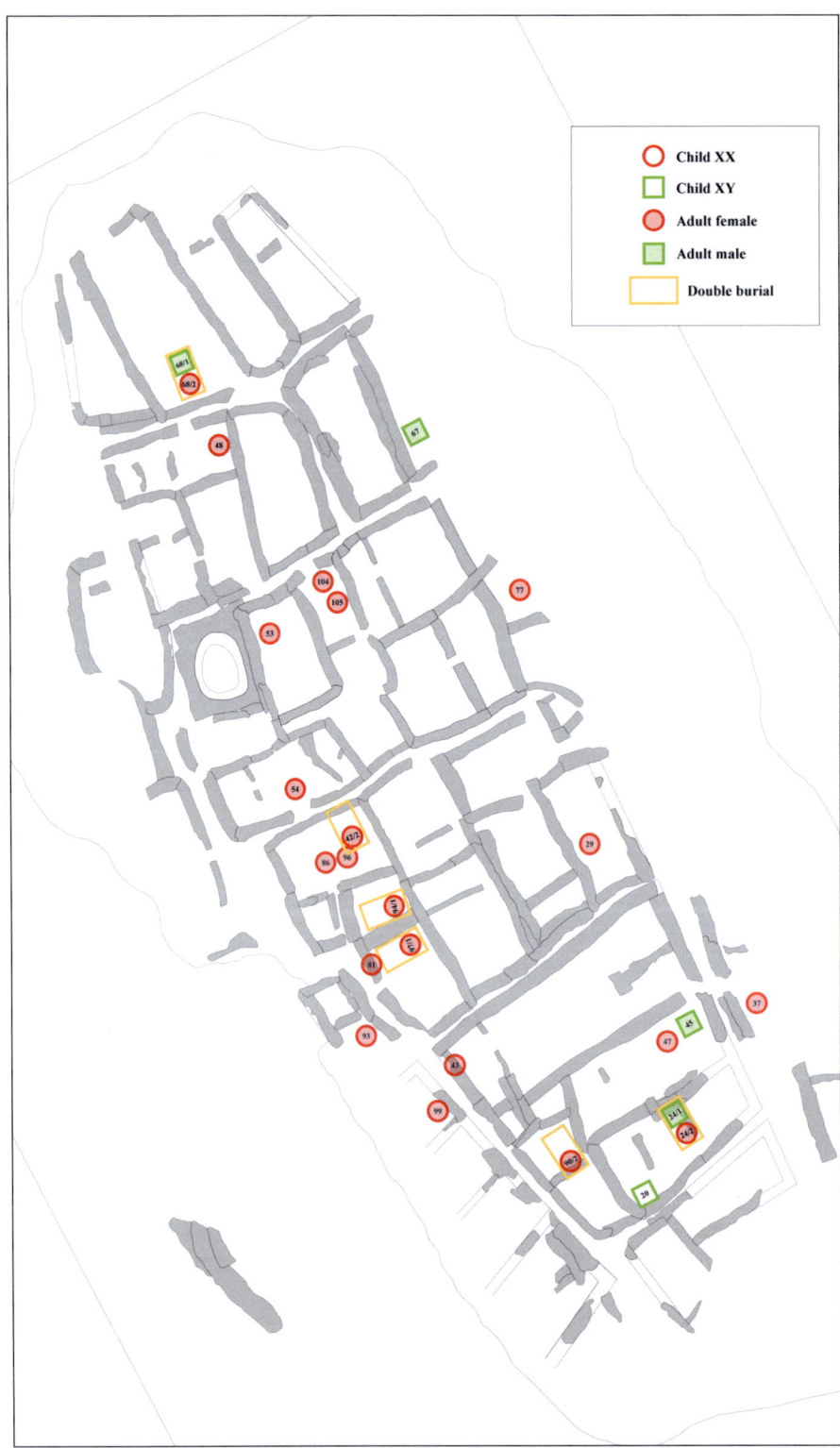

Figure 6.2. La Almoloya layout showing the locations of individuals identified as genetically unrelated through aDNA analysis (source: ASOME-UAB).

A SOCIAL ARCHAEOLOGY OF KINSHIP IN IBERIA AND BEYOND

Figure 6.3. La Almoloya layout showing the locations of individuals identified as genetic relatives through aDNA analysis (source: ASOME-UAB).

Kinship degree	Number of paired relationships	Number of individuals involved	Paired individuals genetically related
first	14	23	AY89-AY88, AY22/1-AY16, AY26/2-AY27/1, AY38/1-AY30/1, AY38/2-AY30/2, AY28-AY80/0, AY80/2-AY42/1, AY80/1-AY42/1, AY38/2-AY30/1, AY8-AY17, AY50-AY27/1, AY22/2-AY16, AY21/2-AY21/1, BA23/2-BA23/1
second	10	19	AY90/1-AY94/2, AY30/2-AY30/1, AY95-AY87, AY75-AY27/1, AY60/1-AY94/2, AY57-AY58, AY50-AY26/2, AY82/1-AY97/2, AY8-AY23, AY14-AY13
third	16	15	AY80/2-AY80/0, AY85/2-AY32, AY32-AY102, AY5-AY25, AY80/0-AY42/1, AY5-AY32, AY90/1-AY46, AY80/1-AY80/0, AY60/1-AY90/1, AY32-AY25, AY28-AY42/1, AY28-AY80/2, AY94/2-AY46, AY85/2-AY5, AY60/2-AY25, AY85/2-AY25

Table 6.2. Synthesis of paired genetic relationships from first to third degree (according to BREADR and READv2).

Might it be that the adults in double graves were couples who had children together? Genetic evidence shows that in three cases, both individuals in the double tombs had first-degree descendants (offspring). Additionally, in more than one-third of the adult double burials, at least one individual buried in double graves had direct descendants in La Almoloya. This points to an affirmative answer.

However, parenthood was not expressed only in double tombs. It is noteworthy that at least five mothers were buried in single tombs or alongside their offspring, but not with their male mating partners. This is the case of the female in the single tomb AY17, who was buried close to her daughter (AY8); burials AY21 and AY85, where females were buried alongside their neonates; and tombs AY3 and AY36, which contained pregnant females that most probably died due to complications during childbirth. From the perspective of gender, it is remarkable that there is no evidence of fathers buried close to their children, either in nearby graves or inside double burials. In fact, there was a double tomb with two XY individuals, a young adult and a neonate, at La Bastida (BA6), but genetics showed they were unrelated (see below). Double tombs containing adults of different sex may represent reproductive couples, but these are not the only burials that entail reproduction. We have also found mothers buried with or near their offspring. Therefore, the ritual expression of reproduction extended beyond double tombs, and the variety of funerary rites likely conveyed additional meanings.

The deliberate nature of adult double tombs is indicated by evidence of planning in the graves themselves. At least in La Bastida and La Almoloya, domestic *pithoi* were reused as burial containers. Thus, the size of these funerary vessels relied on the pottery available for reuse at the time of death. However, cists were newly constructed, allowing for easier customization. At La Almoloya, the size of the adult cists varied depending on whether they were intended for one or two occupants.[19] This suggests that when the first burial took place, plans were already in place for a second person to be laid to rest in the same grave.

---

19 At La Almoloya, 47 cists containing at least one adult have been found during the excavations of the Universitat Autònoma de Barcelona (2013-2023): 26 single and 21 double burials. Despite the overlap in the distributions of the values of inner cist dimensions (single graves: 0.66 ± 0.04 m²; double graves: 0.86 ± 0.07 m²), the difference between the two series reaches the threshold of statistical significance (Mann-Whitney test, U=182.5, z=1.9258, p=0.053).

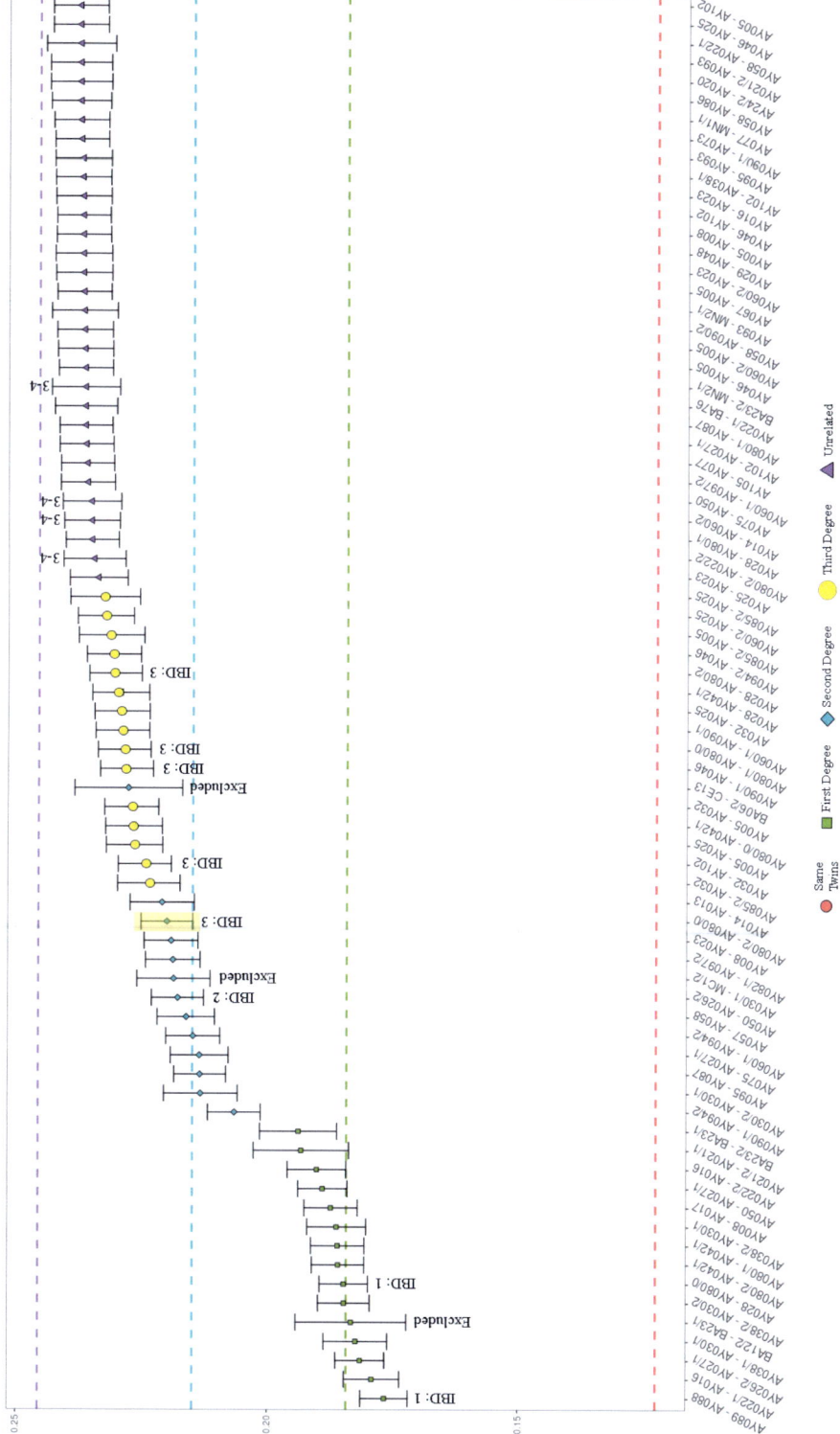

Figure 6.4. The first 74 ordered pairwise-mismatch rates (PMRs) from the Argaric collection (source: BREADR, plot modified to include third-degree relationships). The relationship between AY80/0 and AY80/2 is indicated as second degree by BREADR and READv2, but IBD assigns it to third degree. The IBD assignment is consistent with the pedigree (Villalba-Mouco et al. 2022).

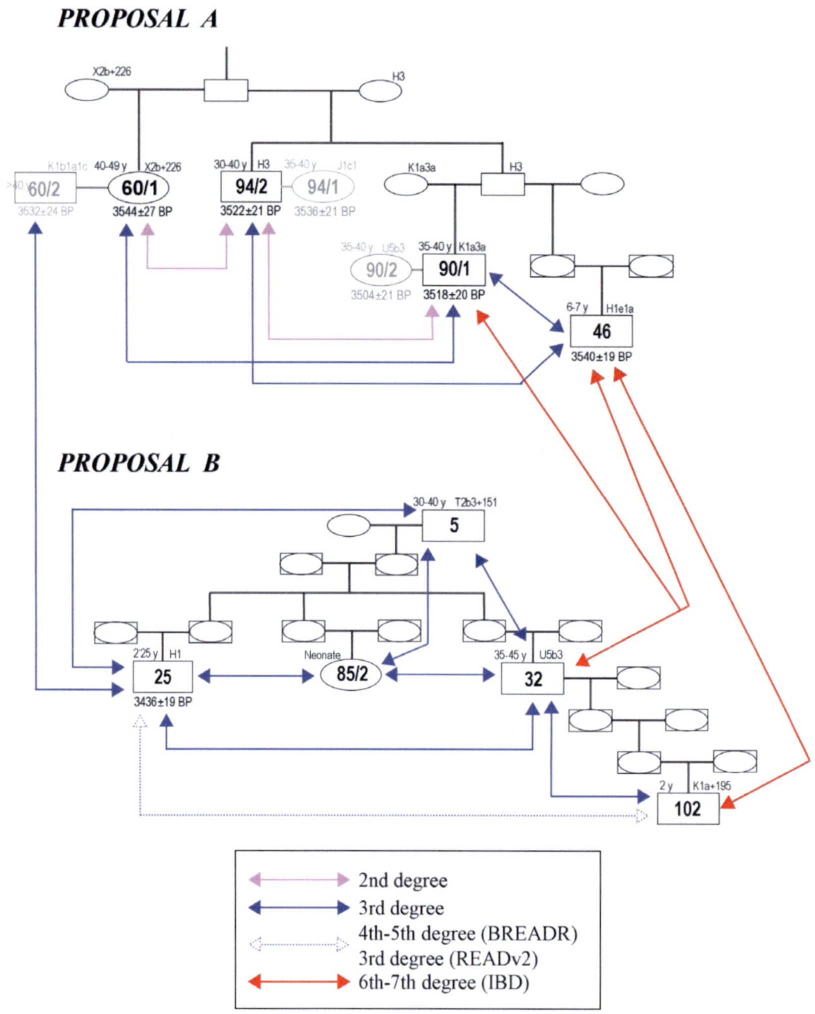

Figure 6.5. Proposals A and B. The A, A′ and B, B′ alternative pedigrees for the sets with the related individuals AY60/1, AY94/2, AY90/1, AY46, AY32, AY102, AY25, AY85/2, and AY5. Proposals A and A′, B and B′ are interchangeable.

By contrast, our findings for the double tombs with an adult and a child are quite different. We have seen that a close biological relationship has been observed in most cases. In Tomb 21 at La Almoloya, which contains the simultaneous burial of an adult female and a neonate,[20] a first-degree genetic relationship (mother-child) is attested. The intertwined positioning of the two skeletons suggests that they died at the same time or quite shortly one after the other, likely due to complications during childbirth (Oliart and Rihuete 2024). A similar context is seen in Tomb 85, which also features a woman embracing a XX neonate against her chest; unfortunately, no genetic data is available for the putative mother.

Thus, in our sample the burial of an adult woman with a child represents a mother-child bond. Straightforward as this interpretation might seem, so far we have just a handful of confirmed cases. Moreover, other forms of social kinship might be at play. We have

---

20   The chromosomal sex of the neonate is XXX (X trisomy) (Villalba-Mouco *et al.* 2021).

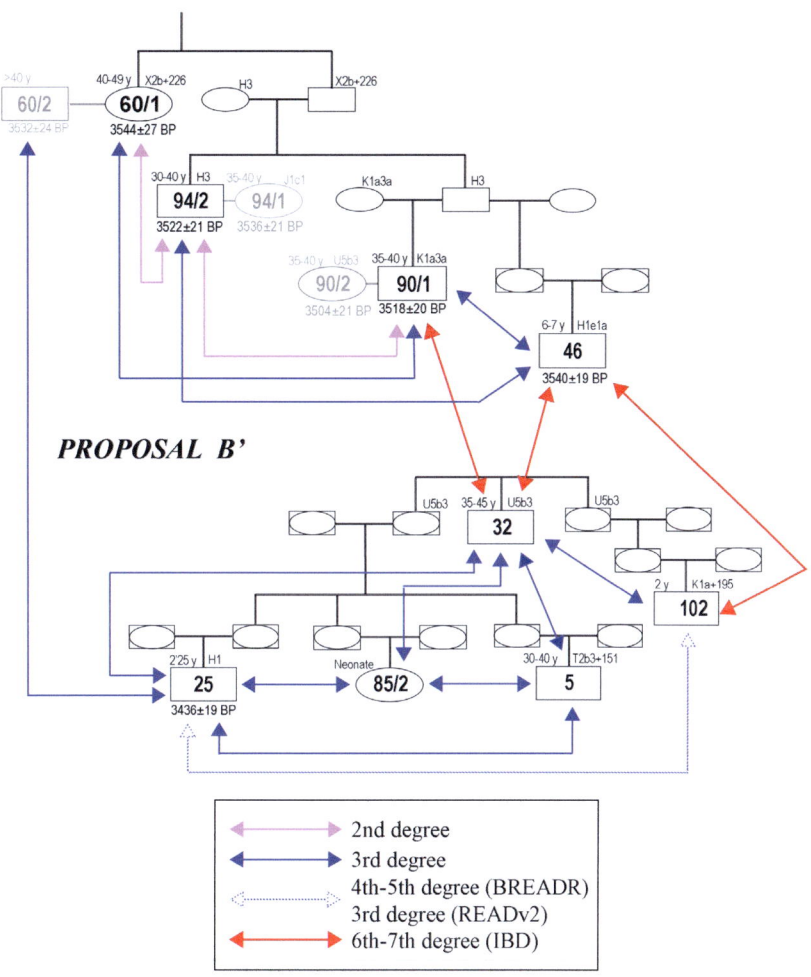

Figure 6.6. Proposals A' and B'. The A, A' and B, B' alternative pedigrees for the sets with the related individuals AY60/1, AY94/2, AY90/1, AY46, AY32, AY102, AY25, AY85/2, and AY5. Proposals A and A', B and B' are interchangeable.

seen the case of BA6 from La Bastida, a tomb where an adult male was buried, followed some time later by an XY neonate who was not biologically related. Here, the funerary treatment might be that of a putative father-son bond,[21] or signify the social recognition of non-consanguineous kinship, such as that of a stepfather and his "adopted" child.

What about the few double graves of children? Are there any genetic links between them? The two cases tested so far reveal consanguinity: in Tomb 23 from La Bastida a girl was buried some time before her brother, while at La Almoloya Tomb 30 the female baby, whose parents were found in Tomb 38, was buried shortly before her half-sister –on her father's side– (Lull *et al.* 2021). Crucially, in the latter case the funerary rite acknowledged their bond despite not sharing the same mother. However, not all siblings were buried

---

21   As the saying goes, *pater semper incertus est* ("the father is always uncertain"), and at the same time, *pater vero est, quem nuptiae demonstrant* ("he is the true father whom the marriage indicates").

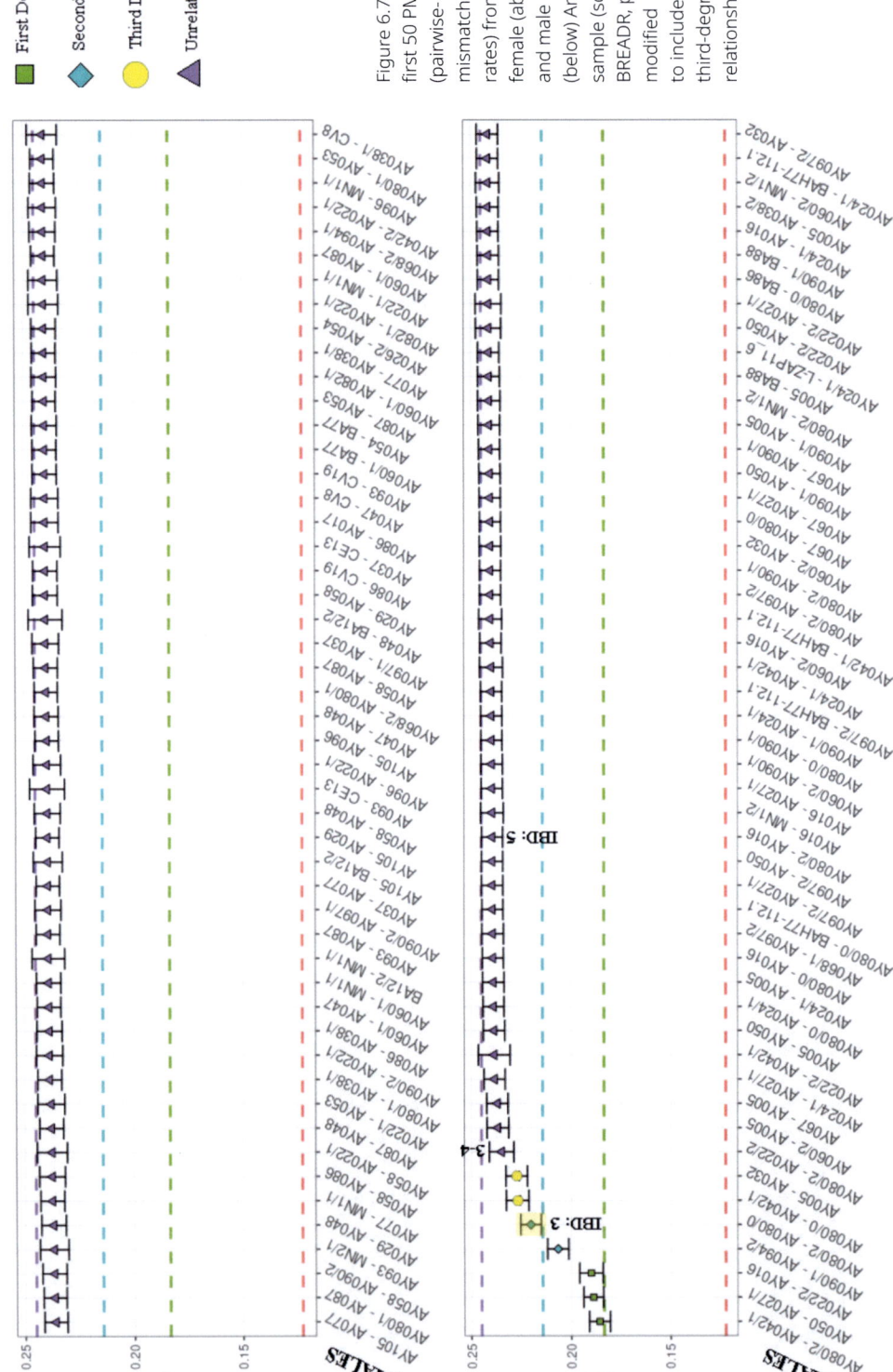

Figure 6.7. The first 50 PMRs (pairwise-mismatch rates) from the female (above) and male (below) Argaric sample (source: BREADR, plot modified to include third-degree relationships).

First Degree
Second Degree
Third Degree
Unrelated

together: first-degree siblings (La Almoloya Tombs 88 and 89) and second-degree or half-siblings (La Almoloya Tombs 13 and 14, as well as Tombs 8 and 25) were also found in separate graves, although quite close to each other.

Our hypothesis on Argaric kinship also outlined matrilocal and matrilineal practices. What insights can we gain from archaeogenetic analysis on its sex/gender system? Up to now, half-siblings in La Almoloya have always been traced through the father's side. Moreover, all the fathers identified at the site were buried with a female,[22] while mothers are found with a male (AY38/1, AY22/1, AY26/2, AY80/1), in a single grave (AY17), or with their children (AY21, AY85). However, the most relevant data for this question is that there are no close biological ties between adult females. No first- nor second-degree relationships have been identified between adult females. The only blood ties among females are those between mothers and their very young daughters. Moreover, no fathers of female adults have been identified. This suggests that the females born in La Almoloya left the settlement once they reached a certain age, probably around puberty. Conversely, females born elsewhere moved to La Almoloya, where they were buried.

Argaric males show more genetic ties at the site (7), even if the closest relationships detected do not include adult brothers (first degree) or half-brothers (second degree). Also crucially, the new analysis of our database with READv2 and BREADR supports our proposal of patrilocality (Villalba-Mouco *et al.* 2022). As mentioned above, adult male siblings are not directly attested at La Almoloya, but the instances of first cousins provide some support for their cohabitation. At the same time, this uncertainty must be considered alongside the meager evidence for the presence of nephews or the descendants of nephews. In addition to the pedigrees reconstructed in Figs. 6.5 and 6.6, a case in point is AY80/0: this male was the great-nephew of AY42/2, whose partial remains were buried near his great-grandparents' double tomb (AY80/1 and AY80/2) (Villalba-Mouco 2022). This suggests that, even if one or more siblings left the settlement, their sons could have come back, as if second-generation male kin returned to their patrilineal home.

To sum up, the latest empirical testing does not support our initial hypothesis about Argaric matrilineal and matrilocal kinship practices. Instead, the results suggest a range of different practices:

a. Widespread female mobility.
b. Restricted male mobility implying patrilocal and patrilineal practices among certain blood-related males.
c. The symbolic significance of adult double tombs containing a female and a male who were mating partners.
d. Social recognition of half-siblings on the father's side.

Once the matrilocal/matrilineal hypothesis has been rejected, does this mean that the traditional view of Argaric kinship is validated? Do the new findings support the claim that the El Argar society was structured around heterosexual monogamous couples under male dominance? Could El Argar represent a patriarchal takeover, which was later naturalized and ultimately shaped the "common sense" of our times?

---

22   These fathers are AY22/2, AY27/1, AY38/2, AY80/2. AY80/0 is an exception to this trend, but his funerary context is also unique, as very few parts of the body were reburied in the place where they were found.

6.6 Towards a new state of the art: new questions, new horizons

New avenues for research have now opened up. In this final section, we will outline some questions that might guide the research agenda.

6.6.1 Does the identification of reproductive couples in double graves support the claim that heterosexual monogamous couples were the backbone of Argaric society?

Indeed, a reasonable inference between the results of genetic data and heterosexual monogamous couples may be drawn. Additional arguments can be made in support of this inference. First, considering the sequence of interments and age at death, both males and females could be buried first, regardless of the time elapsed between the two deaths. This would mean that the social/political alliance between two parties symbolized in double tombs was stable and enduring. Second, the preeminence of this dualistic "political" union over biological ties is further stressed by the instances of half-siblings. These half-siblings resulted from reproductive relationships between one member of the union and a third partner, yet such relationships did not produce triple or multiple adult tombs at La Almoloya –nor, most likely, across the Argaric homeland. In other words, only two reproductive partners were chosen, to the possible exclusion of others. This conclusion need not apply to the whole Argaric territory, because at several western sites, such as Castellón Alto, Cerro de la Encina and Peñalosa, some examples of more than two adults of different sex in the same grave have been attested.[23] But it suggests that "monogamy" may better represent the primary alliance in the ritual expression of interindividual ties than other categories, such as "polygyny". The few but meaningful double adult male tombs might be understood as examples of monogamous alliances as well.

Yet, other evidence disavows a positive answer to this section's question. At La Almoloya, the number of individuals buried in double adult tombs (n=52) is now like those in single graves (n=53),[24] meaning that only half of the adults took part in whatever was represented by the double tombs. In addition, this type of burials decreases over time, from 20 in Phase 2 to just 6 in Phase 3, so that paradoxically these lower numbers coincide with the demographic peak of the El Argar society (also Phase 3). While the rules expressed through double tombs of adults clearly played *a* role in Argaric social organization, they probably were not *the* main organizing principle. More likely other rules governed the distribution of individuals across the Argaric territory.

Exploring the characteristics and scale of these alternative organizational principles is a key challenge for Argaric archaeology today. Notably, one of the most promising findings from kinship analysis concerns genetically unrelated individuals. At La Almoloya, only 41 of 68 individuals with genetic data were involved in first-, second- or third-degree relationships. Even when considering individuals genetically related up to the sixth or seventh degree (according to IBD), the number increases by just one case (AY73). This suggests that 38% of those buried at the site were not biologically related to other residents (Fig. 6.2). Thus, Argaric communities were likely not closed groups of

---

23   This is the case with Tombs 7 and 36 from Castellón Alto (Rubio 2021), Tomb 18 from Cerro de la Encina (Aranda *et al.* 2008) and Peñalosa 2 (Sánchez Romero and Alarcón 2012).

24   These figures correspond to the findings of the fieldwork seasons between 2013 and 2023.

blood relatives but allowed significant individual mobility, as expected in a "complex", "urban" society.

The joint burial of two people of different sexes who had common offspring may suggest an important role of binarism in the symbolic representation of gender. Research into gender expression has also found interesting trends in the treatment and positioning of the corpse, and funerary goods, in which gender intersects with class and age (Lull *et al.* 2005, 2011, 2016). Still, these patterns need further exploration. Moreover, the number of connections between parents buried in double tombs and their children are particularly remarkable, especially since finding descendants in single burials is quite rare. This suggests these unions played a key role in reproduction. Undoubtedly, anthropological analysis of motherhood and archaeogenetics research to identify individuals related by descent will be crucial for future progress.

6.6.2 Certainly, there were political alliances, but between whom and under what rules? Do they necessarily imply exogamous marital alliances between males who controlled females, as suggested by Lévi-Strauss?

This question can only be answered affirmatively if patrilineality and patrilocality are equated with patriarchy, but such correlation lacks strong support. Again, only a limited number of females are found as reproductive partners with an adult male in double graves. Thus, "exogamy" could apply, at best, only to these females. All other females had the right to intramural burial in single graves and received grave goods based on their positions within the wealth hierarchy of El Argar, with the same status of males and likely exceeding them in some spheres (Lull *et al.* 2021). Some of them were mothers and had the right to be buried with or close to their offspring. By contrast, we have not yet found any father in these types of funerary contexts. In short, it seems that many females were fully involved in social life on their own, without archaeological trace of a subordinate position in relation to any male, as we might expect if women were exchanged among patrilineages. It is not justified to attribute widespread female mobility solely to exogamy rules, as most of the women involved may have relocated for reasons that can currently only be broadly understood as political, economic, or both.

Furthermore, the archaeogenetic analysis has uncovered two aspects of the female mobility rule whose implications have yet to be explored in depth:

a. According to the results from La Almoloya, the significant biological distance among all adult women suggests that their places of origin were quite varied and/or that the pathways for managing mobility prevented females with shared biological ancestry from being sent to the same location or returning to their birthplaces.

b. There is evidence for a great deal of diversity in the female lineages at the local level; this stems from the fact that for a single site (La Almoloya, Murcia) the variability in mitochondrial DNA haplogroups attested is virtually the same as that found across all the other sites of Bronze Age Iberia combined together. In fact, at La Almoloya we find almost as many haplogroups as we find adult men,[25] a distribution that squares well with a patrilocal system and an extensive network of female mobility.

---

25 Only 2 out of the 12 mutually exclusive haplogroups identified among the male adult population are repeated (three instances of K1a+195 and two of U5b3).

c. We also observe a wide diversity of mitochondrial haplogroups among females[26] and, moreover, half of these attested haplogroups do not match those found in adult males. This strengthens the idea that the movement of females was not tied to their reproductive role, but rather to other types of practices.

While only a suggestion, the extreme variability in the spatial distribution of female lineages may have stemmed from mechanisms for managing the movement of blood-related women that prevented the repetition of points of origin and destinations in a very effective way. And therefore, these mechanisms would have avoided the association of certain lineages with specific territories. Instead, what emerged was a network that was neither random nor chaotic that spanned the whole Argaric territory and facilitated the transmission and generalization of practices we recognize as "Argaric".

Argaric research faces the challenge of determining whether the variability in female territorial distribution is due to a model of exchange of females among patriarchal partners, or a female-led management of women's mobility, whereby only some women engaged in political and reproductive alliances with territorially established men.

6.7 Conclusion

Recent interdisciplinary archaeological and genetic research suggests that El Argar was comprised of open communities with high mobility. Regarding the widespread movement of women, "female exogamy" at best only partially accounts for it. And while patrilocality has been observed at La Almoloya, only a small number of males were involved. New findings of third-degree relationships have broadened our understanding of kinship, but more than one-third of the individuals analyzed were not biologically related. This suggests that other factors, likely economic and political, played a role in shaping the allocation of Argaric social roles. Finally, we should remember that current genetic data covers only 35% of the individuals found at La Almoloya, so expanding the sample size could impact our interpretations, much like additional radiocarbon dates and newly excavated funerary contexts have in the recent past.

In the same way the matrilocality/matrilineality hypothesis was tested –and ultimately rejected–, it is crucial to question the findings coming mainly from La Almoloya. This means that we need to expand the sampling both in time by including Chalcolithic and Early Argaric sites –and space– by covering more Argaric settlements at comparable scales. Doing so will also help us connecting the changes seen in the second half of the third millennium BCE in southeast Iberia (Micó *et al.* 2024) with the genetic shifts (Villalba-Mouco *et al.* 2021) and kinship patterns observed for El Argar.

Acknowledgments

This research has been supported by the projects "Kinship, population and production in El Argar (2200-1550 cal BCE): a genealogical approach to sexual asymmetries and economic disruption" (MINECO, PID2020-112909GB-I00), "Transformations: the formation of El Argar society (2200-2000 cal BCE) and the making of a symbolic and political order (2000-1550 cal BCE)" (MINECO, PID2023-146504NB-I00), *Grup de Recerca*

---

26 Among female individuals there are 17 haplogroups, of which only 3 are confirmed to be repeated (two instances of X2b+226, two of V and two of K1a1b1).

en Arqueoecologia Social Mediterrània (AGAUR, 2021SGR0525), by Fundación Palarq (III National Prize of Archaeology and Palaeontology) (all authors), a María Zambrano grant (Next Generation-EU) (M.V.) and Programa ICREA Acadèmia (2024ICREA00082-AGAUR) (R.M.). We would like to express our gratitude to Vanessa Villalba-Mouco for her support and guidance in the field of genetics. As always, we are solely responsible for any possible shortcomings.

References

Ainsworth, C. (2015): Sex redefined. *Nature* 518, pp. 288-291. DOI:10.1038/518288a.

Alaçamlı, E., Naidoo, T., Güler, M.N. *et al* (2024): READv2: advanced and user-friendly detection of biological relatedness in archaeogenomics. *Genome Biology* 25(216). DOI:10.1186/s13059-024-03350-3.

Alarcón, E. (2010): *Continuidad y cambio social. Las actividades de mantenimiento en el poblado argárico de Peñalosa (Baños de la Encina, Jaén).* PhD thesis. Granada: Universidad de Granada.

Aranda, G., Molina. F., Fernández, S., Sánchez Romero, M., Al Oumaoui, I., Jiménez-Brobeil, S. and Roca, M.G. (2008): El Poblado y Necrópolis del Cerro de La Encina. Campañas 2003-05. *Cuadernos de Prehistoria y Arqueología de la Universidad de Granada* 18, pp. 21-264. DOI:10.30827/cpag.v18i0.746.

Barad, K. (2007): *Meeting the Universe Halfway. Quantum physics and the entanglement of matter and meaning.* Durham: Duke University Press.

Bennett, J. (2010): *Vibrant Matter. A Political Ecology of Things.* Durham: Duke University Press.

Buikstra, J.E. and Hoshower, L. (1994): Análisis de los restos humanos de la necrópolis de Gatas. In Castro, P.V., Chapman, R.W. *et al.* (eds.), *Proyecto Gatas: Sociedad y economía en el sudeste de España c. 2500-900 cal ANE.* Unpublished archaeological report. Sevilla: Junta de Andalucía, pp. 339-398.

Castro, P., Chapman, R.W., Gili, S., Lull, V., Micó, R., Rihuete, C., Risch, R. and Sanahuja, M.E. (1993-1994): Tiempos sociales de los contextos funerarios argáricos. *Anales de Prehistoria de la Universidad de Murci*a 9-10, pp. 77-107.

Castro, P., Chapman, R., Gili, S., Lull, V., Micó, R., Rihuete, C., Risch, R. and Sanahuja Yll, M.E. (1996): Teoría de las prácticas sociales. *Complutum* Extra 6, pp. 35-48.

Castro, P.V., Gili, S., Lull, V., Micó, R., Rihuete, C., Risch, R. and Sanahuja, M.E. (1998): Teoría de la producción de la vida social. Un análisis de los mecanismos de explotación en el sudeste ibérico. *Boletín de Antropología Americana* 33, pp. 25-77.

Castro, P.V., Gili, S., Lull, V., Micó, R., Rihuete, C., Risch, R. and Sanahuja, M.E. (2001): Teoría de la producción de la vida social. Un análisis de los mecanismos de explotación en el sudeste peninsular (ca. 3000-1550 cal ANE). *Astigi Vetus* 1, pp. 13-54.

Celdrán Beltrán, E., Lull, V., Micó, R., Oliart, C., Rihuete Herrada, C. and Risch, R. (eds.) (2023): *Cerro del Morrón: Un asentamiento en la frontera argárica.* Barcelona: Grup de Recerca en Arqueoecologia Social Mediterrània.

Chapman, R.W. (1990): *Emerging Complexity: The Later Prehistory of South-East Spain, Iberia and the West Mediterranean.* Cambridge: Cambridge University Press.

Childe, V.G. (1958): *The Dawn of European Civilization.* Sixth edition. New York: Alfred Knopf.

Contreras, F., Rodríguez Ariza, M.O., Cámara, J.A. and Moreno, A. (eds.) (1997): *Hace 4000 años. Vida y muerte en dos poblados de la Alta Andalucía.* Sevilla: Empresa Pública de Gestión de Programas Culturales.

Díaz-Zorita Bonilla, M., Aranda Jiménez, G. *et al.* (2024): Female sex bias in Iberian megalithic societies through bioarchaeology, aDNA and proteomics. *Scientific Reports* 14, 21818. DOI: 10.1038/s41598-024-72148-x.

Fausto-Sterling, A. (2000): *Sexing the Body. Gender politics and the construction of sexuality*. New York: Basic Books.

Fox, R. (1983 [1967]): *Kinship and Marriage. An Anthropological Perspective*. Cambridge: Cambridge University Press.

González Marcén, P., Montón, S. and Picazo, M. (2008): Towards an archaeology of maintenance activities. In Montón, S. and Sánchez Romero, M. (eds.), *Engendering Social Dynamics: The Archaeology of Maintenance Activities*. Oxford: British Archaeological Reports, pp. 3-8.

Harman, G. (2018): *Object-Oriented Ontology: A New Theory of Everything*. London: Pelican Books.

Inchaurrandieta, R. (1870): Estudios Pre-Históricos. La Edad del Bronce en la provincia de Murcia. *Boletín-Revista de la Universidad de Madrid* II(13), pp. 806-815.

Jiménez-Brobeil, S., Roca, M. and Al-Oumaoui, I. (2009): Análisis paleoantropológico. In Lizcano, R., Nocete, F., Nocete Calvo, F. and Peramo, A. (eds.), *Las Eras: Proyecto de puesta en valor y uso social del patrimonio arqueológico de Úbeda (Jaén)*. Huelva: Universidad de Huelva.

Klein, E. (1971): *A Comprehensive Etymological Dictionary of the English Language, dealing with the origin of words and their sense development thus illustrating the history of civilization and culture*. Amsterdam: Elsevier.

Latour, B. (1991): *Nous n'avons jamais été modernes. Essai d'anthropologie symétrique*. Paris: La Découverte.

Lewis, C.T. and Short, C. (1879): *A Latin Dictionary. Founded on Andrews' edition of Freund's Latin dictionary*. Oxford: Clarendon Press.

Lull, V. (2000): Argaric society: Death at home. *Antiquity* 285(74), pp. 581-590. DOI: 10.1017/S0003598X00059949.

Lull, V., Celdrán Beltrán, E., Oliart Caravatti, C., Rihuete Herrada, C., Valério, M. and Micó, R. (2024): Categorías arqueológicas de ajuar funerario y clases sociales. In Núñez Calvo, F.J., Mederos Martín, A., Suárez Padilla, J. and Martín Córdoba, E. (coords.), *Entre Málaga y Tiro. Una travesía mediterránea en memoria de la profesora María Eugenia Aubet Semmler*. Málaga: Universidad de Málaga, pp. 641-654.

Lull, V., Micó, R., Rihuete Herrada, C. and Risch, R. (2005): Property relations in the Bronze Age of southwestern Europe: an archaeological analysis of infant burials from El Argar (Almería, Spain). *Proceedings of the Prehistoric Society* 71, pp. 247-268. DOI: 10.1017/S0079497X0000102X.

Lull, V., Micó, R., Rihuete Herrada, C. and Risch, R. (2013): Funerary practices and kinship in an Early Bronze Age society: a Bayesian approach applied to the radiocarbon dating of Argaric double tombs. *Journal of Archaeological Science* 40, pp. 4626-4634. DOI: 10.1016/j.jas.2013.07.008.

Lull, V., Micó, R., Rihuete Herrada, C. and Risch, R. (2016): Argaric Sociology: Sex and Death. *Complutum* 17(1), pp. 31-62. DOI: 10.5209/CMPL.53216.

Lull, V., Rihuete Herrada, C. *et al.* (2021): Emblems and spaces of power during the Argaric Bronze Age at La Almoloya, Murcia. *Antiquity* 95(380), pp. 329-348. DOI: 10.15184/aqy.2021.8.

Mallick, S. and Reich, D. (2023): *The Allen Ancient DNA Resource (AADR): A curated compendium of ancient human genomes. (Harvard Dataverse, V9)*. DOI: 10.7910/DVN/FFIDCW.

Micó, R., Celdrán Beltrán, E., Lomba Maurandi, J., Oliart Caravatti, C., Rihuete Herrada, C. and Valério, M. (2024): Tracing social disruptions over time using radiocarbon datasets: Copper and Early Bronze Ages in Southeast Iberia. *Journal of Archaeological Science: Reports* 58, 104692. DOI: 10.1016/j.jasrep.2024.104692.

Monsalve, A., Sánchez Romero, M. and González, A. (2014): Las comunidades de la Edad del Bronce de La Mancha desde la arqueología y la antropología física: el caso del Cerro de la Encantada (Granátula de Calatrava, Ciudad Real). *Menga. Revista de Prehistoria de Andalucía* 5, pp. 175-197.

Oliart, C. and Rihuete Herrada, C. (2024): Mujeres gestantes, fetos y neonatos en tumbas prehistóricas de la Edad del Bronce argárica. *Actes del XVI Congrés Nacional i Internacional de Paleopatologia (Girona, 2022).* Girona: Museu d'Arqueologia de Catalunya, pp. 47-54.

Pateman, C. (1988): *The sexual contract.* Stanford: Stanford University Press.

Robledo, B., Trancho, G.J. (2003): *Análisis antropológico y condiciones de vida de la población argárica de Cerro del Alcáza*r. Serie Informes Antropológicos. Madrid: Complutense University.

Rohrlach, A., Tuke, J., Popli, D. and Haak, W. (2023): BREADR: An R Package for the Bayesian Estimation of Genetic Relatedness from Low-coverage Genotype Data. *bioRxiv* 2023.04.17.537144. DOI: 10.1101/2023.04.17.537144.

Rubin, G. (1975): The Traffic in Women: Notes on the 'Political Economy' of Sex. In Reiter, R.R. (ed.), *Toward an Anthropology of Women.* New York: Monthly Review Press, pp. 157-210.

Rubio, A. (2021): *Paleopatología en los yacimientos argáricos de la provincia de Granada.* PhD thesis. Granada: Universidad de Granada.

Saini, A. (2020): *Superior: The Return of Race Science.* London: Fourth Estate Books.

Saller, R.P. (1984): *Familia, Domus*, and the Roman Conception of the Family. *Phoenix* 38(4), pp. 336-355.

Sánchez Romero, M. and Alarcón, E. (2012): Lo que los niños nos cuentan: individuos infantiles durante la Edad del Bronce en el sur de la Península Ibérica. In Justel Vicente, D. (ed.), *Niños en la antigüedad. Estudios sobre la infancia en el Mediterráneo antiguo.* Zaragoza: Universidad de Zaragoza, pp. 57-97.

Schubart, H. (2012): *Die Gräber von Fuente Álamo. Ein Beitrag zu den Grabriten und zur Chronologie der El Argar-Kultur.* Wiesbaden: Deutsches Archäologisches Institut.

Siret, H. and Siret, L. (1890): *Las Primeras Edades del Metal en el Sudeste de España.* Barcelona: Tipografía de Henrich y Cía.

TallBear, K. (2013): *Native American DNA. Tribal Belonging and the False Promise of Genetic Science.* Minneapolis: University of Minnesota Press.

Villalba-Mouco, V., Oliart, C. *et al.* (2021): Genomic transformation and social organization during the Copper Age-Bronze Age transition in southern Iberia. *Science Advances* 7, p. eabi7038. DOI: 10.1126/sciadv.abi7038.

Villalba-Mouco, V., *et al.* (2022): Kinship practices in the early state El Argar society from Bronze Age Iberia. *Scientific Reports* 12, p. 22415. DOI: 10.1038/s41598-022-25975-9.

Kinship and Clientele Structures of the Iron Age Southern Iberians (Spain) from the Burial Realm

Carmen Rísquez Cuenca

University Institute for Research in Iberian
Archaeology, University of Jaén, Spain,
crisquez@ujaen.es

Arturo Ruiz Rodríguez

University Institute for Research in Iberian
Archaeology, University of Jaén, Spain,
arruiz@ujaen.es

Abstract

This paper aims to present proposals regarding the kinship patterning among the southern Iberians in the Iron Age, between the seventh and fourth centuries BCE. It discusses the structures that may have governed the organisation of a series of funerary areas, and how they evolved towards other models such as clientelism. We begin with the example of Cerrillo Blanco (Porcuna, Jaén), in the Turdulan region, using the new data provided for its spatial reading by different analytical methods. We hypothesise that we are dealing with an extended family, without ignoring other readings that go beyond the strict literal horizon of kinship. We compare it with La Noria (Fuente Piedra, Málaga), which shows the continuation of the process initiated in the Porcuna tumulus and has allowed forms of dependence to be recognised archaeologically. Finally, the evolution of the funerary landscape in the Oretani-Bastetani-Mentesani area is presented from this same period up to the development of large necropolises. This was a process that led to a single, gentilitial-clientele model, which we can follow with the examples from the necropolises of Cástulo and Cerro del Santuario in Baza (Jaén).

Keywords: *kinship, extended family, clientele, funerary spaces, Iberians.*

7.1 The tumulus of Cerrillo Blanco (Porcuna, Jaén)

7.1.1 Contextualisation and initial considerations

The Early Iron Age funerary monument of Cerrillo Blanco (seventh or early sixth centuries BCE) occupies the upper part of a small hill of marl and sandstone, which gives

In Blanco-González, A. and Alarcón-García, E. (eds.) 2025, *A Social Archaeology of Kinship in Iberia and Beyond. Recent Multistranded Approaches from aDNA to Household Archaeology.* Leiden: Sidestone Press, pp. 139-158.

it a whitish hue that is clearly visible in the landscape. It is located just over 2 km to the northeast of the settlement of Los Alcores (ancient city of *Ipolka*) with which it would have been linked as part of the Turdulan territory (Fig. 7.1). Archaeological excavations in the area have revealed that the site had an extensive sequence beginning in the Late Bronze Age and was maintained as a burial site until the second century BCE (González Navarrete *et al.* 1980; Torrecillas 1985). This recurrence is striking, as it reflects an interest in maintaining links with ancestors, and a collective memory accumulated through the ritualisation of this area.

The tumulus was completely excavated in 1978-1979, although some tombs had already been previously revealed. It was a circular area, 21 m in diameter and delimited by stone slabs. Within it, 25 inhumation burials were documented: 24 single pits and one double burial. The configuration of the monument's space reveals an interest in differentiating this double burial, both in its location –leaving a wide area of respect with the rest of the burials– and its construction –a chamber lined with large orthostats– and its lack of any grave goods (González Navarrete *et al.* 1980) (Fig. 7.2). It should be noted that not all the burials had evidence of grave goods and that these were neither numerous nor ostentatious.

Anthropological studies carried out at that time showed a balanced presence of women (8) and men (9), mostly adults –aged between 20 and 40– with only three male individuals and one female over that age –between 41 and 60 years old– (Torrecillas 1985). The couple from the burial chamber contained one female and one male skeleton (Arteaga 1999: 114) with no reference as to their age at death. The cohort of young people is missing, yet it is important to point out the presence of children (7), representing 30% of the total and subject to the same ritual patterns as the adults.

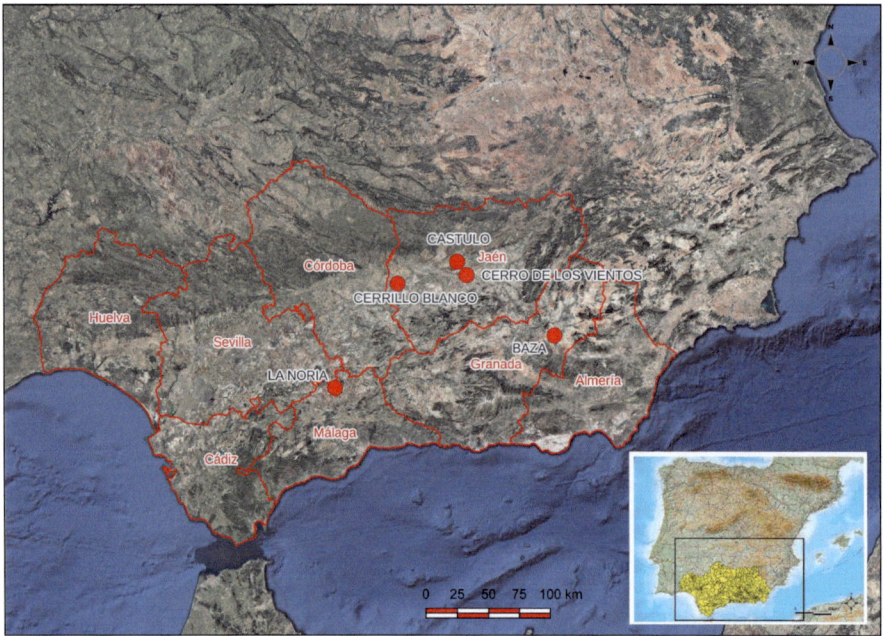

Figure 7.1. Location of the tumulus of Cerrillo Blanco (Jaén, Spain) and sites referred to in the text.

A SOCIAL ARCHAEOLOGY OF KINSHIP IN IBERIA AND BEYOND

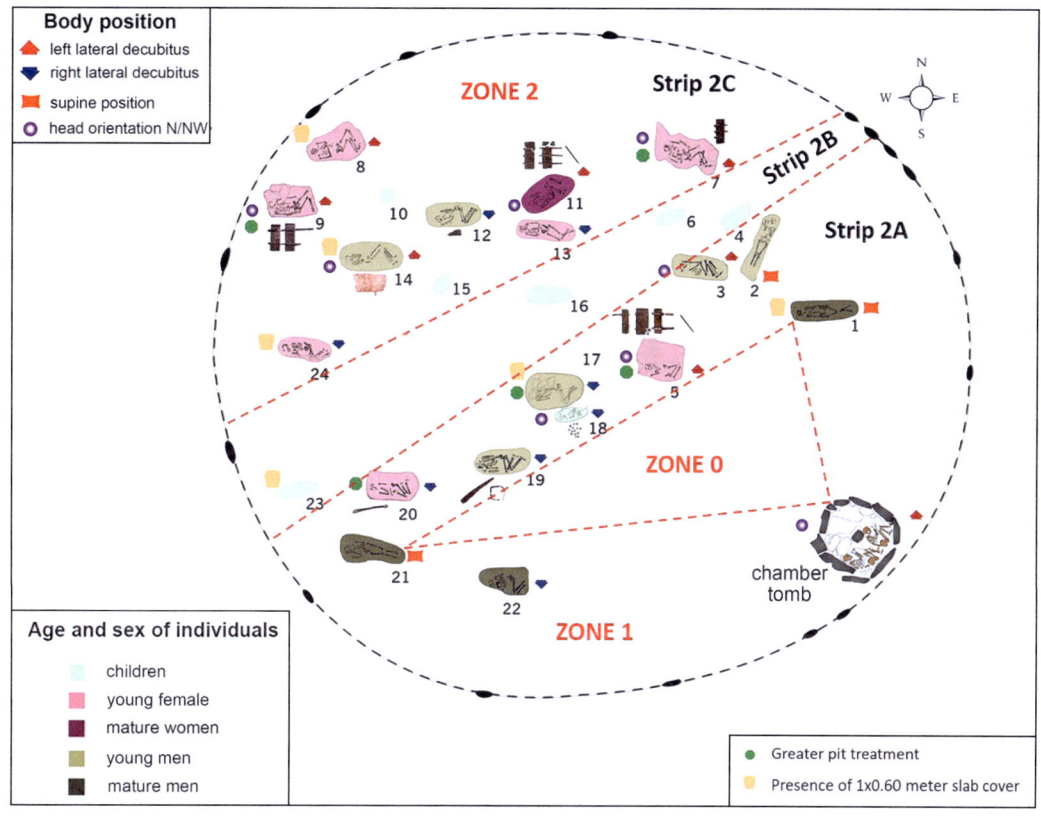

Figure 7.2. Spatial distribution of burials within tumulus of Cerrillo Blanco (Porcuna, Jaén) during the late seventh or early sixth centuries BCE.

We have previously dealt with the burial ritual, bodies depositions and furnishings (Rísquez *et al.* 2022a). This allowed us to verify different patterns in the posture of the corpses: right and left lateral decubitus and supine decubitus, the latter being minority (Tombs 1, 2 and 21). Since only five women, three men and one child had grave goods, gender and age are not the only variables that determine their presence, and further aspects such as social position and filiation must be considered. Burial goods consisted of one or two items related to personal body care and adornment: tweezers, razor blades, combs, fibulae, belt clips or needles. Among the body ornaments, an ivory comb found with the adult male (Tomb 14) is remarkable. Located by his skull, it could have been used for a specific headdress or hairstyle and features the only iconographic representation of the entire assemblage; its narrative focus on the mythical nature, thus distinguishing his possessor, who emulated Mediterranean aristocratic ideology. Similarly noteworthy is the micro-bead necklace associated with the infant individual in Tomb 18, which also indicates a certain status despite premature death. His proximity to the adult male in Tomb 17, together with the intentional disposition of the body, with his head turned to look towards him, led us to propose a possible family link between them. The proximity of both to Burial 5, an adult female with the largest number of belt brooches (three), as well as her location in the tumulus, the closest to the couple buried in the chamber, could allude to a possible social or family affinity. As for the elements related to clothing, we suggested elsewhere (Rísquez *et al.* 2022b) differentiated rules according to the sex of the deceased, since belt brooches and needles were always associated with women; however, only one fibula was found in context and linked to a male (Tomb 19). We must emphasise the connotations of belt brooches as symbols of female collective identity, marking the individualisation of some women, who were laid to rest in the same position as the couple buried in the chamber –left lateral decubitus– (Rísquez 2015).

7.1.2 The new approaches from bioarchaeology and archaeometry

From the above, we restudied these human remains again –except for the missing couple– and the grave goods from some burials, applying various analytical techniques with the aim of better characterising the group buried there.[1] Osteoarchaeological studies (Peláez 2013 and Martín 2022)[2] ratified the results already presented in terms of sex and age diagnoses, adding other factors such as palaeopathological factors and observing stress markers in the adult and mature individuals of this group.

Regarding the different pathologies documented, the metabolic and nutritional ones stand out, such as the presence of *criba orbitalia* in the males from Burials 17 and 22. These are related to diets poor in iron, a deficiency in certain essential nutrients, or generalised infectious processes (Peláez 2013). Enamel hypoplasia affected the same two individuals, as well as the females from Burials 8 and 13. Such episodes occur at early ages, between 4 and 6 years old (Peláez 2013). Maxillo-dental pathologies also stand out, with caries present in 64% of the adult and mature individuals, affecting males to a greater extent. Cervical osteoarthritis was identified in the mature woman in Tomb 11, which may be a sign of degeneration, given her age. In terms of stress markers (Martín 2022), there are individuals whose lives were devoid of major stress factors, such as the women in Tombs 9 and 5, the latter with a consolidated fracture at the level of the wrist, and the male in Tomb 14 (Martín 2022: 17), as well as the male in Tomb 17, who was found to have been free of intense physical exertion during his life. In contrast, there are several individuals –the male from Tomb 2 and the females from Tombs 7 and 20– with periosteal lesions in the tibiae. This indicates tasks that would have required a continuous translocation or an effort involving the lower limbs, exposing them to the possibility of suffering traumatism. Another hypothesis, in the case of the male could be periostitis caused by an infection, judging by damage (Martín 2022: 40).

Carbon (δ^{13}C), nitrogen (δ^{15}N), oxygen (δ^{18}O) and strontium (^{87}Sr/^{86}Sr) isotopic analyses provided insights into the palaeodiet and mobility of the target individuals[3] (Rísquez *et al.* 2022a; Rísquez *et al.* 2025). The uniformity of the δ^{13}C or δ^{15}N values shows that there are no significant differences between the adults of either sex. They present a homogeneous diet based on the C_3-type terrestrial resources –cereals, trees and shrubs, fruit and vegetables from temperate climates– leading to conclude that these people obtained their food from the same ecosystem and consumed the same proportions of animal protein. However, the two sampled sub-adult individuals show higher nitrogen values (1.5‰) than the adults, reflecting that they were still infants or that –even if they had begun to introduce solid food into their diet– the

---

1   The study was carried out as part of the project "Bioarchaeology for the investigation of gender relations in prehistory and protohistory" funded by the Spanish Ministry of Science and Innovation (HAR 2016-80020-p).
2   Studies under the supervision of Inmaculada Alemán (University of Granada).
3   Sample processing and analysis by Marta Díaz-Zorita (University of Tübingen, Germany).

signal of maternal milk consumption in their bone collagen had not had time to be eliminated and therefore the ^{15}N values are enriched (Rísquez *et al.* 2022a: fig. 3). This, together with the osteological determinations and the preserved teeth, allowed us to estimate their age of death at around 3 or 4 years (Rísquez *et al.* 2025). A factor that distinguishes this burial site from other necropolises in the Tartessian area –i.e., the Early Iron Age in Southwestern Iberia– is the high number of early children, and importantly, they received the same funerary practices as adults. This emphasises their importance in the economic and social structure of the group and could be related to the manifestation of the hereditary nature and the remarkable loss that this death represented within the group. As for the analysis of stable oxygen (^{18}O) and strontium (^{87}Sr/^{86}Sr) isotopes for those questions related to mobility, only Individual 11 –the only mature female– presented strontium values incompatible with the group and was interpreted as an outlier (Rísquez *et al.* 2025).

Another issue addressed during the project was the dating (^{14}C AMS) of the human remains.[4] Based on chrono-typology of the occupation finds, this funerary area is conventionally dated to between the mid-seventh and early sixth centuries BCE (González Navarrete *et al.* 1980; Torrecillas 1985; Arteaga 1999; Torres 1999, 2002), which would indicate its use by at least two or three generations. It is remarkable the old chronologies obtained for two radiocarbon-dated burials lacking grave goods, and therefore indicating an earlier date: Tomb 21 dates to between the late eleventh and early tenth centuries BCE and Tomb 1 dates to the ninth century BCE.[5] This would place even greater emphasis on the ritualisation of this area over time, since using a previously occupied site would have legitimised those interred in this tumulus in the seventh century BCE. This is not an exceptional occurrence; the latest studies of the burial mound of Setefilla (Lora del Río, Seville) show that some of the tombs in Tumulus A are dated to *ca.* 1200-1020 cal BCE –although most are earlier– whereas some of the earliest tombs of Tumulus B date to *ca.* 1120-1000 cal BCE (Brandherma and Krueger 2017).

Lastly, aDNA analyses[6] are in progress. Preliminary results indicate that it has been possible to obtain genetic information from six adults (35.2% of the adult individuals) whose molecular sex has been recovered, ratifying the anthropological analyses. Finally, the two infants sampled gave poor results.

7.1.3 The spatial organisation of Cerrillo Blanco

As explained above, two of the radiocarbon-dated burials have a marked chronological difference with the rest, yet it is unlikely that the tumulus was used from the late eleventh to the early sixth centuries BCE. This would mean that there had been a functional pause in its funerary use of more than 150 years, presumably between the previous earlier use and the seventh century BCE. In kinship terms six generations is an acceptable figure for maintaining generational memory, as Ramos (2009) points out following Faron

---

4   Teeth and bone dating of the 17 adult individuals was part of project HAR 2016-80020-p. Sample processing was carried out by Marta Díaz-Zorita and AMS dating analyses (^{14}C) was undertaken at the Ion Beam Physics Radiocarbon Laboratory of ETH University (Zürich, Switzerland).

5   Bayesian modelling and Kernel distribution calculations (Bronk Ramsey 2017) were performed using *OxCal* software (version 4.3.2) by Gonzalo Aranda (University of Granada), Lara Milesi (University of Málaga) and Ray Kidd (Aberdeen University). This study is pending publication.

6   Studies undertaken by Maribel Torres and Alberto Marchal (University of Jaén) and Iñigo Olalde (University of the Basque Country).

in his study of Mapuche lineages. However, we are inclined to believe that although generational memory may have existed, the tumulus was initiated as a functional space from the moment that most of the burials were accumulated in a continuous temporal process. As the artefacts from the grave goods indicate, this took place from the mid-seventh century BCE and ended at the beginning of the following century, with its use corresponding to two or three generations. However, the location of the two earlier tombs must have been known to the re-users of the old burial area and undoubtedly influenced the location of the tumulus. To interpret this potential symbolic function of Tombs 1 and 21, it is essential to consider the role played by the circular chamber tomb. This followed a megalithic tradition: building with stone slabs. The layout of the tumulus floor plan, together with that of the two earlier burials form a triangular space inscribed on the tumulus ground plan, with no ritual use in its interior space. This is regarded as a space of respect, where the axis defined by the two ancient burials served as the boundary of the burials. It is interesting to add that neither of them contained grave goods. As for the chamber burial, it was not possible to take samples for dating, nor did it have grave goods. However, we wonder if it could also have been a Late Bronze Age chamber, or even whether an earlier megalithic burial was refurbished. In contrast to this hypothesis, no earlier megaliths are known in the region, yet its circular floor plan, no corridor and paved floor seem to emulate the old megalithic burials. Moreover, the position of the two corpses is left lateral decubitus and not supine decubitus as in Burials 1 and 21.

As for the spatial arrangement of tombs within the monument, following the axis of Tombs 1 and 21, we find a first group of burials in which males predominate (2, 3, 17 and 19), compared to two females (5 and 20) and one infant (18). Immediately to the northwest, a second space opens that only contains infant burials (4, 6, 16, 15 and 23) occupying a central strip. The last group however shows a predominance of females (7, 8, 9, 11, 13 and 24), only two males (12 and 14) and one infant (10) (Fig. 7.2). From this spatial distribution three major spatial units could be defined: **Zone 0** is an area of respect framed by the triangle defined by the two earlier burials and whose vertex is formed by the chamber tomb with a double burial; **Zone 1** to the west of this triangle, with only one grave (Tomb 22), the only mature male individual outside the two earlier burials, and **Zone 2** to the northwest of the aforementioned axis, subdivided into several strips according to the number of burials. The first is **Strip 2a**, dominated by adult male burials, whereas **Strip 2b**, features infant burials and **Strip 2c**, shows a predominance of adult female burials.

Pending the results of genetic analyses, this spatial patterning leads us to hypothesise that we are dealing with an extended family. This is based on three aspects: a) the existence of two burial groups, interpreted here as an expression of a double ranking; b) the lack of strong differences in the wealth of burial goods; and c) the internal subdivision of Zone 2. As for the inference of a double ranking within the interred social group, Cerrillo Blanco has traditionally been interpreted as an extended family buried around the supposed patriarch of the parental group, who was also buried with his spouse (Roos 1997). This last question is very important, because it adds the ritual visibility of the institution of marriage to the constructive difference of burial pits versus the chamber. In other words, monumentality and double burial characterise and distinguish the first rank, to which is added the representation of the funerary landscape based on its hierarchy with the creation of Zone 0 and the separation of burial pits from the axis defined by the two Late Bronze Age burials. Thus, this boundary line becomes the legitimising factor of the two rankings, sustaining the difference in the lineage's past. A special case is defined in Zone 1, which is separated from the second-ranking funerary group, outside the Zone 0 triangle of respect with Burial 22. It presents the same ritual used in the group of burials from the seventh-sixth centuries BCE in Zone 2. Secondly, from the uniformity of the grave goods no major differences can be ascertained between subgroups. Only nine tombs from Zone 2 contained grave furnishings –mainly clothing and items of personal hygiene limited in number and lacking weapons– suggesting that they did not convey hierarchy. The second-ranking burials had these grave goods, belt brooches are associated with females of any age and only one child burial featured such items. Thirdly, within Zone 2, three spatially distinct strips have been identified, whose indicators of age and gender are significant, as are those of filiation and alliance. In this case, four different spatial units could be defined, excluding the chamber with the couple:

A. **Zone 1.** Only one mature male is documented (Burial 22), buried close to the chamber tomb, something that gave him a certain prestige in the eyes of the group.

B. **Zone 2, Strip 2a**: Composed of adults –four males and two females– except for one child. As this part of Zone 2 is closest to the tombs of the ancestors, it may correspond to an important position. The two female tombs contained grave goods highlighting their position; the furnishings of the woman from Tomb 5 included plaques from three different belts that embodied symbolic codes emphasising their importance, as explained above. Likewise, the infant in Burial 18 contained a small necklace and could be related to the male in Tomb 17. We could hypothesise that those buried in this group are the closest relatives to the couple at the head of the lineage –from the first to the fourth degree of kinship, in the case of an extended group. This group may have followed virilocality/patrilocality principles or in fewer cases uxorilocality/matrilocality principles. However, as we believe that the weight of marriage was very significant, uxorilocality could also be associated with the use of the institution of "son-in-law marriage", although for its practice, the absence of male offspring is frequent and even obligatory, as in the case of Nausicaa in the Odyssey (Leduc 1991).

C. **Zone 2, Strip 2b**: There are four burials (4, 6, 16 and 23) to which Burial 15 could be added on the northern edge of the strip. All of them are infants up to four years old and none contained grave goods (Rísquez *et al.* 2025). This space might have acted as a delimiting buffer between Strips 2a and 2c.

D. **Zone 2, Strip 2c**: Predominantly adults, mainly female (six females, one of them mature, compared to two males) with two child burials. This group of ten burials could be associated with alliance practices of the extended family. The imperceptible marital relationship in this second rank of burials could justify the existence of Strip 2b, where infants who died at an early age were buried, as they had not yet been weaned (Rísquez *et al.* 2025). We should not rule out the possibility that there could be family units such as the one formed by Burials 17 (male), 5 (female) and 18 (infant). This would be a spatial association with no recognisable delimiting signs.

In any case, the existence of an extended grouping does not rule out other readings that go beyond the strict, literal horizon of kinship; it opens up other possibilities of interpretation by valuing the hierarchical factors existing in the funerary landscape, a question we will assess below.

7.2 La Noria necropolis and the archaeological recognition of forms of dependence

Before taking this analysis further, it is of interest here to add the case of the necropolis of La Noria (Fuente Piedra, Málaga). This site has been the subject of two archaeological excavations, a developer-funded in 2006-2008 and by the authors' research group in 2009-2010. La Noria, like Cerrillo Blanco, is part of the Turdulan area and dates from the second half of the sixth century BCE, so it would theoretically be a continuation of the funerary process initiated in the Porcuna tumulus.

It is a necropolis with 35 burials (Ruiz *et al.* 2017) in which, unlike the multi-tomb tumulus of Cerrillo Blanco, several tumuli were individualised, each with a central tomb. The different sizes and typology of the burials also indicate a significant degree of hierarchy (Fig. 7.3). There are at least two funerary rankings among the tumuli and a third of extra-tumulus flat burials. Their spatial distribution was organised by gender, men to the north and women to the south, and, in addition, the distribution of these burials surrounded Tumuli C and E. We should not rule out a fourth ranking also associated with flat graves and consisting of burials in the pits or ditches of Tumuli C, E and H, and one more case at Tumulus B. This group is not widespread and is only documented in four of the ten existing tumuli. There is a single funerary rite: cremation *in bustum*, although the fourth ranking features two cases of burial in urns, whose typology indicates they likely were the last burials (in the late sixth or early fifth centuries BCE). Despite these clear differentiating nuances, it was also common here for exotic or prestige objects to be distributed indiscriminately among the different rankings. Another common trait in all cases is the lack of weapons deposited in the funerary offerings.

Tumuli C is the largest, with a diameter of 21.4 m while the others are no larger than 15 m Both Tumuli C and E are surrounded by flat burials, which act as a buffer zone separating the first ranking from the second. Furthermore, Tumulus C is a double burial and contained grave goods inside the tomb, probably belonging to the second burial, in which of particular note is a bronze libation jug-brazier, characteristic of the Tartessian sphere. A bronze amphora-cauldron ritual set was also documented on its exterior and was also present in the pit of Tumulus E. There were burials in the surrounding pits of these two tumuli, four in Tumulus C and one in Tumulus E. Other offerings included plates, a gold earring and even a necklace of conical bronze sheet beads and vitreous paste rings, in the former, and bovid remains in the latter. The central burial in Tumulus C contained the remains of two females, a young woman buried first, and a mature woman buried later. The central tomb of Tumulus E contained no human remains, although the complete funerary rite had been carried out, probably as a cenotaph. It could be interpreted as the funerary representation of a married couple in two different tumuli, despite the absence of the body of the man, whose representational function was assumed by the woman with symbols such as the libation set. In fact, the two tumuli had been connected by a corridor linking their peripheral ditches to emphasise the profound relationship between occupants and absentees. If this ensemble forms the first ranking, the second ranking would have been made up of the remaining eight tumuli. We do not observe external rites using the amphora and bronze cauldron as in the previous burials, nor offerings in the ditches, although in some cases, such as in H and B, burials are recorded in their pits. The third ranking consists of flat burials located between the first and second and the fourth ranking, those in the pits or ditches of the tumulus. Anthropological studies have allowed us to distinguish a group

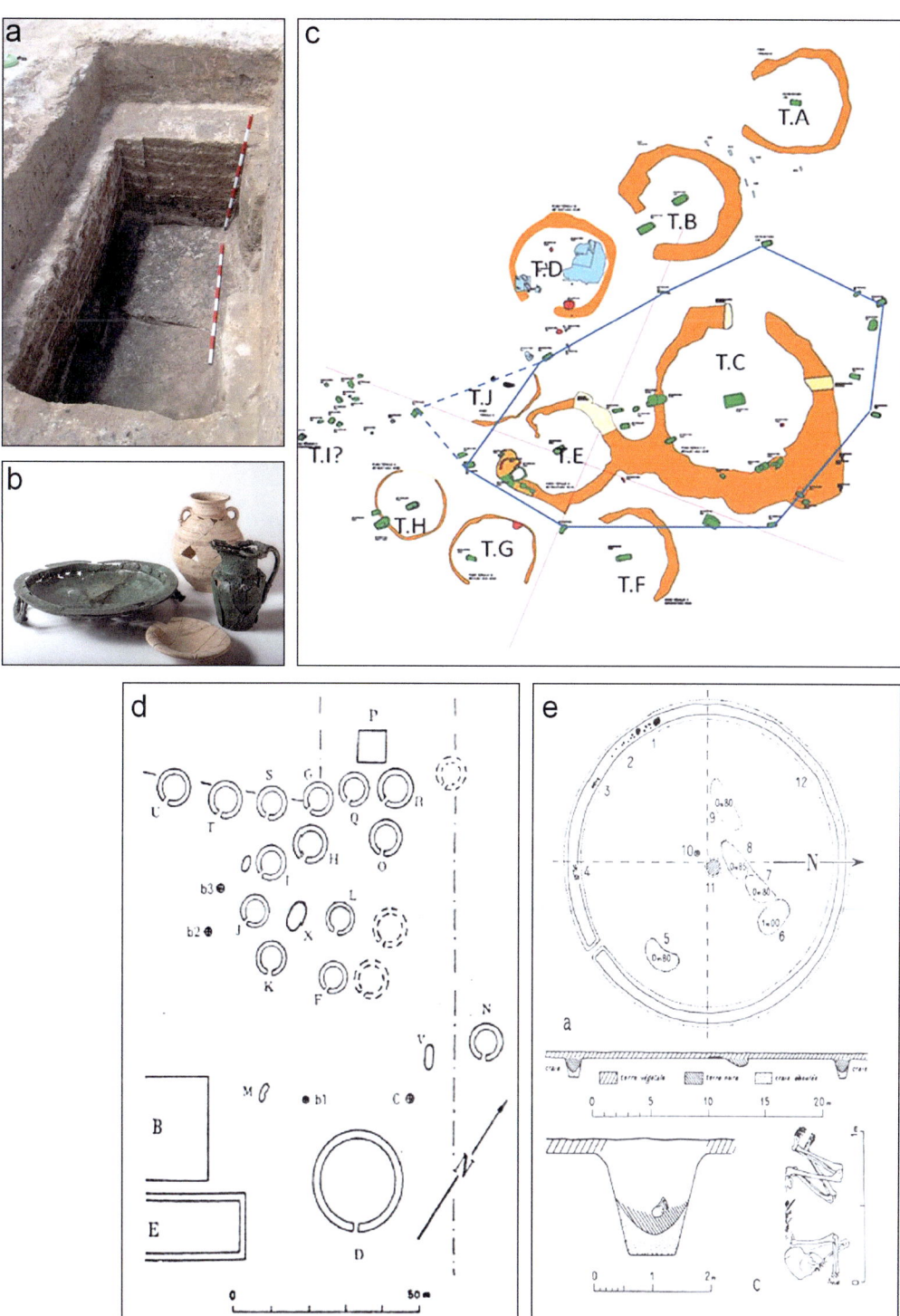

Figure 7.3. Necropolis of La Noria (Fuente Piedra, Málaga, Spain), second half of the sixth century BCE: double burial of Tumulus C of La Noria (A); bronze libation jug-brazier set of La Noria (B); plan of La Noria (C). Necropolis of Aulnay aux Planches (France), plan of phase III (Hallsttat D) (D) and plan and detail of Tumulus D (E) (after Daubigney 1984: figs. 7-8).

of female burials to the south of the first ranking tumulus. Although we do not have the bone remains from the first excavation to confirm this, we cannot rule out the possibility that there were male burials to the north, as some male tombs were excavated in the second phase within this area. In this group of burials, some prestige objects were found, different from the symbolic finds documented in the first two rankings: small perfume containers such as a vitreous paste *aryballos* found in the tomb of a woman (255) and a faience *aryballos* in the form of a pomegranate in Tomb H233-239, also of a woman, located in the pit of Tumulus H. It is important to note the presence in third and fourth ranking burials of this type of personal hygiene items. The circulation of these items is reminiscent of the Cerrillo Blanco burials, where they were also deposited in the second ranking, if we recall the case of the Orientalising comb.

A comparison with a necropolis whose funerary structure is surprisingly like that of La Noria, with nine flat urn cremation tombs, eighteen tumuli with circular ditches and a single central burial (Brisson and Hatt 1955; Chertier 1976), may be of interest for the determination of parental or other practices. This is Phase III (Hallsttat D) of the necropolis of Aulnay aux Planches (France) (Fig. 7.3). As at La Noria, Tumulus D is in an eccentric position and has a diameter of 22 m, much larger than the rest, which at most reach 8 m. It includes, in addition to the central cremation, six peripheral burials arranged in the outer ditch –two inhumations (adult and infant) and cremations. In other words, it also has four rankings as observed at La Noria. The interpretation initially proposed (Brisson and Hatt 1955) was that it was the funerary expression of the conquest of the territory by a Proto-Celtic group, the tumuli being the tombs of the invaders and the flat tombs those of the conquered. However, this proposal was simplistic and implausible. As has been advocated for other cases in Champagne (Chertier 1976), a reuse of the tumulus was proposed, with burials that followed the aura generated by the ancients, considered to be tutelary. However, in the absence of definitive chronological evidence for such an action, a third reading based on the principle of "coetaneity", and dependence (Daubigney 1984) has made headway. In fact, the peripheral burials are found –as would be expected of dependent people– in tombs without grave goods or with very few objects. Therefore, the researcher (Daubigney 1979, 1984) did not rule out that in Tumulus D there are variants that can be explained by social hierarchy, rather than a sign of the evolution of funerary rites. This adds up to an obvious fundamental observation: That the tumulus was intended to protect and monumentalise their ancestors, and its construction implied an ideological intervention of the community that tied the connection between tumulus and aristocracy. In this direction Daubigney went a step further by putting forward the hypothesis of human sacrifice at the two burial sites documented in the pit of Tumulus D. This question led him to argue that there are indicators of the existence of *ambacti*. This was a type of dependent person documented during the Roman conquest of Gaul that Jullian defined as "companions in war" of the Gallic aristocracy. It was a superior class in the hierarchy of clients, reminiscent of "a court of vassals" (Jullian 1908), the number of which was indicative of the power of the aristocrat/lord. In a reference in the *Gallic War* (BG 3, 22. 1-4), it is recounted that the king of the Sosiates (Aquitanians), Adiatunnus, took with him six hundred *devoti*, called *solduri*, who accompanied their leader in life and death. In return, these ultra-faithful followers dressed in the same clothes and led the same kind of life as their leader, but if their leader died, they sacrificed themselves with him. The *devotio*, the driving force behind the behaviour of this type of hard clientele, is the expression of a religious obligation whose meaning went beyond the Latin *fides* between client and patron, which basically expresses attachment and devotion. Testart and Baray (2007) oppose this approach to the *ambactus*. For them it is in fact a type of soft clientele that was not necessarily based on a sworn component of loyalty, unlike the Aquitanian *solduri*, the Greek *terapontes*, the Germanic *comites* or the Celtiberian *devoti*. This led them to define the latter cases as a form of vassalage different to the Roman formula of *fides*. In our opinion, although it is an interesting debate, the reflection opened by Dauvigney in his reading on Aulnay-les Planches, as in the case of reuse, has not documented any information that would allow a positive evaluation of human sacrifices.

At this point we must return to La Noria, as the presence of women in the pit of Tumulus C, or even the urn burial of a couple in the pit of Tumulus E, calls into question Daubigney's proposal that the dead deposited in the tumulus ditch were *devoti*, a clearly masculine institution, and in our case it is not even clear that it existed among the Iberians of the south. In any case, the existence of dependence indicators at La Noria is not questioned, due to the common use of the tumulus that, according to Daubigney, represents an aristocratic ideological form. This does not question the fact that the first two rankings of tumuli constitute the group of members of an extended family. In our view, the key is their relationship to the burials in the flat or third ranking tombs, as it is not clear whether the nature of the group differences with respect to the tumulus is based on kinship (Kirchhoff 1959) or on political dependency factors such as clientship (Testart and Baray 2007). In any case, these are soft hierarchical dependency factors, as we show in the proposed triangular diagram based on the triple relationship: kinship relations, class relations and arms relations (Fig. 7.4). We state that these would have been the mildest models of dependence in reference fundamentally to the clientele, as there is no presence of weapons in the funerary offerings, there are no ritual practices of sacralisation that would have mediated social relations and because we know of no extreme decisions that would have affected the lives of the clients, i.e., the self-sacrifices that would characterise *devotio*. The members of this group also had access to a series of prestige objects that could define the sphere of gifts given by the patron to the client. This fact further differentiates the client from the vassal and highlights the left side of the diagram, where the dependency practices in free people were developed. The other path towards more hierarchical forms –the conical clan– does not require the breaking of kinship ties. It could be that this third ranking of burials is of one or more extended families subordinate to the main extended family of the patriarchal husband and wife, generationally descended from the founding ancestors of the clan. Nevertheless, the difference between this third ranking of flat graves and the burials in the peripheral ditches of the tumulus is not clear. What is clear however is that their intentional proximity to their burials individually identifies them with the people buried in each tumulus.

Closing the cycle of reflection and returning to Cerrillo Blanco, what could be the readings of the case if not an extended family, could there be answers like those suggested at La Noria? Firstly, it is difficult to recognise a clientele funerary structure. This is due to the difficulty imposed by the limited existence of two rankings, since, even if we would like to see it in Zone 2, there are barely indicators that differentiate kin from clientele and this in the framework of a reduced hierarchy exclusively limited to spatial positioning and the possession of supposed gifts. Furthermore, if it were a case of clientele, even a soft version of the relationship in a two-rank structure it would eliminate the existence of relatives

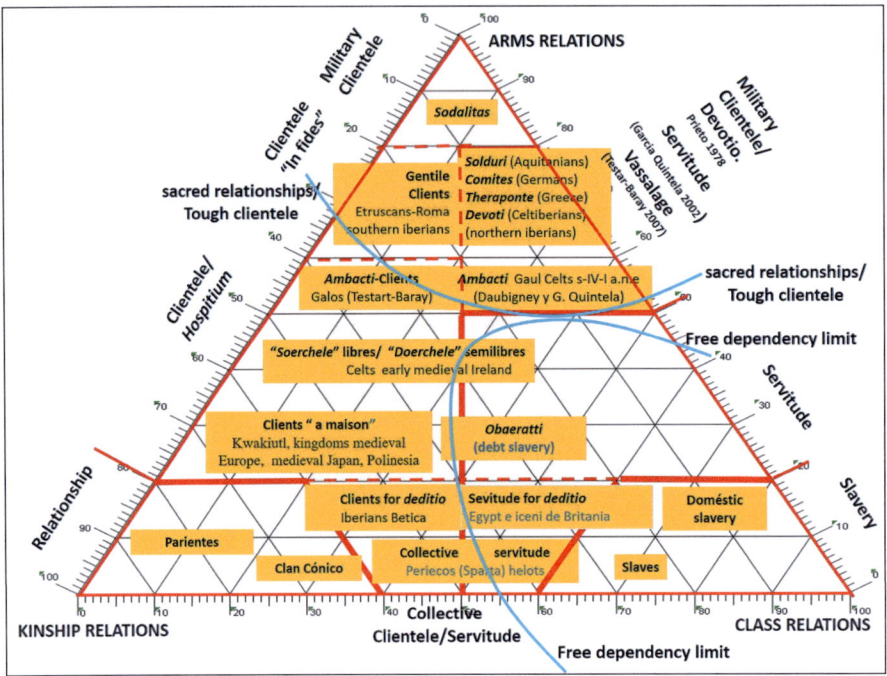

Figure 7.4. Triangular diagram based on the triple relationship of kinship relations, class relations and arms relations.

in the funerary space and would mean that the tomb of the main couple would have reached a high degree of social and political hierarchy by that early moment. This cannot be absolutely ruled out, but it is difficult to accept. The case is different if the interpretations are limited to the differences between the strips (2a and 2c) of burials in Zone 2, linking their reading to a conical clan model in which two differentiated extended families were hierarchically organised. In their spatial location, the strips would have reproduced their degree of parental proximity to the patriarchs and therefore to the common ancestor of the clan. In any case, and although the reading is historically coherent, it is possible to note imbalances in the gender relationship between one group and the other, which would not easily justify the existence of two structurally similar family units.

7.3 The Oretani-Bastetani area. The funerary landscape of the gentilic lineage

In the eastern part of the province of Jaén and the north of Granada, several cases of burials allow us to reconstruct the patterns of the funerary landscape developed in the same chronological arc as before, but in the Iberian nuclear zone. This is the Oretani-Bastetani-Mentesani area. We should start with some cases from the Late Bronze Age, prior to the Iberian period. In Haza de Trillo (Guadiana Menor), Mergelina (1943-44) excavated a shaft burial with five inhumations and grave goods of six bronze bracelets. Of similar characteristics and chronology are Tombs 32 and 33 in the Cerro del Santuario necropolis in Baza (Granada), the latter from the fourth century BCE. The two linked tombs (Presedo 1982) contained four burials each and grave goods of bronze bracelets. Another case, also dated to the eighth century BCE, was documented at La Guardia (Jaén), an *in bustum* cremation burial with grave goods of handwork pottery and a small square-shaped bronze bracelet (Pereira 2001). Overall, these burials from the eighth century BCE show the coexistence of two funerary rites, although in territorial terms the inhumations point towards the easternmost part of the area, the Bastetani, and the cremations to the westernmost part, where the Mentesani are defined. The latter borders the Turdulan area that opens towards the west after the frontier between the present-day provinces of Jaén and Córdoba (Quesada 2008; De Hoz 2015; Ruiz and Molinos 2023). We will look at the evolution of this initial funerary landscape through the sequence of two cases. Cerro de los Vientos, Puente del Obispo (Jaén), on the banks of the Guadalquivir in the Mentesani area, is dated to the second half of the seventh century BCE (Lechuga and Soto 2017). There, a funerary structure was documented; it was dug into the ground and had a hall that led to two niches with cremated corpses. In one, the urn contained the remains of an adult female, while the other had those of a sub-adult individual, whose sex could not be determined, although the presence of an iron spearhead could indicate a male individual. Opposite the niche of the woman was a third niche with the entrance covered by a slab. It presented no funerary use, which suggests, although there is no funerary rite to confirm it as at La Noria, that it was a cenotaph. We can conclude that it could have been a family with the absence of the father, to which could be added a sub-adult woman who was placed individually in another burial nearby. In the surrounding area there was also a structure associated with commensal rites. It contained post-fired painted pots of the Real type and a large wheel-thrown vessel of the caliciform type, probably used for wine. It is important to emphasise that the tomb of the adult woman appears to be associated with that of the young man, as heir to the family line. Among its grave goods, the latter included the most technologically innovative elements: iron and wheel-thrown pottery. In the Bastetani area, the second case to be analysed is the hypogeum of Hornos de Peal (Molinos and Ruiz 2007) dated, like La Noria, to the second half of the sixth century BCE. The burial, cut into the rock, consisted of an elongated chamber where two urns were deposited with the cremated remains of a man and a woman. The grave goods were not particularly rich, although weapons were documented: a spearhead and an arrowhead of the Macalon type. However, the constructive structure of the tomb indicates the importance of the couple buried there, as the hill in which the hypogeum was carved out had been sculpted to resemble a cylindrical-shaped tumulus, reserving next to the chamber a right-angled ramp that ascended to the top of the tumulus on top of which the *ustrinum* was built. The entire upper surface was treated with red ochre, a technique that made it stand out visually as a territorial landmark. The monumentality of the structure was enhanced by a baetyl erected at the chamber door. In this case there were no burials around it, however, at that time one km away a small nucleus was being formed on the hill of El Ahorcado de Toya. A century later this would give rise to an important necropolis that was unfortunately plundered at the beginning of the twentieth century. It was the site of the Toya Chamber, the most architecturally important princely tomb in the entire Iberian cultural area (Rísquez *et al.* 2022c).

From the fifth century BCE onwards, the small nuclei developed, changing the funerary landscape until they became large necropolises. In addition to their size, they developed two characteristics that were present in the historical antecedents. On the one hand, the importance of the nuclear family and, on the other, the presence of weapons in the grave goods. The new model of the agglomerated necropolis was closely connected to the autonomous political development of the *oppida*, as that process relegated the old ethnic

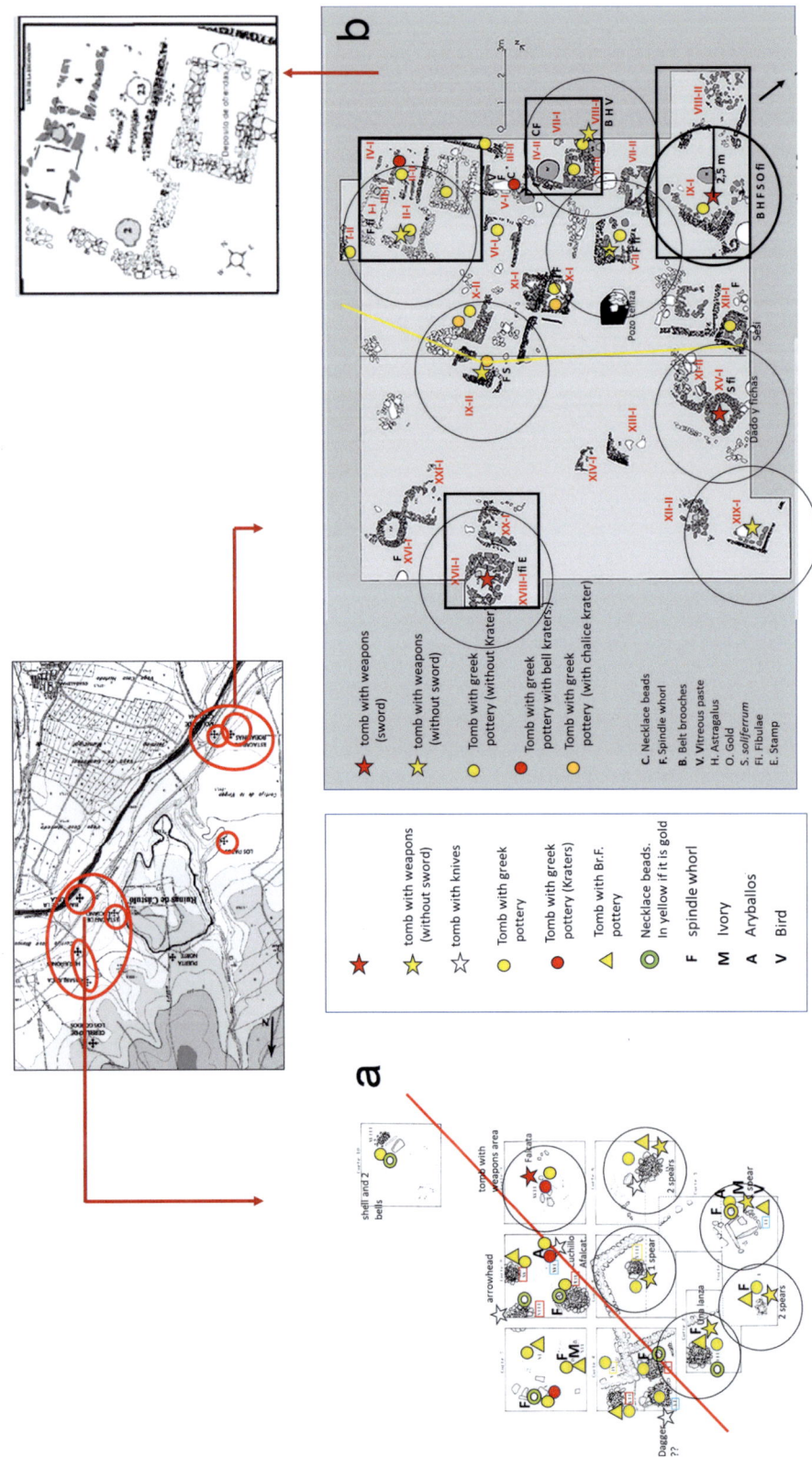

Figure 7.5. Necropolis of Cástulo (Linares, Jaén, Spain) fifth-fourth centuries BCE: A) Higuerones-Baños de la Muela, a dispersed necropolis with chambered tombs; B) Estacar de Robarinas, an agglomerated burial space, with the largest tombs on stepped platforms (after Ruiz and Molinos 2022).

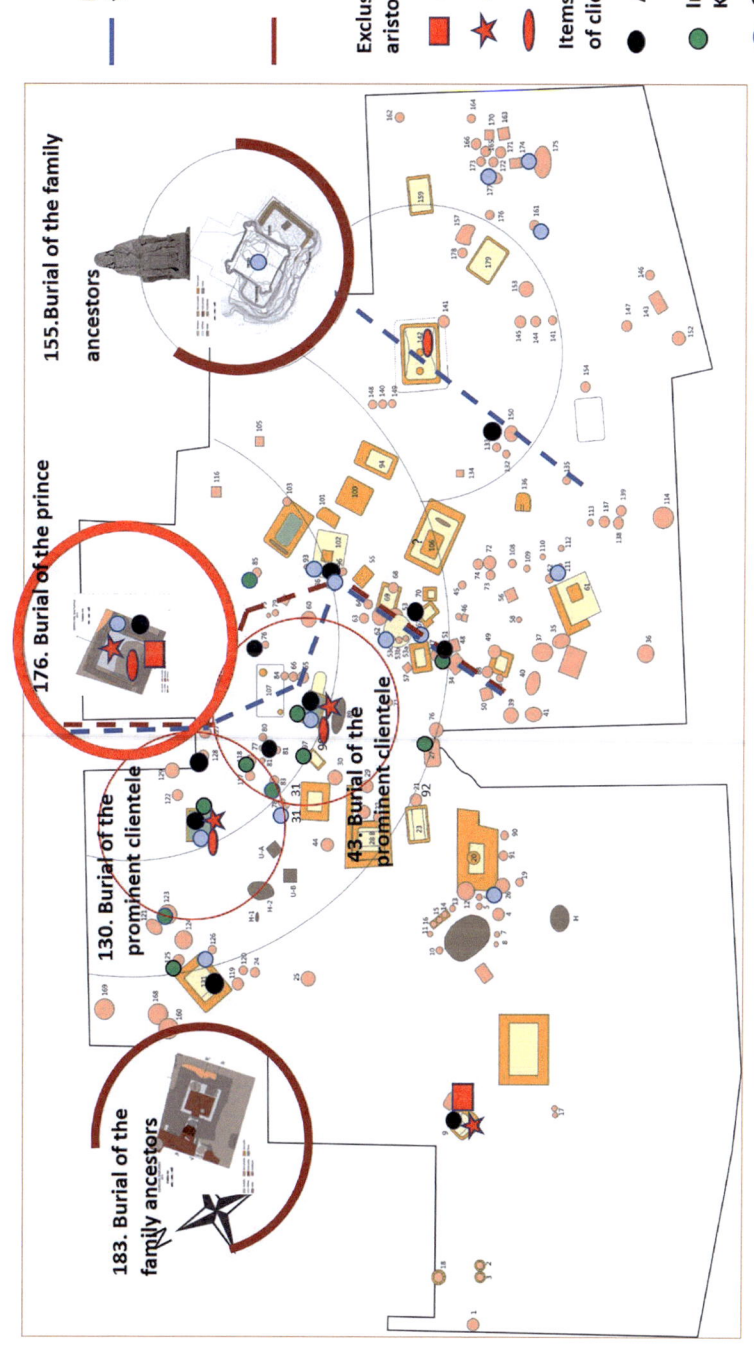

Figure 7.6. Necropolis of the Cerro del Santuario (Baza, Granada, Spain) as a gentilic funerary landscape.

Boundary of high-neck flared cremation urns

Boundary of Kalathos cremation urns

Exclusive items from aristocratic grave goods

Cart-wheel

Attic Kraters

Bronze brazier

Items type gifts given of clientele grave goods

Attic Kilices

Imitations of Attic column Kraters

Soliferreun

155. Burial of the family ancestors

176. Burial of the prince

130. Burial of the prominent clientele

43. Burial of the prominent clientele

183. Burial of the family ancestors

units to mere cultural entities and led to the development of the gentilic[7] lineage political structure in the *oppida*, as an alternative political mechanism to the old kinship structures. The process was neither rapid nor absolute. A recent review of the fifth-fourth-century BCE necropolis of Cástulo (Ruiz and Molinos 2022) led to conclude that at least two lineages lived side by side in that large Iberian Oretani *oppidum* (Fig. 7.5). One was buried in Higuerones-Baños de la Muela, a dispersed necropolis with chambered tombs, while the other was buried in the agglomerated space of the necropolis of Estacar de Robarinas, with the largest tombs on stepped platforms. These were not the only formal differences between the two lineages. Although they shared the bases of the traditional Bastetani funerary model, the distribution of the tombs in the small necropolis of Baños de la Muela has allowed us to observe two differentiated areas, according to the presence of weapons in the burials, which suggests a spatial arrangement marked by gender. The case of Higuerones-Baños de la Muela confirms that in Cástulo the patriarchal model could still have been active at the beginning of the fourth century BCE, although it had evolved in two directions; on the one hand, by segregating the funerary space of the prince-patriarchs, absent in Baños de la Muela and probably buried in the tumuli of Higuerones.

On the other hand, this process was undertaken by arming the extended families, as segments of the patriarchal lineages. The other lineage of Castulo, buried in Estacar de Robarinas, represented a new type of society, identified in the large gentilitial-clientele necropolis (Ruiz and Molinos 2022). The model had its paradigm in the necropolis of the Sanctuary of Baza (Fig. 7.6), which gives visibility to the three basic generations that define a gentilic funerary landscape: that of the anonymous or ancient ancestors, whose exact parental chain cannot be reconstructed (Tombs 32 and 33 from the eighth century BCE); the generation of the family ancestors (Tomb 155 of the Lady of Baza); and the first generation of the nuclear families of the princes, in nuclear family pantheons (Burial 176). The burials of the clientele, born in an *in fides* pact, as proposed by Mario Torelli (1996), are added to the sphere of this last generation. This allowed the aristocrat, by integrating them into the funerary space of the lineage, to expand his private army and his social dependence groups. These groups are distinguished from the consanguineous members of the lineage, both by the segregated spatial distribution in the funerary space of the necropolis and by the type of urn and the gifts given by the patron –imitations of Attic column craters, Attic cylices or weapons such as the *soliferrum*. An additional characteristic of these necropolises is the distribution of burials by mixed micro-groups, nuclear families, as noted at Estacar de Robarinas, where these groups of burials are defined as spatial units around the tomb with weapons (Ruiz and Molinos 2022). This supports Peroni's reading in defining the gentilitial structure as opposed to the consanguineous lineage model, as a structure of hierarchical nuclear families (Peroni 2004).

The process in the Turdulan area was unified in funerary terms at the end of the fifth century BCE, with the Bastetani-Mentesani forming a single model, the *gentilitium*, until its crisis at the end of the fourth century BCE, when the necropolises were abandoned in many *oppida*. This was probably due to the crisis of the gentilitial model in the face of other developing political models such as the "House Society". These structures that had a greater impact on territorial forms and cognatic alliance relations, as well as on the emergence of new free social strata such as citizenship and state political systems as in the kingdom of Cástulo.

---

7   A word derived from the Roman *gens*, meaning an agnatic patrilineage (note by the editors).

References

Arteaga, O. (1999): La delimitación del territorio entre Cástulo y Obulco. In Salvatierra, V. and Rísquez, C. (eds.), *De las sociedades agrícolas a la Hispania romana. Jornadas históricas del Alto Guadalquivir, Quesada (1991-1995)*. Jaén: Universidad de Jaén, pp. 95-141.

Brandherm, D. and Krueger, M. (2017): Primeras determinaciones radiocarbónicas de la necrópolis de Setefilla (Lora del Río) y el inicio del periodo orientalizante en Andalucía occidental. *Trabajos de Prehistoria* 74(2), pp. 296-318. DOI: 10.3989/tp.2017.12196.

Brisson, A. and Hatt, J. (1955): Cimetières gaulois et gallo-romains à enclos en Champagne. I. Le cimetière de l'Homme mort à Ecury-le-Repos (Marne). *Revue Archéologique de l'Est* 6, p. 313-333.

Chertier, B. (1976): *Les nécropoles de la civilisation des champs d'urnes dans la région des marais de Saint-Gond (Marne)*. Paris: Gallia préhistoire, supplément 8.

Daubigney, A. (1979): Reconnaissance des formes de la dépendance gauloise. *Dialogues d'Histoire Ancienne* 5, pp. 145-189.

Daubigney, A. (1984): Tombes et signes hiérarchiques en Champagne protohistorique: problèmes. In Daubigney, A. (dir.), *Archéologie et rapports sociaux en Gaule: Protohistoire et Antiquité*. Besançon: Université de Besançon, pp. 123-154.

Díaz-Zorita, M., Aranda, G., Bocherens, H., Escudero, J., Sánchez Romero, M., Lozano, A., Alarcón, E. and Milesi, L. (2019): Multi-isotopic diet analysis of south-eastern Iberian megalithic populations: The cemeteries of El Barranquete and Panoría. *Archaeological and Anthropological Sciences* 11(2), pp. 3681-3698. DOI: 10.1007/s12520-018-0769-5.

González Navarrete, J.A., Arteaga, O. and Unguetti, C. (1980): La necrópolis del Cerrillo Blanco y el poblado de Los Alcores (Porcuna, Jaén). *Noticiario Arqueológico Hispánico* 10, pp. 185-217.

Jullian, C. (1908): *Histoire de la Gaule. II La Gaule indépendante*. Paris: Hachette.

Lechuga, M.A., Moreno, M.I. and Soto, M. (2022): El área funeraria del Cerro de los Vientos (Puente del Obispo, Baeza, Jaén). Un ejemplo de hibridación de las élites locales durante el periodo Orientalizante en el Alto Guadalquivir. In Rísquez, C., Rueda, C. and Herranz, A.B. (eds.), *El reflejo del poder en la muerte. La cámara sepulcral de Toya*. Jaén: Universidad de Jaén, pp. 187-210.

Leduc, C. (1991): ¿Cómo darla en matrimonio? La novia en Grecia, siglos IX-IV a.C. In Schmitt Pantel, P. (ed.), *Historia de las mujeres en Occidente, I. La Antigüedad*. Madrid: Taurus, pp. 251-316.

Kirchhoff, P. (1959): The principles of clanship in human society. In Fried, M. (ed.), *Readings in Anthropology*. New York: Crowell, pp. 259-270.

Martín, D. (2022): *Análisis de entesopatías. Necrópolis de Cerrillo Blanco (Porcuna, Jaén)*. BA dissertation. Granada: Universidad de Granada.

Molinos, M. and Ruiz, A. (2007): *El hipogeo ibero del Cerrillo de la Compañía de Horno (Peal de Becerro, Jaén)*. Sevilla: Junta de Andalucía.

Peláez, M. (2013): *Estudio descriptivo de la necrópolis tartésica de Cerrillo Blanco (Porcuna, Jaén)*. MA dissertation. Granada: Universidad de Granada.

Pereira, J., Chapa, T. and Madrigal, A. (2001): Reflexiones en torno al mundo funerario de la alta Andalucía durante la transición Bronce Final-Hierro I. *Spal* 10, pp. 249-273. DOI: 10.12795/spal.2001.i10.17.

Peroni, R. (2004): Culti, comunità tribali e gentilizie, caste guerriere e figure di eroi e principi nel secondo millennio in Italia tra Europa centrale ed Egeo. In Marzatico, F. and Gleirscher, P. (eds.), *Guerrieri, principie ed eroi. Fra el Danuvio e il Po, dalla Preistoria all'Alto Medievo.* Trento, pp. 161-173. DOI: 10.3989/tp.2023.12328.

Ramos, A.M. (2010): *Los pliegues del linaje. Memorias y políticas mapuches-tehuelches en contextos de desplazamiento.* Buenos Aires: Eudeba.

Rísquez, C. (2015): La arqueología ibérica y los estudios de género en Andalucía. *Menga* 6, pp. 61-91.

Rísquez, C., Díaz-Zorita, M., Rueda, C., Herranz, A.B., Torres, M., Alemán, I. and Peláez, M. (2022a): Compartir el espacio en la muerte. El túmulo funerario de Cerrillo Blanco (Porcuna, Jaén). In Rísquez, C., Rueda, C. and Herranz, A.B. (eds.), *El reflejo del poder en la muerte. La cámara sepulcral de Toya.* Jaén: Universidad de Jaén, pp. 137-162.

Rísquez, C., Rueda, C. and Herranz, A.B. (2022b): Objetos de vestir y adornos personales en la construcción de identidades femeninas. De los orígenes a la consolidación del modelo aristocrático ibérico en el alto Guadalquivir. In Graells, R., Lorrio. J.A. and Camacho, P. (eds.), *Problemas de cultura material. Ornamentos y elementos del vestuario en el arco litoral mediterráneo-atlántico de la Península Ibérica durante la Edad del Hierro (ss. X-V A.C.).* Alicante: Universidad de Alicante, pp. 157-172.

Rísquez, C., Rueda, C. and Herranz, A.B. (eds.) (2022c): *El reflejo del poder en la muerte. La cámara sepulcral de Toya.* Jaén: Universidad de Jaén.

Rísquez, C., Herranz, A.B., Rueda, C., Díaz-Zorita, M. and Alemán, I. (2025): Morir demasiado pronto. Enterramientos infantiles en el túmulo funerario de Cerrillo Blanco (Porcuna, Jaén). In Sánchez Romero, M., Alarcón, E. and Rivera, A. (eds.), *Pequeños cuerpos con grandes biografías. Una mirada a la infancia desde la (bio) arqueología.* Granada: Comares, pp. 233-262.

Roos, A. M. (1997): *La sociedad de clases, la propiedad privada y el estado en Tartesos. Una visión de su proceso histórico desde la arqueología del proyecto Porcuna.* PhD dissertation. Granada: Universidad de Granada.

Rueda, C., García, A., Ortega, C. and Rísquez, C. (2008): El ámbito infantil en los espacios de culto de Cástulo (Jaén, España). In Gusi, F., Muriel, S. and Olaria, C. (eds.), *Nasciturus, infans, puerulus vobis mater terra: la muerte en la infancia.* Castellón: Servei d'Investigacions Arqueològiques i Prehistòriques, pp. 473-496.

Ruiz, A. (2023): Panorama de los iberos del sur. A propósito del territorio sudoccidental de los iberos. *Palaeohispánica* 23, pp. 93-114.

Ruiz, A., Molinos, M., Cano, M.F., Montes, E. and Ortuño, E. (2017): El túmulo C de la necrópolis de La Noria (Fuente de Piedra) ¿La tumba de las dos mujeres? In Ruiz, A. and Molinos, M. (eds.), *Catálogo de la Exposición La Dama, el Príncipe, el Héroe y la Diosa.* Sevilla: Junta de Andalucía, pp. 119-124.

Ruiz, A. and Molinos, M. (2018): Genealogía, matrimonio y residencia en el proceso político de los iberos del alto Guadalquivir. In Rodríguez Díaz, A., Pavón Soldevila, I. and Duque Espino, D.M. (eds.), *Más allá de las casas. Familias, linajes y comunidades en la protohistoria peninsular.* Cáceres: Universidad de Extremadura, pp. 41-72.

Ruiz, A. and Molinos, M. (2022): La secuencia genealógica de los linajes iberos a través de los paisajes de la muerte: de Baza a Cástulo. In Rísquez, C., Rueda, C. and Herranz, A.B. (eds.), *El reflejo del poder en la muerte. La cámara sepulcral de Toya.* Jaén: Universidad de Jaén, pp. 41-92.

Testart, A. and Baray, L. (2007): Ambactes et soldures. Figures gauloises du compagnon-nage guerrier. In Lecrivain, V. (ed.), *Clientèle guerrière, clientèle foncière et clientèle électorale. Histoire et anthropologie.* Dijon: Presses Universitaires de Dijon, pp. 51-84.

Torelli, M. (1988): Le popolazioni dell'Italia antica: società e forme del potere. *Storia de Roma,* I, *Roma in Italia.* Roma: Einaudi, pp. 53-74.

Torrecillas, J.F. (1985): *La necrópolis de época tartésica de Cerrillo Blanco (Porcuna, Jaén).* Jaén: Instituto de Estudios Giennenses.

Torres, M. (1999): *Sociedad y mundo funerario en Tartessos.* Madrid: Real Academia de la Historia.

Torres, M. (2002): *Tartessos.* Madrid: Real Academia de la Historia.

Archaeogenetics Beyond Kinship: The Iron Age Intramural Child Burials of Northern Iberia

Roberto Risch

Department of Prehistory, Universitat Autònoma de Barcelona, Spain, Robert.Risch@uab.cat

Patxuka de Miguel Ibáñez

University Institute for Research in Archaeology and Historical Heritage (INAPH), University of Alicante, Spain

Luka Papac

Department of Archaeogenetics, Max Planck Institute for Evolutionary Anthropology, Germany

Adam B. Rohrlach

Department of Archaeogenetics, Max Planck Institute for Evolutionary Anthropology, Germany

Marcello Peres

Department of Prehistory, Universitat Autònoma de Barcelona, Spain

Javier Armendáriz

Department of Humanities and Education, Public University of Navarra, Spain

Stephan Schiffels

Department of Archaeogenetics, Max Planck Institute for Evolutionary Anthropology, Germany

Abstract

This chapter addresses the long-lasting custom of placing infant (perinatal and natal) burials inside dwellings in the northern half of Iberia during protohistory. The authors use archaeogenomics and statistical tools to identify consanguineous relationships ("biological kinship") between a sample of analyzed infant individuals from several Early Iron Age villages in the Ebro basin. Genomic results indicate that these individuals had been selected and buried, presumably, for their unusual features, such as trisomies, Down and Edwards syndromes or for being twins. This case study demonstrates how archaeogenetics combined with highly detailed contextual information can provide hypotheses on the reasons for different burial treatments and how genetic data can be used to shed new light on the social groupings attached to residential buildings.

Keywords: *settlement archaeology, perinatal burial, chromosomal trisomy, paleodemography, Early Iron Age.*

In Blanco-González, A. and Alarcón-García, E. (eds.) 2025, *A Social Archaeology of Kinship in Iberia and Beyond. Recent Multistranded Approaches from aDNA to Household Archaeology.* Leiden: Sidestone Press, pp. 159-178.

8.1 Introduction

Recent developments in archaeogenetics have enabled to identify and quantify close biological relationships between people living in the past. In the beginning, studies of ancient DNA (henceforth aDNA) focused on a few individuals from each cemetery, region and period with the aim to grasp the "big picture" of social mobility, migrations and stability (Meller *et al.* 2017; Reich 2018; Krause 2022). The reduction in the analytical costs and the extension of the scope of the questions asked within the newly emerging discipline of archaeogenetics has allowed the genetic characterisation of whole communities buried together, when preservation and resources allow it. This information on purely biological relations –but provided with a spatial, temporal, and material (e.g., grave goods) reference in the archaeological context– has in turn impacted kinship studies, one of the main research fields of social and cultural anthropology. As ethnographic research and theoretical discussions have generated and critically scrutinised an exceptional corpus of information, the meaning of kinship has been extended to a growing array of relations among humans, but also between humans and animals, plants and rocks (Frieman 2023; TallBear 2018). In contrast, the information on biological ascend or descent of individuals provided by the sequencing of ancient genomes, neither rely on perception of ethnographers nor their theoretical perspectives but exclusively depends on the degree of conservation of aDNA. The first- and second-degree relationships between two persons can be reliably determined with statistical tools like READ –when >20,000 genome-wide polymorphisms are preserved (Kuhn *et al.* 2018)– and relations up to the tenth degree are detectable with the identical by descent (IBD) procedure –if >500,000 genome-wide polymorphisms are retrieved (Ringbauer 2024). Before written records existed, such a web of biological relations went far beyond what any human was or is capable of recalling concerning their descent. This fundamental difference in the way that data is generated in anthropology and archaeogenetics and what needs to be stressed with determination in view of some of the discussions maintained in anthropology and, to a lesser extent, archaeology, which has contributed comparatively little to our understanding of kinship in prehistory (Ensor 2013: 272-298).

A recent attempt to bring both research lines closer has shown that differences probably lay more in the part of (past) social reality addressed by each field, than in an underlying epistemological difference between social and natural sciences (Meller *et al.* 2023). The information used by social and cultural anthropology derives from observations and conversations done mostly by western priests, colonial functionaries or academics on others. Most of these observations and conversations cannot be replicated, as recent developments under the capitalist mode of production have changed all societies on the planet or even made them disappear (e.g., Wolf 1982; Davis 2001). Where anthropology has re-visited old field studies and scrutinised previous views on kinship, conclusions reached were often different from the original anthropological studies (e.g., Hutchinson 2000).

Archaeology, on the other hand, tries to replicate this information through the analysis of the material remains left by societies which can no longer be observed nor directly questioned. Inferences concerning kinship are usually based on the size and internal organisation of houses (e.g., Flannery 1973; Kuijt 2002) or on the spatial/stratigraphic arrangements between individuals of different age and sex buried in the same place or monument (e.g., Hager and Boz 2012; Lull *et al.* 2016; Cveček and Schwall 2022). Yet, neither house architecture nor funerary practices are always or only ruled by what anthropology has defined in multiple ways as kinship. This observation allows for interpretations from many other realms of social organisation, such as interaction with the environment, productive capacity, wealth differences, demography, ideology, etc. It follows that any reading of settlement and funerary records in terms of past kinship relations can only be tentative.

Finally, as already mentioned, genetic studies can only provide information about who was and who was not biologically related among the individuals buried in a cemetery. Unfortunately, an increasing number of studies confirm that many cemeteries only included a fraction of the original community, with the whereabouts of the rest remaining unknown. Physical anthropology has provided abundant examples where the reconstructed demographic profiles are flawed in terms of sex, age or funerary ritual (e.g., Angel 1971; Boz and Hager 2013). The reconstruction of pedigrees has recently started to quantify the proportion of the population inferred to have existed but not identified in a cemetery (Rivolat *et al.* 2023). The discussion on the reasons and criteria applied to grant subsets of the community archaeologically identifiable burials and the others not, has remained largely speculative.

The funerary practice, rather than being fragmented –as often stated– provides insight into performative and ritual practices of social groups of variable size, apparently bound to social conceptions concerning life and death. These rituals are a staged (self)representation of a community in front of the event of death and ancestry. Which persons and objects were interred where, and how, is the result of specific decisions taken in the past, rather than a more or less random result of material preservation and archaeological recovery. While a social or cultural anthropological perspective might be tempted to think that the funerary record is a representative but incomplete image of the whole living society, it is actually an intentional and meaningful selection of individuals, materials, places, and temporalities used to assert, maintain, or transform social, economic, and political relations. Consequently, whoever is buried in a certain way and whoever is not expresses a social selection and the intention to ascertain (or overcome) specific social rules. Whatever bioanthropological, archaeological or genetic results are retrieved from funerary contexts, they are inevitably affected by the decisions made by past societies about who was buried or dealt with where and in which way (Risch *et al.* 2023).

One archaeological realm where this performative complexity can be captured exceptionally well is intramural burial practices. The close proximity between human remains and the persons living in the dwellings already attaches a special importance to the dead in the ideological, political, economic or other practices of these communities. A selection of corpses is practically inevitable, as the size of houses already sets physical limits on the number of possible interments, if tombs are not reopened and regularly emptied of corpses (Kuijt 2008). When the whole community cannot be entered, special rules need to be established concerning who deserves attention; rules which were definitely anchored in the social relations of the living. Understanding the rules of intramural interment should provide us direct insight into the social relations of the persons living, working or simply having access to these social spaces. While bioarchaeology has been crucial to identifying rules of interment in terms of sex, age, pathologies, and nutrition, and while archaeological observations have retrieved spatial and structural regularities, archaeogenetics allows us to establish if the buried individuals where biologically related or not. Whichever array of relations existed in the past, sexual relations leading to descendants definitely formed part of them and merit special attention given their crucial importance for the physical as well as political reproduction of the community. Consequently, archaeogenetics can provide information to understand our prehistoric past, which goes beyond the realm of the relation between biological relations, sex and kinship (Risch *et al.* 2023).

Resorting to recent aDNA results obtained from intramural Iron Age inhumations of the North of the Iberian Peninsula (Papac *et al.* 2023), the present study aims to highlight the heuristic gains achieved by this type of studies. This case shows how archaeogenetics can provide, for the first time, a hypothesis on the reasons followed by a community to separate their deceased into different groups. But we also want to highlight how genetic data can be used to understand the social group attached to architectonic buildings, conventionally interpreted in archaeology as houses and settlements, and the temporal duration of these relations.

8.2 Intramural burials in Iron Age Iberia

While cremation became a widespread funerary practice in large parts of Europe and the Mediterranean by the end of the Bronze Age, communities in northern Iberia and along the Mediterranean coast –from Languedoc to Murcia– reserved intramural burial for a selected number of children, most of which died at perinatal age (Fig. 8.1). Absolute dates and stratigraphic observation indicate that infants started to be interred in settlements in this large territory around 1000 BCE and continued to be practiced under Roman rule (Moya *et al.* 2005: 44).

So far, around 770 intramural burials of young children have been identified in nearly 90 Final Bronze Age and Iron Age settlements, mostly in the north-eastern part of the Iberian Peninsula (Fig. 8.1). Although numbers differ notably between sites, over half of the known settlements have only produced evidence of one or two intramural burials. According to the available funerary evidence, the rest of the population was cremated and buried in cemeteries outside the settlements. The discourse surrounding the meaning of this very restricted intramural burial rite has developed over a long time in Iberian archaeology. Sacrificial rituals have been suggested for a long time (Gusi 1970), but these are not supported by available anthropological evidence. Alternative explanations have related the burials to offerings during the construction or remodelling of buildings (Guérin and Martínez 1987-1988; Dedet and Schwaller 1990). Intramural interments have also been considered as a selection of children who died of natural causes (Armendáriz and de Miguel 2006: 41; Lorrio *et al.* 2010). Age was clearly a selection criterion as most of skeletons that have been studied anthropologically died around 40 weeks of intrauterine life, probably during or shortly after birth (Risch and Carbonell 1986; Guérin and Martínez 1987-1988; Gómez and Oliver 1989; Gracia *et al.* 1989; Maluquer de Motes *et al.* 1990; Agustí *et al* 2000; Armendáriz and de Miguel 2006; de Miguel 2009; Lorrio *et al.* 2010; Carnicero-Cáceres and Torres-Martínez 2021). Double and triple burials, which are well-documented in a dozen settlements, have been interpreted by some authors as twins or triplets, who died more or less simultaneously at a perinatal age (Agustí *et al.* 2000; Armendáriz and de Miguel 2006; de Miguel 2009; Lorrio *et al.* 2010; Subirà and Molist 2016; Blasco and Montón 2019). No specific burial patterns can be recognised in terms of the spatial distribution of the skeletons within the settlements. Neither the number of burials nor the rooms and buildings where they are placed in seem to follow any obvious rule.

The apparent randomness concerning the number of tombs, the number of inhumations per tomb, and their location inside the settlements suggests that the events or circumstances leading to intramural burial practices were unforeseeable or erratic, as well as selective in view of the limited number of skeletons found in living spaces.

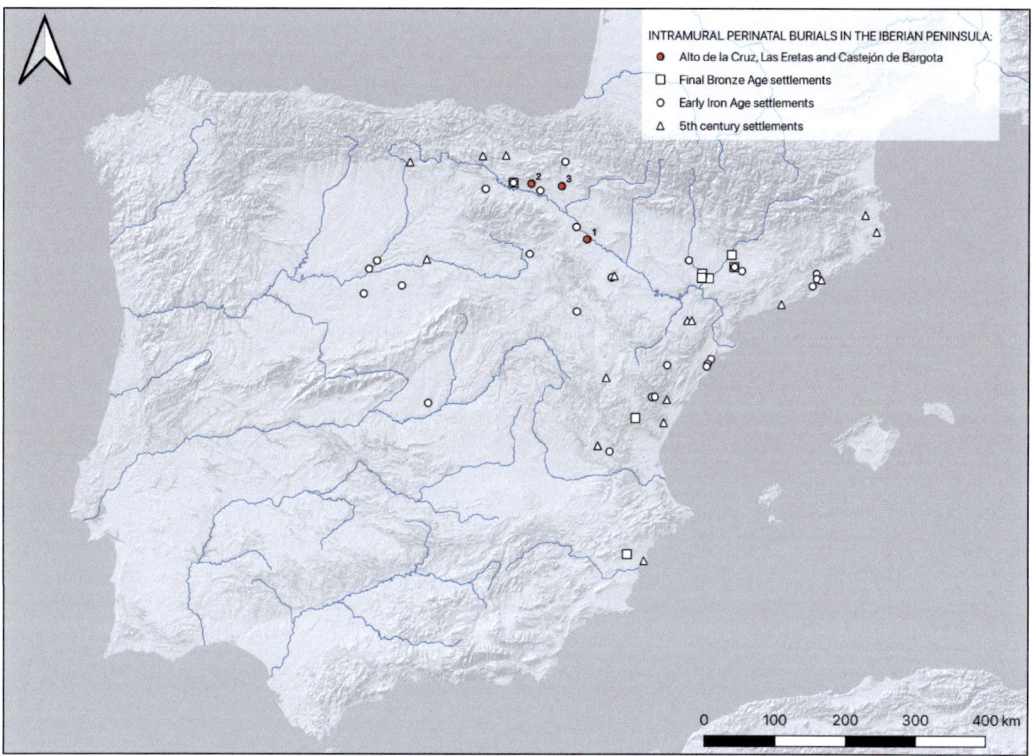

Figure. 8.1. Map of Late Bronze Age and Early Iron Age settlements with intramural perinatal burials in the Iberian Peninsula: 1. Alto de la Cruz; 2. Las Eretas; 3. Castejón de Bargota.

8.3 The analysed burials and their archaeological context

So far, the genomes of 37 intramurally buried children have been studied, found in three Early Iron Age (henceforth EIA) settlements of Navarra (Spain), dated between *ca.* 800-450 BCE. They include 29 individuals from the well-known fortified settlement of Alto de la Cruz, excavated extensively during the second half of the twentieth century (Fig. 8.2). Another six individuals come from a very similar, though smaller, settlement of Las Eretas, and two more individuals were found at El Castejón, all excavated in the last decades (Armendáriz and de Miguel 2006; Castiella *et al.* 2009). The three sites are fortified villages with blocks of rectangular houses typical of a new occupation pattern emerging in the northeast of the Iberian Peninsula during the last stage of the Late Bronze Age (Maluquer de Motes 1958; García *et al.* 1994). While Alto de la Cruz and Las Eretas are good examples of lowland villages located at the bottom of valleys next to the Ebro and Arga rivers, respectively, El Castejón de Bargota is a fortified hilltop settlement that occupies one of the heights in the southern foothills of the Cantabrian mountains.

8.4 Insight into funerary ritual and social organisation provided by genetics

From the 37 EIA children of the three settlements, the petrous part of the temporal bone was sampled and analysed for aDNA. Enough data was produced from 35 children (>40,000 1,240k single nucleotide polymorphisms, henceforth SNPs) for downstream autosomal,

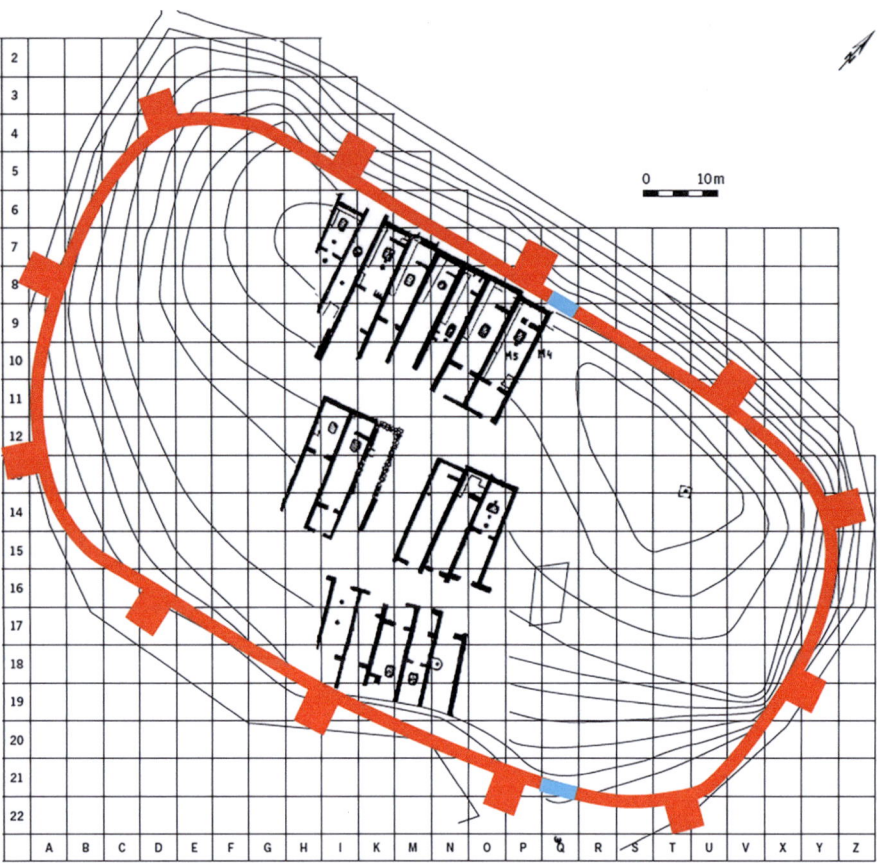

Figure. 8.2. General plan of Alto de la Cruz (village PIIb) with the authors' proposal for the reconstruction of its fortification; in blue, documented parts of the wall; in red, proposed layout (modified after Maluquer de Motes 1954: pl. 1).

Y-chromosomal, and mitochondrial analyses. READ (Kuhn *et al.* 2018) was used in the first place to identify close biological relatedness. Recently, IBD (i.e. identity by descendent) analyses have been performed on those individuals with >500.000 SNPs recovered. This information allows us to identify relations which can reach up to the tenth degree. The patterns of consanguinity and effective sizes of the population from which individuals are sampled were determined by runs of homozygosity (henceforth ROH) in genomes of various haplotype lengths using HapROH (Ringbauer *et al.* 2021). Finally, the individuals studied here were also subject to a recent screening project involving 9,855 ancient samples with genetic data for chromosomal trisomies (Rohrlach *et al.* 2024). This combination of genetic and statistical approaches has allowed us to provide insight into the biological conditions of the children and the social relations that might have led to their burial within the settlements.

8.4.1 What was different or "special" about the few children buried inside the houses?

Taking into account that child mortality in late prehistory probably affected between one quarter and one third of the population (Alesan *et al.* 1999; Séguy and Buchet 2013; McFadden *et al.* 2022), the few intramural burials recovered in the Iron Age settlements of northern Iberia must have fulfilled certain criteria to be spared from cremation and

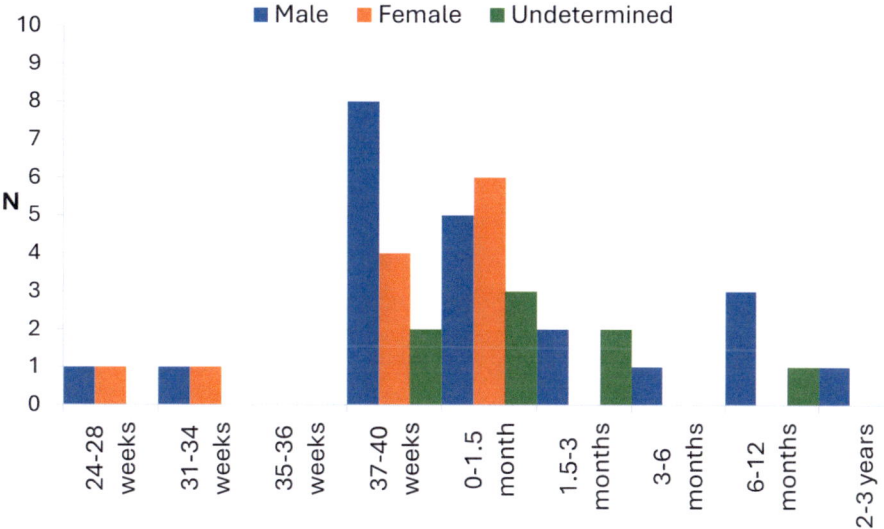

Figure 8.3. Age and sex of the anthropologically and genetically studied buried children from Alto de la Cruz, Las Eretas and El Castejón (Navarra, Spain).

kept inside the architectonically circumscribed spaces where part of the social practices of these communities took place.

Age of death clearly seems to have been one of these criteria (Fig. 8.3). Two thirds of the buried children had died at the end of gestation (37-40 weeks), possibly because of birth complications, or immediately afterwards. During the first month of extra-uterine life, deaths are frequently related to the after-effects of a complicated birth, infectious disease, often of respiratory and digestive type, and to genetic and congenital alterations that limit the development of the babies, leading to their death. Very few of the buried children (~10%) had not reached the age of foetal maturation (37 weeks). This distribution of age and sex offers a demographic profile which, despite possible limitations, may be representative of child mortality in EIA populations. More difficult to explain are the children over six months of age, some of whom featured clear asymmetry between dental and bone development.

The results of genetic sex determination reject the possibility that children were selected according to their biological sex (Fig. 8.3). Despite more male than female intramural child inhumations being recovered, the ratio does not deviate statistically significantly from 1:1 (p=0.24, binomial test). A clear disproportion is only observed among the children buried above three months of extrauterine life, all of which could be identified as male when genetic analysis was successful (7 of 10 individuals). It is unclear if this mirrors the higher male than female mortality commonly observed among children (Drevenstedt *et al.* 2008; Pongou 2012).

More significant is the exceptionally high number of identical twins identified through READ among the 35 sequenced children. The first case was identified in Las Eretas and the second in El Castejón. In both cases, twins were buried together under the respective house floors, suggesting they died simultaneously at a perinatal age (Fig. 8.4). It is noteworthy that the two individuals originating from another possible double burial were not related in a first or second degree, implying that not all multiple EIA infant burials need to be seen as twins, siblings or half-siblings. However, this conclusion needs to be taken with caution, as no information on the context of both skeletons, found in 1949, has been published.

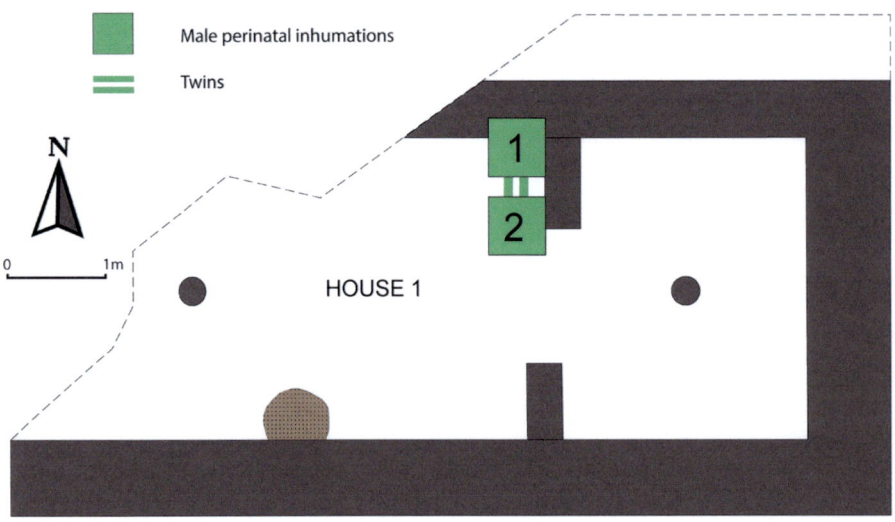

Figure 8.4. Location of a grave with identical twins inside House 1 of El Castejón.

We also identified two cases of first-degree relatives, and the fact that all four individuals are newborns and share mitochondrial and Y-chromosome haplogroups implies they must be siblings. Five cases of second-degree relatives were also found. It follows that some couples and some biologically related groups buried more than one of their children inside the settlements, while others did not. This excludes, for example, that an exclusive intramural burial right was followed when the first child of a couple was born dead.

More unexpected has been the identification of four cases of children with chromosomal trisomy, three from Alto de Cruz and one form Las Eretas. The screening of 9,855 ancient genomes generated at the Max-Planck Institute for Evolutionary Anthropology only revealed two more prehistoric cases, which were also buried inside settlements (Rohrlach *et al.* 2024). Only two more cases of trisomy in prehistory have been published so far (Cassidy *et al.* 2020; Anastasiadou *et al.* 2024). All of them have been identified as having Down syndrome (trisomy 21), whereas one child from Alto de la Cruz is the first and, so far, the only case of Edwards syndrome (trisomy 18) known from prehistory. The high frequency of children with trisomy in Alto de la Cruz and Las Eretas suggests that these children were perceived as "special" and worthy of keeping near the living.

In sum, the exceptionally high number of twins (4/35) and cases of chromosomal trisomies (4/35) identified through genetics provide a first hint concerning the funerary criteria applied to a small number of children. In this sense, the intramural burial right appears as an expression of a certain perception of illness and way to approach it.

8.4.2 How large were the communities related to EIA settlements?

The observed patterns of ROH imply that the dwellings of the Iron Age maintained relations with a population of several thousand, which in general allowed them to maintain low levels of consanguinity. The cases of trismy 18 and 21 identified through statistical screening of ancient genomes offered a new way to approach paleodemography. As these genetic disorders seem to have occurred at the same rates in past human populations as they do today, their appearance in a community implies a specific number of births (Rohrlach *et al.* 2024). Today, the rates of prevalence of Down and Edwards are 1:705 and 1:3,226, respectively, when considering live births, stillborn and terminated pregnancies. These mean values were used to simulate the total population required to produce the cases of trisomy observed for Alto de la Cruz and Las Eretas. This simulation took into account: a) that neither settlement was completely excavated, nor all samples yielded aDNA; b) that the corrected life expectancy at birth (e^0) in the EIA could have been 22.9±1.37 years; and c) considered the size and duration of the settlement (see Papac *et al.* 2023 for details). According to the results of five million simulations, the cases of trisomy 21 expected to have been buried in Alto de la Cruz would have required between 9,249-16,947 births over the course of 400 years of EIA occupation. Running the same calculations with trisomy 18, a similar range of births was reached (11,643-21,334). In the case of Las Eretas, with an occupation of 250 years, between 6,085-10,929 births would have been required to result in the number of children with trisomy 21 expected to have been buried in the settlement.

If this cumulative census population size (D) over the settlement's lifetime is transformed into actual population numbers, using the well-known paleodemographic formula proposed by Acsádi and Neneskéri (1970), and adding the maximum and minimum figures derived from the trisomy 18 and 21 cases, the population burying these children in Alto de la Cruz (P) would need to reach between 530-1,221 persons. The population placing their dead trisomy 21 children in Las Eretas can be simulated between 557-1,001 persons. Though not quantifiable in the same way, these large population sizes are also supported by the high numbers of identified twins.

8.4.3 How large were the communities performing intramural burials?

Genetic analysis of intramural burials, especially in the case of children, can also inform about the social composition of the persons who had access to these spaces. We consciously avoid using the term "living" as such actualism of household organisation might not be as universal as Western ways of dwelling might suggest. The EIA communities of Navarra might have organised build spaces in a very different way, as we will see.

Few of the archaeological labels of the anthropological remains recovered between 1948 and 1957 provide sufficient information to relate them, at least tentatively, to some of the individuals mentioned in the excavation reports (Taracena and Gil 1951; Gil 1953; Maluquer de Motes 1958). But in at least two of the more recently excavated spaces of Alto de la Cruz and one from Las Eretas we were able to sequence groups of children buried under the same floor level and to define their degree of biological relatedness. The most complete bioarchaeological record has been provided by Building 87/8 of settlement phase IIa, where all five children could be sequenced (Fig. 8.5). Two brothers (CRU026-CRU027) were buried at a certain distance from each other in the northwestern part of the *ca.* 90 m² space. A set of sisters (CRU022-CRU023) was placed more to the southwest, again at a certain distance from each other. These sisters are among the five highly consanguineous individuals identified by the ROH pattern (Papac *et al.* 2023: fig. 18). Their long stretches of ROH ($\geq$ 33 cM) totalling $\geq$ 62 cM in length, suggest that they are the offspring of parents who were related to one another at approximately the third-degree (possibly first-cousins). A third female infant (CRU025) was placed between both pairs, but no first- or second-degree relation with both pair of siblings could be identified (Fig. 8.5). It is of interest that the three children

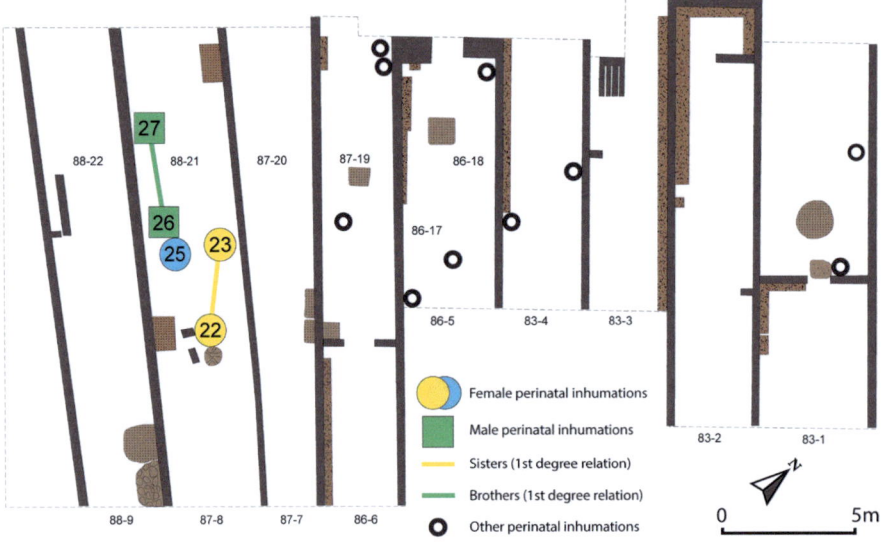

Figure 8.5. Alto de la Cruz settlement phase IIa and first-degree relationships between perinatal inhumations identified in building H87/8.

buried closely together in the central part of the room had no close biological relations between them, while the siblings of two of them are located at a certain distance from this central group (Fig. 8.5). The newly generated IBD data confirm that the two brothers and the two sisters belong to nearly mutually exclusive pedigrees. Both siblings are related in a third or more distant degree to other children of the settlement, but neither of these groups show relations between them.

A similar situation is observed in the building immediately below 87/8, which was destroyed by intense fire at the end of settlement Phase IIIb. Two of the three perinatal burials excavated in the central part of Room 88/21 could be analysed with READ (CRU028, CRU029), showing that a first- and second-degree relation can be excluded (Papac *et al.* 2023: fig. 18). The new IBD data confirm that both individuals also belong to mutually exclusive pedigrees. The boy CRU029 is related (>20cM) to at least five more children, including the two brothers CRU026-CRU027 and the girl CRU025, buried in the successive building of settlement phase IIa, but none of these individuals has a similarly close relation to the young boy CRU028, who instead was distantly related (>12cM) to CRU024, a case of trisomy 21 also dated to phase IIIb, and to the twins ERE001-ERE002 found in Las Eretas. The pedigrees of CRU028 and CRU029 can only be related through CRU024, which is related (several stretches >20cM) to CRU006, which in turn is more distantly linked to CRU029 (only one stretch >20cM).

The same situation is also observed in Las Eretas, where three of the four children buried in Building 2 were successfully sequenced (Fig. 8.6). Meanwhile, ERE004, a boy with Down syndrome and the individual with the highest ROH values in the whole sample (Papac *et al.* 2023: fig. 18), and the girl ERE005 had a second-degree relation, probably implying they were half-siblings rather than "uncle/aunt-nice/nephew". As they share the same mt haplogroup (Papac *et al.* 2023: tab. 6), the first scenario would imply that both were children from the same mother, but of different fathers. The second would

Figure 8.6. Buildings and infant burials of Las Eretas. Besides the twins ERE001 and ERE002, only inhumations ERE005 and ERE006 revealed a second-degree relationship.

suggest a matrilineal line of descent. Both individuals were distantly related to twins ERE001-ERE002 of the neighbouring building (Fig. 8.6). In any case, the biological relations between the two pairs were neither through a paternal nor through a maternal line, as mitochondrial and Y-chromosome haplogroups are different. On the other side, boy ERE003, which was not buried at a large distance from the two previous closely related individuals, shows no match with any individual from Las Eretas or Alto de la Cruz.

In sum, in all rectangular buildings where we could sequence two or more individuals excluding twins, the present results do not support the identification of individual dwellings with nuclear families, as conventional archaeological models tend to assume (e.g., Maluquer de Motes 1958). Rather, the EIA buildings seem to have been managed by extended groups formed by several couples and their offspring. In all three analysed buildings, the offspring implied markedly different pedigrees, both in ascending and descending lines. These pedigrees were only remotely related, as in fact were most inhabitants in both settlements. The spatial closeness of the children's burials does not necessarily preclude biological closeness, except in the case of twins which were always buried together. In Building 87-8 of Alto de la Cruz, three different couples even had the right to bury (some of) their children in the same space and in close proximity to each other.

Given the short duration of each settlement phase, it is unlikely that these genetic and spatial patterns are the result of successive occupation of the buildings by different groups. Such a scenario, with extended groups attached to the buildings, also finds support in the archaeological record of the successive and exceptionally well-preserved Rooms H88/21 and H87/8. Unfortunately, very few remains were found in the interior of Building 2 of Las Eretas, which was neither destroyed by fire nor suddenly abandoned. However, the contents

	Surface	Hearths	Kilns	Grinding Slabs	Pottery	Burials
Alto de la Cruz, Building 87/8 (level IIa)	93 m²	2	-	2	9	5
Alto de la Cruz, Building 88/21 (level IIIb)	Unknown, but probably ca. 90 m²	1	1	3	17	3
Las Eretas, Building 2	71.5 m²	1	0	1	3	4
Las Eretas, Building 4	99.5 m²	2	2	3	12	not excavated

Table 8.1. Food processing structures, grinding tools and small-sized pottery (collared cups and bowls) found in buildings with multiple intramural burials from Alto de la Cruz and in well-preserved buildings from Las Eretas.

of Room 4 provides a better glimpse of what could have been the common inventory of an EIA house at Las Eretas (Tab. 8.1). The number of hearths, grinding stones, and pottery vessels have been used in archaeology as indicators of the number of persons involved in the productive and consumptive activities of a social space. Particularly tight is the relation observed in different ethnographical contexts between grinding stones and adult women, given the physical constraints of cereal grinding, which usually is carried out daily for several hours by each woman (e.g., Hayden 1987; Horsfal 1987; Gronenborn 1994, also based on our own fieldwork in rural northern Ghana and Mali). Quern stones are usually a lifetime possession of women. Only where markedly different types of cereals –such as wheat and millet– are ground in similar quantities, can a women use two different grinding slabs (Nixon-Darkus *et al.* 2024), a situation which is not supported by the botanical record of Alto de la Cruz (Cubero Corpas 1990). Demographic inferences based on grinding tools must also take into account that in Navarra only the stationary grinding slabs seem to have been made of stone (Maluquer *et al.* 1990), while the grinders are missing in the lithic record and, probably, were made of wood, as confirmed experimentally and through use wear analysis in other parts of Iberia (Delgado-Raack and Risch 2016). In relation to pottery, only small vessels have been considered, as they were probably related to the individual consumptions of food or beverages.

According to these parameters, the food processing structures, tools and pottery recovered in the buildings of Alto de la Cruz and Las Eretas conform better to a pattern produced by extended groups than by nuclear families (Tab. 8.1). Large buildings had at least two grinding slabs, two firing structures, and a large number of small pottery vessels. The presence of two identical hearths placed in different parts of the square room is particularly revealing in Building 87/2 of Alto de la Cruz and in Building 4 of Las Eretas. Instead, in other rooms, cooking hearths were clearly shared by the extended groups, suggesting a high degree of cooperation. The large amount of pottery of different sizes and shapes present in the buildings also supports the occupation by a large number of individuals rather than of a nuclear family. If the small vessels and bowls are considered to be representative of the number of (juvenile and adult) individuals consuming in a certain space, the larger houses can be expected to have provided food to more than nine persons (Tab. 8.1). All 12 vessels from Building 4 of Las Eretas come from the back of the room. According to their find context, they seem to have been stored in a pantry

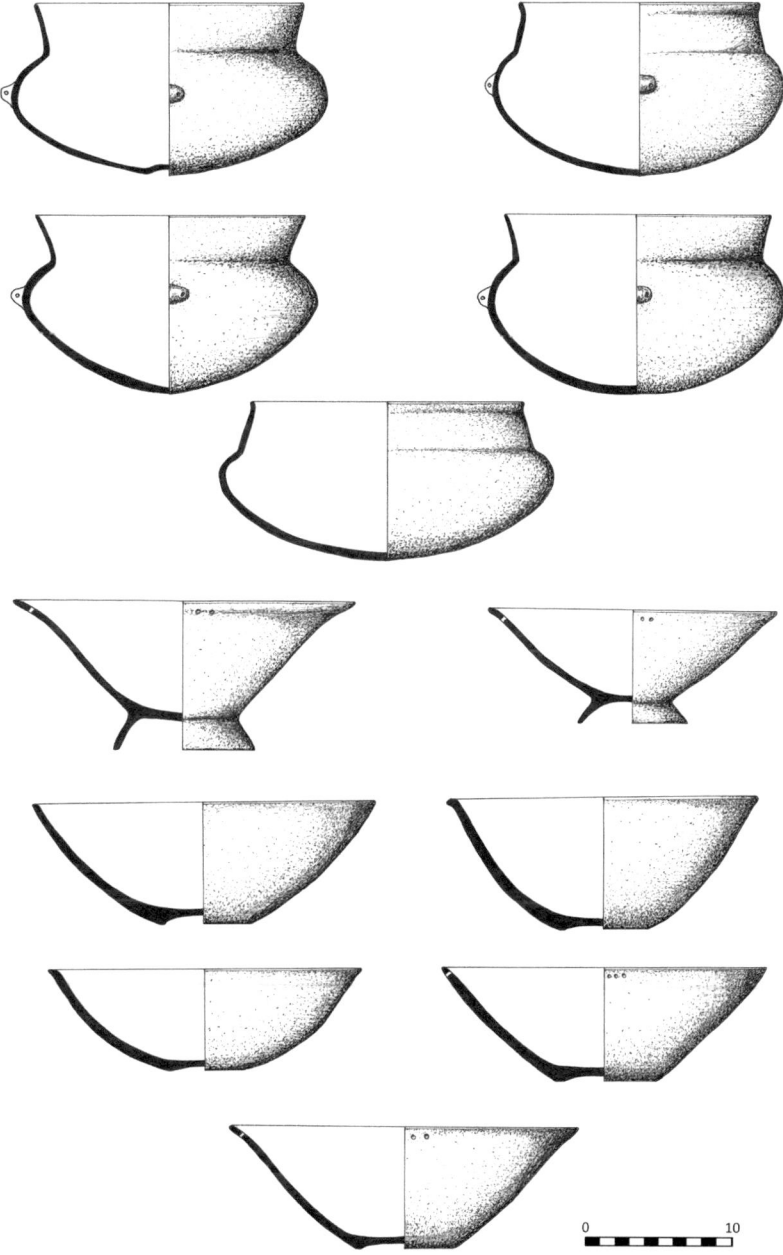

Figure 8.7. Small collared cups and bowls found in the rear of Building 4 of Las Eretas, probably stored in a pantry.

or cupboard, rather than used in daily activities (Fig. 8.7). Five vessels are simple bowls while another five classify as collared cups. They would have allowed serving different types of foods and beverages to five additional persons, who were always not present in the building (Fig. 8.7). A special standfood and another two bowls have exceptionally fine walls and might have been used for special purposes. Only Building 3 of Las Eretas seems to have relied on one hearth, one grinding slab and a few pots; but its size is significantly smaller. It must also be mentioned that at least in Las Eretas, modern excavations could show that grinding slabs were also placed and used in open spaces outside the buildings. In any case, the larger and better-preserved constructions have provided an exceptional variety and quantity of daily means of production and consumption, making the engagement of an extended group more likely than a nuclear family.

8.4.4 Did the EIA communities of northern Iberia follow patri-, matrilineal or ambilineal descent?

Another interesting result of IBD is the observation that pairs of individuals having second- to fourth-degree relations (nine or more shared sequences of >20cM) often share the same mt haplogroups (7 out of 11). Three more pairs of individuals sharing only one to six sequences of >20cM also have the same mt haplogroup. Such a pattern suggests that some of the children were related through the maternal line, while others –slightly less– were through the paternal line. This is even more likely, taking into account that the buried children could not have had descendents, and that some second- fourth-degree related pairs belong to different settlement phases and/or were found in different buildings, making it unlikely that they were the children of the same woman. Apparently, rules of matrilocality and "male exogamy" were followed by these EIA communities, although patrilineality and patrilocality also seem to have existed, as the related cases with different mtDNA support.

8.5 Conclusions

The substantial progress in our understanding of prehistoric societies achieved through the close collaboration between genetics and archaeology becomes manifest recalling, for example, the conclusions reached by the excavator of Alto de la Cruz concerning family life in the settlement's houses. It was published over 60 years ago, but echoes the views held in relation to many regions and periods in prehistory: "The type of house and the subsequent tradition throughout the Ebro basin and in most of the Peninsula suggests that we are dealing with a monogamous society. The houses present such a unity that it is difficult to combine the coexistence under the same roof of several women with their respective children" (Maluquer de Motes 1958: 142, translated by the authors).

The first genetic results provided by intramural and mostly perinatal burials of EIA Northern Iberia oblige us to reconsider such assumptions perpetuated as conventional wisdom in dominant narratives. In the first place, we have confirmed that the funerary ritual devoted to a small group of children observed strict criteria concerning age, but not sex. Sacrificial rituals can be excluded given the lack of paleopathological evidence of violent actions but is also contradicted by the observation that several of the buried children, including one case of Down syndrome (CRU024), had not reached the age of foetal maturation and could not have survived childbirth or were stillbirths anyway, making sacrifice impossible. Our current hypothesis, which will need to be confirmed in other settlements, is that the perinatal children buried under the living floors of the living were selected for their unusual traits or were born under special circumstances perceived by the community at their birth. Twins and children with trisomies are clearly overrepresented, considering that our sample only included 35 individuals. Clearly, this group of children was socially valued as being different enough to remain close to the spaces of everyday life, instead of being cremated and buried in urnfield cemeteries outside the settlements. The aspect of health or the specific circumstances in which a child was born and –shortly afterwards– died have not been imagined as possible motivations behind these ritual practices and might modify recent reflections of social anthropology on the meaning of intramural child burials in other prehistoric contexts (e.g., Kuijt 2008; Cveček and Schwall 2022).

The archaeogenetic study of Alto de la Cruz and Las Eretas has also provided new strategies to address crucial aspects of social organisation such as community size and continuity. The number of persons involved in this special child burial rite, in the settlements as well as in the individual buildings, was substantially larger than thought in all previous models, which were ultimately derived from ethnological observations. It can be expected that this ideologically connected community was also responsible for the economic and political organisation of the settlements. The size of the populations attached to 0.5-0.75 ha large settlements, such as Alto de la Cruz and Las Eretas, could have reached up to 1,000 people, making it likely that they settled beyond the fortified areas. The sharing of buildings by biologically unrelated groups, otherwise residing in a wider area, suggests a complex economic and political use of the walled space, which challenges the usual household-based models envisaged by archaeology.

A further challenge to current views is the fact that the very regularly built square houses –which is a recurrent trait of Iron Age architecture in Iberia– seems to have been used by two or more groups of mutually exclusive descent lines. The regularity observed in the material inventories of the houses has often been used as evidence for the lack of social inequality and economic exploitation in the EIA settlements (Ruiz Zapatero and Fernández Martínez 1985; Armendáriz 2008: 175-181). This co-habitation, co-working, and co-worshiping of their children, involving a considerable number of people who shared the same spaces, fireplaces, raw materials, and tools, might have been a successful mechanism to avoid social asymmetries.

Although further archaeogenetic studies are necessary to support this result, the high number of second- to fourth-degree relations sharing the same mt haplogroup suggests that settlement continuity, also after moments of destruction, was assured mostly –though not exclusively– through the different maternal lines of each community. Also in this respect, the EIA of northern Iberia seems different to the Bronze Age communities, as well as to the traditional model of Western societies.

Acknowledgements

This project has received funding from the European Research Council (ERC) under the European Union's Horizon 2020 research and innovation programme (grant agreement 851511). It has also been supported by the Spanish Ministry for Science and Innovation (PID2020-112909GB-100) and by the research project "Convergence and interaction between complex Bronze Age societies" from the Academia program of the *Institució Catalana de Recerca i Estudis Avançats* (ICREA) of the Catalan Government. One of the authors, Marcello Peres is the beneficiary of a Margarita Salas Grant from the Spanish Ministry of Universities, under the framework of the European Union "NextGenerationEU" program.

References

Acsádi, G. and Nemeskéri, J. (1970): *History of Human Life span and Mortality*. Budapest: Kiadó.

Anastasiadou, K., Silva, M., Booth, T., Speidel, L., Audsley, T., Barrington, C., Buckberry, J., Fernandes, D., Ford, B., Gibson, M., Gilardet, A., Glocke, I., Keefe, K., Kelly, M., Masters, M., McCabe, J., McIntyre, L., Ponce, P., Rowland, S., Ruiz Ventura, J., Swali, P., Tait, F., Walker, D., Webb, H., Williams, M., Witkin, A., Holst, M., Loe, L., Armit, I., Schulting, R. and Skoglund, P. (2024): Detection of chromosomal aneuploidy in ancient genomes. *Communications Biology* 7(14). DOI: 10.1038/s42003-023-05642-z.

Angel, L. (1971): Early Neolithic skeletons from Çatal Höyük. Demography and pathology. *Anatolian Studies* 21, pp. 77-98.

Armendáriz Martija, J. (2008): *De aldeas a ciudades. El poblamiento durante el primer milenio a.C. en Navarra*. Pamplona: Gobierno de Navarra.

Armendáriz Martija, J. and de Miguel Ibáñez, M.P. (2006): Los enterramientos infantiles del poblado de Las Eretas (Berbinzana). Estudio paleoantropológico. *Trabajos de Arqueología Navarra* 19, pp. 5-44.

Agustí, B., Alonso, N., Junyent E., Lafuente Á. and López, J.B. (2000): Una inhumación múltiple de perinatales en la fortaleza de Els Vilars (Arbeca, Lleida) y las prácticas de enterramiento en hábitat durante la primera Edad del Hierro en el valle del Segre (Cataluña). In Dedet, B., Gruat, P. and Marchand, G. (eds.), *Archéologie de la mort. Archéologie de la tombe au Premier Âge du Fer. Actes du XXIe Colloque International de l'Association Française pour l'Étude de l'Âge du Fer (Conques-Montrozier, 8-11 mai 1997)*. Lattes, pp. 305-324.

Blasco Sancho, M.F. and Montón Broto, F.J. (2019): Enterramientos perinatales de la Primera Edad del Hierro en el poblado de La Codera (Alcolea de Cinca, Huesca). *Bolskan* 27, pp. 41-54.

Boz, B. and Hager, L. (2013): Living above the Dead: Intramural Burial practices at Çatalhöyük. In Hodder, I. (ed.), *Humans and Landscapes of Çatalhöyük: Reports from the 2000-2008 Seasons*. Ankara: Cotsen Institute of Archaeology, pp. 413-440.

Carnicero-Cáceres, S. and Torres-Martínez, J.F. (2021): Child burials in domestic contexts at an Iron Age hillfort: The *oppidum* of Monte Bernorio (Villarén, Palencia). *Munibe Antropologia-Arkeologia* 72, pp. 123-140.

Cassidy, L.M., Maoldúin, R.Ó., Kador, T., Lynch, A., Jones, C., Woodman, P.C., Murphy, E., Ramsey, G., Dowd, M., Noonan, A., Campbell, C., Jones, E., Mattiangeli, V. and Bradley, D. (2000): A dynastic elite in monumental Neolithic society. *Nature* 582, pp. 384-388. DOI: 10.1038/s41586-020-2378-6.

Cubero Corpas, C. (1990): Análisis paleocarpológicos de muestras del Alto de la Cruz. In Maluquer, J., Gracia, F. and Munilla, G. (eds.), *Alto de la Cruz (Cortes, Navarra). Campañas, 1986-1988*. Pamplona: Institución Príncipe de Viana, pp. 199-217.

Cveček, S. and Schwall, C. (2022): Ghost Children: Delayed Personhood and Culture-specific Models of Infancy in Western Anatolia. *Praehistorische Zeitschrift* 97(2), pp. 544-570. DOI: 10.1515/pz-2022-2044.

Davis, M. (2001): *Late Victorian Holocausts: El Niño Famines and the Making of the Third World*. London: Verso.

Delgado-Raack, S. and Risch, R. (2016): Bronze Age cereal processing in Southern Iberia: A material approach to the production and use of grinding equipment. *Journal of Lithic Studies* 3(3), pp. 125-145. DOI: 10.2218/jls.v3i3.1650.

De Miguel Ibáñez, M.P. (2009): Las inhumaciones perinatales de El Castejón de Bargota (Navarra). *Cuadernos de Arqueología de la Universidad de Navarra* 17, pp. 215-248. DOI: 10.15581/012.17.27720.

Drevenstedt G.L., Crimmins, E.M., Vasunilashorn, S. and Finch, C.E. (2008): The rise and fall of excess male infant mortality. *Proceedings of the National Academy of Sciences USA* 105(13), pp. 5016-5021. DOI: 10.1073/pnas.080022110.

Ensor, B.E. (2013): *The Archaeology of Kinship: Advancing Interpretation and Contributions to Theory.* Tucson: The University of Arizona Press.

Flannery, K.V. (1973). The origins of agriculture. *Annual Review of Anthropology* 2, pp. 271-310.

Frieman, C.J. (2023): Kin and connection: Bodies and relations in archaeology and ancient genetics. In Meller, H., Krause, J., Haak, W. and Risch, R. (eds.), *kinship, sex, and biological relatedness. The contribution of archaeogenetics to the understanding of social and biological relations.* Tagungen des Landesmuseums für Vorgeschichte Halle, 28. Heidelberg: Propylaeum, pp. 43-50. DOI: 10.11588/propylaeum.1280.c17994.

García López, E. (1994): Un modelo de análisis de evolución arquitectónica e interpretación social. El asentamiento del Bronce Final-Primera Edad del Hierro del Alto de la Cruz (Cortes, Navarra). *Pyrenae* 25, pp. 93-110.

García López, E., Gracia Alonso, F. and Munilla Cabrillana, G. (1994): Cortes de Navarra. Transición Bronce Final-Hierro en el valle medio del Ebro. *Revista de Arqueología* 160, pp. 14-21.

Gil Farrés, O. (1953): Excavaciones en Navarra III. Campañas realizadas en el Alto de La Cruz de Cortes de Navarra, entre 1950 y 1952. *Príncipe de Viana* 14, pp. 9-46.

Gómez Bellard, F. and Oliver Foix, A. (1989): Nuevos enterramientos infantiles ibéricos de inhumación en Castellón. *Quaderns de Prehistòria i Arqueologia de Castelló* 14, pp. 51-62.

Gracia, F., Munilla Cabrillana, G. and Mercadal i Fernández, O. (1989): Enterramientos infantiles en el poblado ibérico de la Moleta del Remei (Alcanar, Montsià). *Quaderns de Prehistòria i Arqueologia de Castelló* 14, pp. 133-159.

Gronenborn, D. (1994): Ethnoarchäologische Untersuchungen zur rezenten Herstellung und Nutzung von Mahlsteinen in Nordost-Nigeria. *Experimentelle Archäologie Bilanz 1994*, pp. 45-55.

Guérin, P. and Martínez Valle, R. (1987-1988): Inhumaciones infantiles en poblados ibéricos del área valenciana. *Saguntum* 21, pp. 231-265.

Gusi, F. (1970): Enterramientos infantiles ibéricos en vivienda. *Pyrenae* 6, pp. 65-72.

Hager, L. and Boz, B. (2012): Death and its relationship to life: Neolithic burials from Building 3 and Space 87 at Çatalhöyük, Turkey. In Tringham, R. and Stevanovic, M. (eds.), *Last House on the Hill: BACH Area Reports from Catalhoyuk, Turkey.* Bristol: Cotsen Institute of Archaeology, pp. 297-330. DOI: 10.2307/j.ctvdmwx1j.22.

Hayden, B. (ed.) (1987): *Lithic studies among the contemporary Highland Maya.* Tucson: University of Arizona Press. DOI: 10.1126/science.240.4855.1083.

Horsfall, G.A. (1987): Design theory and grinding stones. In Hayden, B. (ed.), *Lithic studies among the contemporary Highland Maya.* Tucson: University of Arizona Press, pp. 332-377.

Hutchinson, S.E. (2000). Identity and substance: The broadening bases of relatedness among the Nuer of southern Sudan. In Carsten, J. (ed.), *Cultures of relatedness: New approaches to the study of kinship.* Cambridge: Cambridge University Press, pp. 55-72.

Krause, J. and Trappe, T. (2022), *A Short History of Humanity: A New History of Old Europe.* New York: Random House.

Kuijt, I. (2000). People and space in early agricultural villages: Exploring daily lives, community size and architecture in the Late Pre-Pottery Neolithic. *Journal of Anthropological Archaeology* 19, pp. 75-102. DOI: 10.1006/jaar.1999.0352.

Kuijt, I. (2008): The Regeneration of Life. Neolithic Structures of Symbolic Remembering and Forgetting. *Current Anthropology* 49, pp. 171-197. DOI: 10.1086/526097.

Lorrio Alvarado, A.J., de Miguel Ibáñez, M.P., Moneo Rodríguez, T. and Sánchez de Prado, M.D. (2010): Enterramientos infantiles en el *oppidum* de El Molón (Camporrobles, Valencia). *Cuadernos de Arqueología de la Universidad de Navarra* 18, pp. 210-262. DOI: 10.15581/012.18.4375.

Lull, V., Micó, R., Rihuete Herrada, C., and Risch, R. (2016): Argaric sociology: Sex and death. *Complutum* 27(1), pp. 31-62. DOI: 10.5209/CMPL.53216.

Maluquer de Motes, J. (1954): *El yacimiento hallstático de Cortes de Navarra. Estudio Crítico 1.* Pamplona.

Maluquer de Motes, J. (1958): *El yacimiento hallstático de Cortes de Navarra. Estudio Crítico 2.* Pamplona.

Maluquer de Motes, J.; Munilla Cabrillana, G. and Gracia Alonso, F. (1990): *Alto de la Cruz (Cortes, Navarra). Campañas 1986-1988.* Pamplona.

Meller, H., Daim, F., Krause, J. and Risch, R. (eds.) (2017): *Migration and integration from Prehistory to the Middle Ages.* Tagungen des Landesmuseums für Vorgeschichte Halle. Heidelberg: Propylaeum.

Meller, H., Haak, W., Krause, J. and Risch, R. (2023): *Kinship, sex, and biological relatedness: The contribution of archaeogenetics to the understanding of social and biological relations.* Tagungen des Landesmuseums für Vorgeschichte Halle, 28. Heidelberg: Propylaeum. DOI: 10.11588/propylaeum.1280.

McFadden, C., Muir, B. and Oxenham, M.F. (2022). Determinants of infant mortality and representation in bioarchaeological samples: A review. *American Journal of Biological Anthropology* 177(2), pp. 196-206. DOI: 10.1002/ajpa.24406.

Moya, A., López, J.B., Lafuente, Á., Rey, J., Tartera, E., Vidal, A. and Equip Vincamet (2005): El Grup del Segre-Cinca II (1250-950 cal. a.n.e.) a les terres del Baix Cinca: el poblat clos de Vincamet (Fraga, Osca). *Revista d'Arqueologia de Ponent* 15, pp. 13-58.

Papac, L., de Miguel Ibáñez, P., Rohrlach, A.B., Armendáriz, J., Peres, M., Lamnidis, T.C., Mötsch, A., Schiffels, S. and Risch, R. (2023): Intramural child burials in Iron Age Navarra: How ancient DNA can contribute to household archaeology. In Meller, H., Krause, J., Haak, W. and Risch, R. (eds.), *Kinship, sex, and biological relatedness: The contribution of archaeogenetics to the understanding of social and biological relations.* Tagungen des Landesmuseums für Vorgeschichte Halle 28, Heidelberg: Propylaeum. pp. 263-295. DOI: 10.11588/propylaeum.1280.c18012.

Penske, S., Orschied, J., Rohrlach, A.B., Meller, H., Haak, W., Schunke, T. and Risch, R. (2023): Life and work. A possible 'house community' at the Early Bronze Age settlement of Schiepzig in Central Germany. In Meller, H., Krause, J., Haak, W. and Risch, R. (eds.), *Kinship, sex, and biological relatedness: The contribution of archaeogenetics to the understanding of social and biological relations.* Tagungen des Landesmuseums für Vorgeschichte Halle, 28. Heidelberg: Propylaeum, pp. 183-194. DOI: 10.11588/propylaeum.1280.c18007.

Pongou, R. (2012): Why is infant mortality higher in boys than in girls? A new hypothesis based on preconception environment and evidence from a large sample of twins. *Demography* 50(2), pp. 421-444. DOI: 10.1007/s13524-012-0161-5.

Reich, D. (2018): *Who We Are and How We Got Here: Ancient DNA and the New Science of the Human Past*. Oxford: Oxford University Press.

Risch, R. and Carbonell, J. (1986): Los enterramientos ibéricos y romanos de Darró (Vilanova i la Geltrú). Estudio osteológico. *Butlletí de la Biblioteca-Museu Balaguer (1983-1985)*, pp. 19-43.

Risch, R., Haak, W., Krause, J. and Meller, H. (2023): Kinship, sex, and biological relatedness. The contribution of archaeogenetics to the understanding of social and biological relations. In Meller, H., Krause, J., Haak, W. and Risch, R. (eds.), *Kinship, sex, and biological relatedness: The contribution of archaeogenetics to the understanding of social and biological relations*. Tagungen des Landesmuseums für Vorgeschichte Halle, 28. Heidelberg: Propylaeum, pp. 9-25. DOI: 10.11588/propylaeum.1280.c18044.

Rivollat, M., Rohrlach, A.B., Ringbauer, H., Childebayeva, A., Barquera, R., Szolek, A., Le Roy, M., Colleran, H., Tuke, J., Aron, F., Pemonge, M.H., Späth, E., Télouk, P., Rey, L., Goude, G., Balter, V., Krause, J., Rottier, S., Deguilloux, M.F. and Haak, W. (2023): Extensive pedigrees reveal the social organization of a Neolithic community. *Nature* 620, pp. 600-606. DOI: 10.1038/s41586-023-06350-8.

Rohrlach, A.B., Rivollat, M., de Miguel Ibáñez, P., Moilanen, U., Liira, A.M., Teixeira, J.C., Roca-Rada, X., Armendáriz-Martija, J., Boyadzhiev, K., Boyadzhiev, Y., Llamas, B., Tiliakou, A., Mötsch, A., Tuke, J., Prevedorou, E.A., Polychronakou-Sgouritsa, N., Buikstra, J., Onkamo, P., Stockhammer, P.W., Heyne, H.O., Lemke, J.R., Risch, R., Schiffels, S., Krause, J., Haak, W. and Prüfer, K. (2024): Cases of trisomy 21 and trisomy 18 among historic and prehistoric individuals discovered from ancient DNA. *Nature Communications* 15, p. 1294. DOI: 10.1038/s41467-024-45438-1.

Ruiz Zapatero, G. and Fernández Martínez, V.M. (1985): Cortes de Navarra: un modelo económico de la 1ª Edad del Hierro en el Noreste de la Península Ibérica. *XVII Congreso Nacional de Arqueología*. Madrid, pp. 371-392.

Séguy, I. and Buchet, L. (2013) (eds.): *Handbook of Paleodemography*. Cham: Springer. DOI: 10.1007/978-3-319-01533-8.

Subirà, M.E., Ruiz, J. and Molist, N. (2016): Anàlisi diacrònica de la necròpolis de l'església de Sant Miquel d'Olèrdola a partir de l'estudi antropològic. In Esteve, X., Miró, C., Molist, M. and Sabaté, G. (eds.), *Jornades d'Arqueologia del Penedès 2011 (Vilafranca del Penedès, 21-22 octubre 2011)*. Vilafranca del Penedès, pp. 335-360.

Taracena, B. and Gil Farrés, O. (1951): Excavaciones en Navarra. *Príncipe de Viana* 12, pp. 211-234.

TallBear, K. (2018): Making Love and Relations Beyond Settler Sex and Family. In Clarke, A. and Haraway, D. (eds.), *Making kin, not population: Reconceiving generations*. Chicago: Prickly Paradigm Press, pp. 145-209.

Wolf, E.R. (1982): *Europe and the People without History*. Berkeley: University of California Press.

Part III

ADDRESSING KINSHIP FROM HOUSEHOLD ARCHAEOLOGY

A Kinship-informed Comparative and Worldwide Survey of the Multiple Residential Group

Antonio Blanco-González

Department of Prehistory, Ancient History and Archaeology,
University of Salamanca, Spain, ablancoglez@usal.es

Abstract

This contribution addresses a basic and ubiquitous collective social agent in antiquity between the conjugal family –which was rarely a social unit in itself– and the settlement or local community. The paper delves into such an intermediate and kin-driven social grouping, often materialised as a neighbourhood or multi-house aggregate. The text proposes an approach from the standpoint of household archaeology, centring on kinship, with the aim of challenging unquestioned claims and to endorse the target formulas of conviviality with supportive and highly detailed historical information from diverse sources. Such residential group cannot be interpreted as an "extended family" or a "cultural trait", nor always be reduced to a complete household or corporate unilineal group. The chapter first discusses its underpinnings and interpretative limits and then surveys the literature of two major household archaeological schools in a comparative fashion to focus on a necessarily restrictive selection of well-known and representative case studies. Out of the suite of combinations of residence and descent options, the text concentrates on those multi-functional and self-sufficient composite residential groups involving either subaltern small conjugal dwellings or elite oversized ancillary subunits: mostly virilocal and patrilineal, and some bilocal and bilateral cases, often confused. This sample ranges from decentralised pre-/protohistoric Near Eastern and Mediterranean organisations to Mesoamerican and South American historical state-based polities. Such an exercise highlights key underlying commonalities of this collective social actor in varied settings across time and space, readdresses misguided points in current archaeological literature and suggests prospects for multi-stranded research integrating the domestic and funerary realms.

Keywords: *household archaeology, archaeology of kinship, Mediterranean and Near Eastern archaeology, Mesoamerican and Andean archaeology.*

In Blanco-González, A. and Alarcón-García, E. (eds) 2025, *A Social Archaeology of Kinship in Iberia and Beyond. Recent Multistranded Approaches from aDNA to Household Archaeology.* Leiden: Sidestone Press, pp. 181-208.

9.1 Introduction

After decades of disinterest and growing confusion in the archaeological literature on kinship, some contributions in the last decade have undoubtedly provided an impetus for a solid development of kinship archaeology with a clear social focus (Ensor 2013a, 2013b, 2021; Souvatzi 2017). Having addressed the theoretical update of the field, the clarification of misunderstandings and the critical revision of cross-cultural inference procedures (Ensor 2021: 80-146) the time is ripe for a new generation of kinship studies. However, the current scholarly picture is complex, with disparate views that often embrace contradictory and incommensurate premises, ways of reasoning and scholarly languages and procedures. Old-fashioned culture-historical clichés on this issue are pervasively alive and kicking. Positivism and scientism predominate, and many archaeologists assume that without biologically-oriented lines of evidence –i.e., genetics and isotope datasets– it is not possible to deal with past –and especially prehistoric– kinship. As for micro-scale accounts, the human protagonists whose social life can be archaeologically traced remain invisible and dim. Work focused on the domestic realm opts mostly for the formal description and classification of architecture (e.g., Bilge 2019; Moore 2021; Nevett 2023), often with sophisticated methods such as space syntax, network, lighting or access-analyses (e.g., Gilboa *et al.* 2014; Smith *et al.* 2019; Susnow and Goshen 2021; but see Bermejo Tirado this volume). This trend is coherently framed within a traditional archaeological baseline which keeps envisaging architectural arrangements as "index fossils" or "cultural traits" indicative of "acculturation" in culture-historical terms. The underlying understanding is that the domestic realm is a passive container or a monolithic and immobile background stage for a materially disconnected and superficial social performance (see Souvatzi 2008 for a smart critique). Moreover, such accounts are either shy and reluctant about the social identification of their protagonists or rely on vague, interpretively neutral or unengaging terms. When implicitly acknowledging the importance of kinship, most concepts and premises used in the archaeological literature are rudimentary, pivoting around the false divide between "nuclear" and "extended families" as opposing realities (e.g., Apostolaki 2015; Huebner 2017; Bilge 2019). Consequently, social archaeological accounts attentive to kinship and rejecting bio-determinism are minority. Lastly, some scholars keep expressing their mistrust on the accessibility of kinship among prehistoric and illiterate societies (i.e., most of the human past): "(...) the role of social kinship in prehistory –even whether the concept can be used there at all– has to remain uncertain at best because no texts (...) remain of those societies but only archaeological sources (...) which are rarely suitable for the reconstruction of immaterial social relations" (Risch *et al.* 2023: 11).

Among the factors that may have contributed to the underdevelopment and widespread reticence about the viability of archaeological approaches to kinship is the difficulty of integrating anthropological and cross-cultural knowledge into historical analysis. Within European academia, all-encompassing and analogy-based generalisations may be easily regarded as a functionalist and top-down "external imposition" on archaeology as a historical –i.e., idiosyncratic– discipline dealing with cases irreducible to general rules (cf., Celdrán and coauthors and Risch and collaborators this volume). However, the validity of such abstract patterns and strong evidential reasoning is statistically supported and has always been illustrated by and tested with pertinent case-studies, such as egalitarian European Neolithic groups (Souvatzi 2017; Ensor 2021), hierarchical organisations in North America (Ensor 2013a) or state-classist ones in Mesoamerica (2013b). In addition, the theoretical debates of the last decades on alternative forms of organisation to kinship (Chesson 2003; González-Ruibal 2006; González-Ruibal and Ruiz-Gálvez 2016) have also left shady aspects that should be critically re-evaluated from recent methodological contributions on kinship (Ensor 2013a, 2013b, 2021; Souvatzi 2017).

Drawing on the above problems and drawbacks, this essay intends to mitigate reluctance among sceptic archaeologists on the pertinence and feasibility of studying prehistoric kinship. The analytical approach of this contribution is that of household archaeology informed by current literature and inferential methods on kinship (e.g., Ensor 2013a, 2021; Souvatzi 2017; Ensor, Souvatzi, Grau Mira and Bermejo Tirado this volume). To counter current uneasiness, the chapter will engage these updated anthropologically-oriented perspectives with historical insight on kinship –including clear-cut demographic variables and long-lived institutions– in well-known ranked societies of the Old World –with a few words on New World cases– featuring highly detailed information from rich and eloquent burial and household archaeology, including textual references. The following sections justify the chosen inferential methodology, discussing and making explicit the premises, as well as the concepts that guide and the limitations that constrain this evidential reasoning. This method concentrates on collective social agents driven by kinship principles and focuses on a selective series of material correlates of their lifestyles and social practices. The composite or joint residential group is tackled here as the physical and immaterial entanglement that afforded, prompted and curtailed sociability led by kin-based social groupings in a disparate and wide range of prehistoric and early historic societies worldwide. Hopefully, this chapter will contribute to better understand what a social approach to kinship from archaeological methods has to offer.

9.2 Household archaeology and kinship

This contribution addresses agrarian and premodern societies, so to disregard kinship would be a misguided approach (Schloen 2001: 70-71; Huebner 2017: 5). Kinship concerns the central core of what portrays society and humanity: social relations. For Sahlins (2013) it is mutuality of being or intersubjective belonging: relatives are intrinsic to one another's existence, both in life and death. Today most social scientists are unanimous in considering that kinship, except in our late capitalist society, is an ineluctable cross-cutting principle that underpins and is at the intersection of everything. Thus, kinship impinges upon: biological procreation, sex, gender and relational identities; ritual and social life; intra-community and remotely kept rights and obligations; production, consumption, succession to titles, ranks or offices; labour and property, transmission and access to material and immaterial assets, resources, alliances and heritage; technology, know-how and craftsmanship; exchange; residential mobility, settlement and burial customs; and political organisation (Godelier 2011; Ensor 2013; Sahlins 2013). This paper adheres to Schloen's (2001: 110) acknowledgement that "aspects of social relationships that a modern Westerner might separate under the headings of 'family', 'friendship' and 'patronage' are usually all expressed in terms of kinship." It also embraces Godelier's (2011) understanding of kinship as the intersection of relations of social mutuality and closeness, enacted though the life-long process of becoming relatives, mostly via sharing and caring practices. Therefore, the focus here is on social practices and cultural decisions –the target of social archaeology– whereas biological progeny/genealogical relatedness is considered a secondary contributing factor –often an inadvertent by-product in history– whose importance is contingent and thus can be widely variable (Ensor 2021: 11).

This chapter tackles the domestic realm and collective social actors to shed valuable light on the in-house immaterial relationships, which are archaeologically reachable (contra Risch *et al.* 2023: 11). Our archaeological approach to kinship relies on anthropologically supported household archaeology drawing on statistically tested cross-cultural patterns (Ensor 2013a, 2013b, 2021, this volume; Souvatzi 2017, this volume). It aims at transcending mainstream dualist and exclusionary functional and morphological classifications of household archaeology. This is ineluctable when addressing ancient realities, whose household materiality can rarely be addressed from a net split of cultural values and segregated activity areas. By contrast, the social life of our target communities was characterised by the simultaneous overlapping and criss-crossing of complementary aspects –such as production, consumption and discard, and material and ritual/symbolic maintenance activities– at several levels and with disparate participants (Goody 1976; Netting *et al.* 1984; Albertz *et al.* 2014).

This approach also aims to refocus attention on collective social groupings organised through kinship as the true historical actors of our research. Indeed, long before the advent of the modern subject and individual personhood in seventeenth-century CE Europe, prehistoric social life was overwhelmingly characterised by relational identities, collaborative action and collective agency (Hernando 2017). The residential group was therefore a crucial arena for the forging of relational identities and therefore also kin making. As Ensor (2021: 12) puts it: "We must orient the study of prehistoric kinship toward the varied strategies to form corporate kin groups." Moreover, in the current climate of interpretive shyness on the kin protagonists of archaeological narratives, this pathway can provide us with powerful conceptual tools, capable of enriching our observations, and therefore they deserve to be fully explored.

The chapter avoids using in interpretive terms several common concepts. Firstly, "family" is circumvented here because it is a polysemic, too connoted and materially elusive idea. Ensor (2013a, 2013b, 2021) or Souvatzi (2027, this volume) have repeatedly warned that the biological "nuclear family" is not a social unit, and many close biological relatives do not dwell together (see below). Archaeologists often identify as "extended families" the intermediate groups between the basic procreation and caring cell, and the quarter or neighbourhood. Yet Ensor (2013a: 40, 311) restricts the "extended family" to the group of blood relatives, often not co-residents, lacking involvement in decision-making or the management of everyday errands. In short, neither the biological "nuclear family" nor the "extended family" are social units and are therefore irrelevant for the purposes of social organisation. Besides this, household archaeology literature normally and implicitly conflates the group of cohabitants with the decision-makers and estate's owners (e.g., Netting *et al.* 1984; Nash 2009; Smith 2010; Pacifico and Truex 2019; Moore 2021). By contrast, this chapter acknowledges Ensor's (2013a: 39-41) distinction between several concepts which are neither reducible nor interchangeable: a) the "household" or physical estate, including the dwelling, subsidiary domestic facilities and agrarian infrastructures; b) the "residential group" or the community of kin relatives –married adults and unmarried offspring– who co-inhabit a household and share dwelling and maintenance activities, whose remnants are traceable by archaeology; and c) the larger "household group" sharing property and decision-making, some of whose members may not reside in the household. Only in the case of bilateral filiation combined with bilocal residence both groupings (b and c) are one and the same, since membership is predicated upon residence instead of descent. For

Ensor (2013a: 47-49, 2013b: 45-46) such a corporate social unit is a *residential-household group*, exclusive of House Societies (*sociétés à maison*). Finally, in a restrictive and strict sense (Ensor 2013a, 2013b, 2021), residential groups practicing patrilineality cannot be equated with the larger corporate patrilineages –typically made of numerous patrilineal extended segments– because women co-members are dispersed through patrilocality (aka gynaecomobility). As for patri-/virilocal residential groups, these include nonmembers (brothers' wives) of the extended corporate patrilineal group. Drawing on these nuances and warnings, the text will refer to "households" –meaning the physical estates– yet will not deal with proper corporate "household groups" (save Houses or *maisons*); it will focus instead on the social practices and archaeological markers of "residential groups", that is, kin-based groups of co-dwellers and co-workers defined by socially and practice-based built links other than biological progeny.

9.3 A Social archaeology of the co-residential group

9.3.1 The analysis of intermediate-scale collective social units

The sociological (Harris 1982, 2018; Kertzer 1991) and archaeological (Hayden and Cannon 1982; Schloen 2001; Chesson 2003; González-Ruibal 2006; Manzanilla 2009; Smith 2010; Grau Mira 2013; Pacifico and Truex 2019; Ensor 2021) study of the neighbourhood and its occupants has increased steadily. This configuration of social life was widespread and very successful in pre-modern agrarian societies worldwide. Advanced prehistoric trajectories in various world foci witnessed the consolidation of increasingly complex agrosystems of deferred surplus yields, and shortly afterwards market-oriented production and heavy investment in infrastructure and heritage. These circumstances very often gave rise to joint or multiple residential groups; that is, integrated and self-contained modalities of communal conviviality, production, consumption and the transmission of heritage whose organisational principles and materiality are cross-culturally comparable (Manzanilla 1996; Schloen 2001; Ur 2014; González-Ruibal and Ruiz-Gálvez 2016; Harris 2018; Moore 2021; Susnow and Goshen 2021; Blanco González and Grau Mira in press). Moreover, even though gender roles are to be discussed on a case-by-case basis, iconography, textual sources, household and burial records, and bioarchaeology abundantly demonstrate that ranked agrarian communities often adopted lifestyles based on gender differences and the predominance of the male gender. Under these material and organisational historical coordinates, men overwhelmingly monopolised authority (patriarchy) and governed key household issues, such as decision-making, social roles, postmarital residence (which entailed attaching the offspring to a corporate group and working for it), succession to offices or inheritance. This is why the text will assume the presence of patriarchs as a reasonable hypothesis when several such bodies of evidence indicate so (cf., Schloen 2001; Ruiz-Gálvez 2013).

The kinship arrangements that characterize the casuistry of joint or composite residential groups is defined by diverse combinations of descent and postmarital residence practices. Firstly, in terms of descent we can encounter two main principles associated with these intermediate-sized collectives (Ensor 2013a, 2021): a) unilineal descent, whereby filiation is traced by only one gendered line –either matrilineal or patrilineal– and whose large corporate groups (lineages nested within higher-order clans) are made up of multiple noncorporate joint patri-/matrilineal segments, and b) bilateral-cognatic networking, whereby any genealogical or kindred relations are traced for whatever purposes, and whose extended residential groups are small corporate units. Secondly, as for residence we may encounter joint collectives practicing unilocality or bilocality. Among unilocality, patrilocality predominates over matrilocality in cross-cultural terms. Patrilocality means that the newly married couple lives with the groom's father, while the more inclusive concept of virilocality is the cohabitation of the spouses with the husband's brothers or parallel male cousins –socially equivalent to brothers– and their families, whereas married daughters/sisters/female cousins live elsewhere (Ensor 2013a: 64-67, 2021: 20). For the sake of simplicity, this chapter conflates patri- and virilocality –since this division makes little sense in terms of practices– and therefore will only mention "virilocal groups". This essay will not deal with matrilocality, which is also a viable option among multi-family aggregates, yet far less common among the selected case-studies. Regarding bilocality, this is a flexible residence strategy associated solely to bilateral principles, whereby couples or individuals negotiate where to live and who to work with using any kindred network (Ensor 2021: 208). Descent cannot be predicted from residence, and therefore virilocality does not exclusively entail patrilineality (Ensor 2021: 83, table 5.2). Actually, there is only a 65% of worldwide association between patrilocal and patrilineal groups –which according to Ensor (2021: 83) is a weak correlation in cross-cultural terms– whereas virilocality is associated to bilateral practices in 30% of cases.

To cover a manageable sample in the limited space available, this paper only considers multi-family aggregates of virilocal and bilocal cases. Indeed, the literature on historical

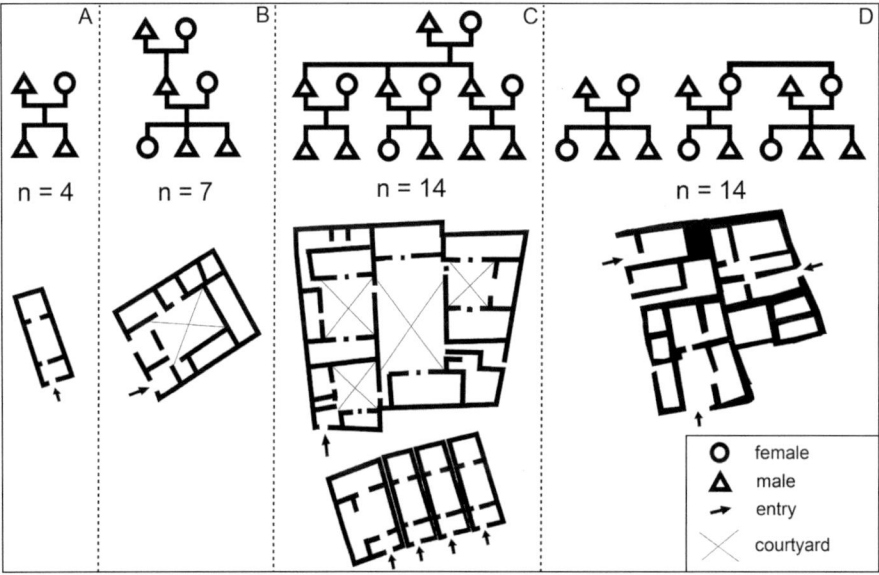

Figure 9.1. Residential units mentioned in the text (top), lines convey practice-based kinship links instead of biological relatedness, with number of co-residents (n), and layouts of their archaeological correlates (bottom): A) single-family subaltern conjugal dwelling; B) small joint patrilocal and patrilineal residential group; C) multi-family and multi-generational virilocal group accommodating diverse unilineal options (e.g., patrilineage segment, parallel male cousins, brotherhood, patronymic association); D) bilateral and bilocal residential group or House (modified from Ortega Ortega 2024: fig. 9.6 and Blanco González and Grau Mira in press).

virilocal households is a rich source of inspiration –food for thought– for feeding our social and archaeological imagination on such groupings. From a shallow survey of this inexhaustible ethnohistorical and archaeological corpus (e.g., Kertzer 1991: 158-159; Todd 1991: 303-471; Schloen 2001: 117-122; Ensor 2013a, 2013b, 2021; Ruiz-Gálvez 2013; González-Ruibal and Ruiz-Gálvez 2016; Souvatzi 2017, etc.) a series of commonalities can be recurrently encountered in space and time. Importantly, this is not about extrapolating Western-premodern institutions as universals applicable elsewhere in a direct historical analogy-like fashion. Resorting to this framework can be useful only in comparative and illustrative terms yet never anachronically projecting the context-dependent conditions of an observed sample on further situations. Nor is this about the fossilization of normative "cultural traits" or ideal "abstract types". The approximation embraced here deals with unstable and transitory stages within the ever-changing cycles of household development (Goody 1956; Netting *et al.* 1984; Smith 1984) and demography is a crucial variable. We can draw from a well-researched historical setting as an enlightening body of evidence providing necessary insight into particular demographic variables. Thus, in the Eastern Mediterranean second and first millennia BCE, Schloen (2001: 117-183) characterises the vegetative dynamics of such societies based on their low life expectancy at birth, high infant mortality and early childbearing age. All in all, based on the mentioned literature and especially on Schloen's (2001) and Ensor's (2013a, 2013b, 2021) works, the following variants of residential units and subunits may be systematized (Fig. 9.1):

a. The noncorporate, single-family, conjugal, elementary or unstable cell (*famille instable*), composed of up to two generations (Fig. 9.1A). This subunit is not reducible to our Western-modern monogamous, kindred-centred and heteronormative nuclear family –i.e., the parents and their biological/foster offspring– which is a historical artifact far from universal. Its inhabitants may follow virilocal, natolocal, neolocal or bilocal residence (Ensor 2021: 83-87). In the above spatio-temporal terms, it is the modal or most repeated phase in any domestic cycle, since it ranges from the young domestic unit –with immature offspring until their nubile age– to the senile, divorced or widowed. Bilaterality often imposes the early emancipation of sons to set up their own home far from the natal one (neolocality) and conditions their role as equal heirs, becoming extinct as a social group upon the death of the patriarch. This is the basic convivial cell of virilocal and bilocal arrangements addressed here, yet this is not a socio-economic unit by itself –except in the case of neolocality.

b. The multiple, composite, complex or extended residential group, as a multi-generational –up to three generations– and multi-family collective social unit –based on the co-residence of several conjugal subunits sharing livelihood and daily activities– which is an ephemeral phase in paleodemographic conditions (Fig. 9.1B and 9.1D). It can take several configurations. When practising strict patrilocality, we can find the joint family (*famille communautaire*), in which all the married sons –or male cousins– remain in their father/patriarch's house, sharing equal inheritance rights. Under the bilocal stem family (*familia troncal, famille souche, famiglia ceppo*) only one son lives and works alongside the father and inherits an indivisible heritage –i.e., individual male natolocality (Ensor 2021: 210). Brotherhood (*hermandad, frérèche*) is the ephemeral cohabitation of the male siblings/cousins and their families after the patriarch's death. A ubiquitous virilocal and patrilineal variant is the patronymic association or loosely defined and variably sized grouping often inhabiting blocks of adjoining conjugal dwellings (Fig. 9.1B-D). Their members consider themselves to be related according to a wide range of kin-like interdependence and even subalternity ties (e.g., patronage/ clientship) entailing cooperating and sharing their daily lives, bear a patronym and claim descent from a common eponymous ancestor and founder of the kin grouping (Kertzer 1991; Schloen 2001: 71). The bilateral joint residential group may practice virilocality or bilocality, in the latter forming a true corporate group –i.e., a *residential-household group* or Lévi-Straussian House institution (*casa, maison*) (Fig. 9.1D).

According to mentioned demographic estimates (Goody 1956; Schloen 2001), the coincidence of three generations would be an oddity and most grandchildren would be born after the death of their grandfather. Thus, two-thirds of the households would exhibit a conjugal organisation, while only one-third would be living in the multi-generational virilocal phase. The most common occurrence would therefore be the single-family conjugal unit, with a range of 3.5-5 persons. Several such subunits would be grouped into segments of corporate units –numerous unilineal patrilineages, neighbourhoods or patronymic associations, rare ambilineal ramages and small bilateral complex groupings– of some 7-12 members. Finally, some of these sections or factions might be nested within widely variable higher-order groupings, such as unilineal and virilocal patriclans or bilateral and bilocal *residential-household groups* (Houses, *casas, maisons*) of up to a few hundred members (Schloen 2001; Ruiz-Gálvez 2013; González-Ruibal and Ruiz-Gálvez 2016; Ensor 2021).

9.3.2 Archaeological indicators of the joint residential group

The basic premise pursued here is that on-site architectural spatiality and settlement layout can either afford, facilitate and channel or inhibit and hamper social relations. Thus, buildings and transit/common spaces contribute to physically fixing bonds of assistance and mutual support or, on the contrary, delineate boundaries and divisions emphasizing their absence and independence (cf., Ensor 2013a, 2013b, 2021; Ensor, Souvatzi, Grau Mira and Bermejo Tirado this volume). Domestic architecture is understood here "(…) as a nexus of informal social interactions among neighbours, who express their solidarity –both within households and between households– in the language of kinship." (Schloen 2001: 110).

Based on the above understanding, the lines of evidence used henceforth conform in a consistent and statistically significant way with cross-cultural patterns (Ensor 2013a, 2013b, 2021, this volume). First, the area of exclusive residential use of a dwelling –i.e., deducting non-living storage or other purposes surface– informs on the format of the residential unit and the type of postmarital residence practised (Ensor 2013a: 63-68, 2013b, 2021: 116-126). In the case of two-storey buildings, the first floor often accommodates the more private living quarters. Thus, small single-hearth dwellings (<60 m^2) are single-family buildings occupied by conjugal units, each with its fireplace. By contrast, large (>100 m^2) multi-hearth buildings regularly host several married women with first-degree ties –i.e., mother and daughters (matrilocality) or sisters (uxorilocality)– and their respective families living under the same roof. Intermediate-sized and multi-hearth dwellings (60-80 m^2) have recently been associated with semi-matrilocal or small uxorilocal units –e.g., two married sisters and their conjugal families living together– using Central European Neolithic longhouses (Ensor 2021: 156-162). Secondly, the placement of dwellings combined with the size of their living area allow confidently diagnosing descent. Thus, formal or pre-planned arrangements of conjugal dwellings around or along a focal space to which its entrances open conform with unilineal descent. By contrast, improvised, accretive, casual or random aggregates of houses lacking a prearranged unitary focal locus, with doors facing opposite directions suggest bilateral filiation, that is, more flexible and opportunistic dynamics practised by kindred networks instead of descent groups (Ensor 2013a: 109-137, 2021: 126-135, fig. 6.1 and this volume). Bilocal stem family groups can be identified as long-lived units after its superimposed remains (Ensor or Souvatzi this volume). Finally, an essential diagnostic pattern for our argument is that virilocality is regularly associated with agglomerated conjugal dwellings sharing the concerted opening of their entrances. Depending on descent, this can be done: a) formally locating dwellings in relation to a nodal, communal space –around an open-air central patio, or along an elongated inner corridor –*pastas* or *prostas* in Greek houses– a roofed central hallway, an external lane, or blind alley (Schloen 2001; Bilge 2019; Nevett 2023)– suggesting unilineal descent; or b) arranging houses concentrically but informally, thus indicating bilateral practices (Ensor 2013a: 63-68, 2021: 20, 25, 123-135, fig. 6.1). Beyond the statistically supported distinction of unilineal and bilateral practices (Ensor 2013a: 141-158, 2021: 126-131) material markers may not be consistent enough to rule out among likely options. Thus, virilocal aggregates of siblings with or without the father (*frérèche*), or more fluid virilocal patterns –e.g., patronymic associations or neighbourhoods of patrons with their clients– are materially indistinguishable.

Ensor (2013a, 2021: 134) cautions that there may be exceptions to these cross-cultural patterns and restricts their application to sedentary and non-aristocratic groupings. Indeed, noble cases –as some revised below– often exceed the dimensions of the statistically significant tested samples. Besides this, the gathering of conjugal dwellings indicative of noble composite virilocal joint-family complexes may host an array of residential groups presided by the patriarch/main house (Fig. 9.1). These may range from extended monogamous groupings –with kinsfolk and domestic retinue with their families– to polygamous (polygynous) households. Again, we have yet no reliable and well tested archaeological proxies to discern between such scenarios. However, this essay prioritises the big picture of the forest over the trees and therefore contends that, for the purpose of hypothesis building, even outlying, eccentric or noble cases can also be relatively and contextually assessed from these cross-cultural patterns as subordinate or ancillary sub-units within the whole joint household.

The archaeological analysis of associations of interdependent conjugal houses has developed considerably (Hayden and Cannon 1983; Schloen 2001; Smith 2010; Ensor 2013a, 2013b, 2021; Grau Mira 2013; Souvatzi 2017; Pacifico and Truex 2019). However, interpretation in terms of kinship is often eluded. Among the virilocal and bilocal casuistry, we may find varying degrees of physical closeness and architectural propinquity between the constituent conjugal subunits sharing spaces, facilities, implements and activities. Often these joint households segregate their convivial space from neighbouring ones, by means of physical boundaries –fences, walls, *periboloi*– with a single entrance, thus forming closed neighbourhoods or quarters. Schloen (2001: 110) defines a quarter as "a cluster of households closely linked by ties of kinship, factional alliance or patron-client relationship, all of which are expressed in familial terms." In this respect, Todd (2001: 68) points out: "the existence of an individualised ensemble of small houses (…) in a hamlet or in an enclosure reveals the existence of a higher-order unit encompassing the nuclear families". The historical literature identifies such collectives living in multi-dwelling aggregates as multi-functional and self-sufficient assemblages of buildings, always presided over by the patriarch's house (Schloen 2001: 108). This complex household qualitatively represents a communal social subject different from the sum of

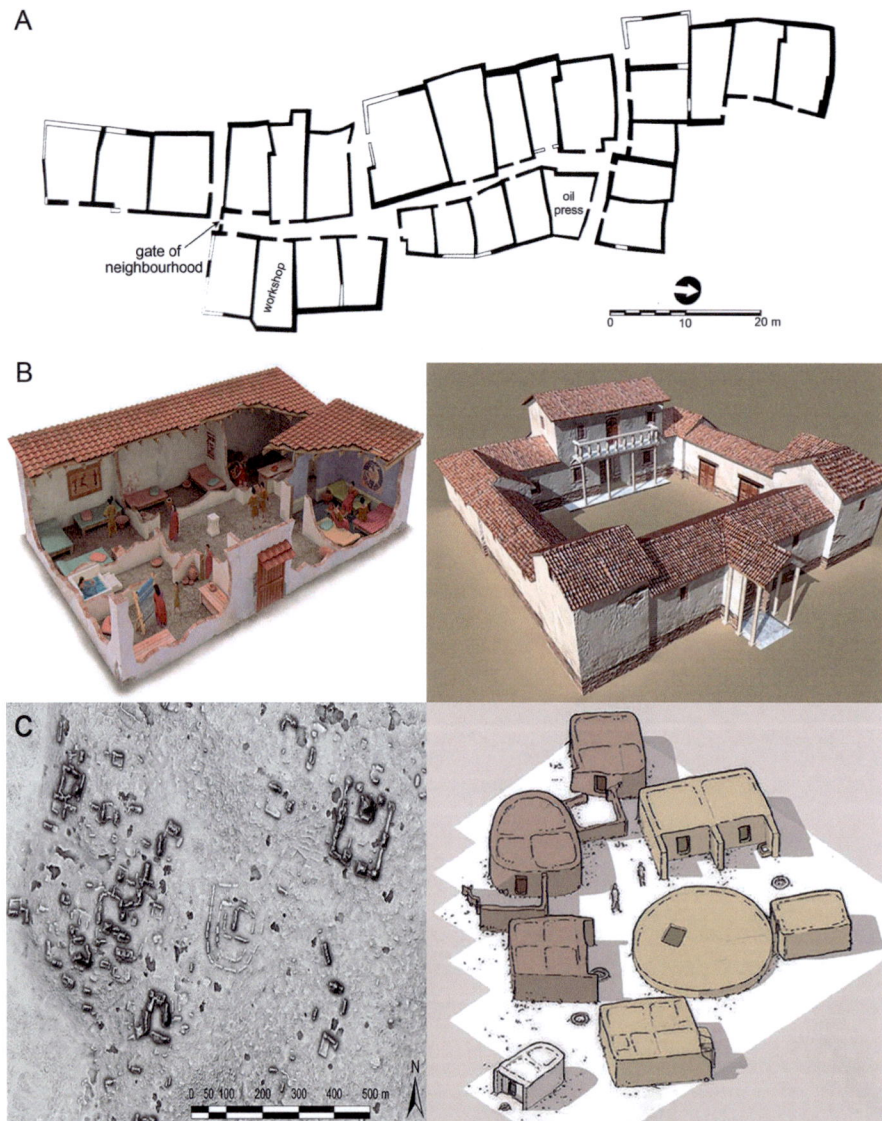

Figure 9.2. The three virilocal scenarios considered in the text: A) conjugal dwellings with abutting walls formally arranged in elongated blocks indicative of small patrilineage segments (e.g., patronymic associations) at Tell en-Nasbeh (Palestine) (modified from Schloen 2001: fig. 13); B) "courtyard households" suggestive of patrilineages, Greek *oikos* (left) and Roman *villa* (right) (Wikimedia Commons); C) random and informally clustered conjugal dwellings symptomatic of virilocal and bilateral groupings, from a LiDAR image of a Mayan site (left) (Wikimedia Commons) and from Jerf el Amar (Syria) *ca.* 9000 BCE (right), with houses opening onto a central patio with a circular underground silo (modified from Stordeur 2019).

A SOCIAL ARCHAEOLOGY OF KINSHIP IN IBERIA AND BEYOND

its constituent subunits, which –in the absence of neolocality– are not socio-economic units on their own (Ensor 2013a, 2021; Pacifico and Truex 2019). As a working hypothesis lacking statistical testing yet, the broad archaeological casuistry could be summarized here into two major co-resident scenarios (Fig. 9.2):

a. The self-contained, and formally designed multi-cellular architectural aggregate, formed by adjoined modular single-cell wards inhabited by conjugal subunits, clustered together in an agglutinative or organic way. This domestic materiality is the product of a foundational social agreement with clear demographic, resource, workforce, mutual support and inheritance provisions. Its architectural design conforms with a nodal space for collective use, with two main variations: a1) a block or quarter with the conjugal dwellings juxtaposed sharing partitioner walls and aligned along a frontage space –an alley or street– (Fig. 9.2A) (e.g., Schloen 2001; Grau Mira 2013, this volume); and a2) a closed, single-entrance and standalone building with adjoined en-suite rooms centripetally arranged opening from an inner rectilinear communal patio/lightwell –the so-called "courtyard house" (Nevett 2023) (Fig. 9.2B)– or a central hallway (Ur 2014). Both solutions contain inbuilt shared facilities for people –warehouses, presses, stockyards, workshops, ovens, wells– and livestock –stables– and sometimes also a dynastic mausoleum, domestic shrine or sanctuary. The compact buildings can be located free-standing in a thinly populated rural environment or in a low-density urban landscape (hacienda, mansion) or can be integrated into a densely populated urban fabric (insula or city block/district). Monumental buildings typically include connecting transit spaces –inner open-air patios, roofed hallways or corridors leading to the rooms– as conspicuous indicators of a well-off household. Based on their pre-planned and organic design, this contribution suggests that these modalities were dwelled by virilocal groupings practicing unilineal descent. They may have included kinsfolk, retainers, dependents or clients, all of them considered the patriarch's relatives.

b. The informal or loosely scattered compound –because of its cumulative growth, lacking a unitary and preconceived blueprint– of several conjugal dwellings, with a centripetal disaggregated spatial design, haphazardly clustered around an open focal space –a courtyard or patio– with ancillary buildings for exclusive use (Fig. 9.2C). Its conjugal subunits are linked through cognatic practices entailing fluid and casual negotiations of kindred membership based on co-residence instead of descent. They share the plot of land and the social, symbolic and material capital of the corporate composite social group, formed through the flexible and opportunistic selection of ego-centred kindreds, either small ambilineal ramages or bilateral groupings (Ensor 2013b, 2021).

9.4 Surveying the literature on joint aggregates of houses

Considering the above cautionary notes and acknowledging the limits of the working hypotheses, this epigraph explores a selection of examples. Previous contributions from a kinship-informed household archaeology approach have analysed abundant bilateral extended households, such as Mesoamerican Mayan (Ensor 2013b), Atlantic (González-Ruibal 2006) and Mediterranean (González-Ruibal and Ruiz-Gálvez 2016; Grau Mira 2013 and this volume; Blanco-González and Grau Mira in press) protohistoric examples, and European Neolithic cases (Souvatzi 2017; Ensor 2021). This section pays special attention to the multi-functional and joint-family composite household hosting extended unilineal residential groups –i.e., co-resident members and non-members of patrilineages. It also mentions bilocal and bilateral *residential-household groups* as counterexamples that can contribute to defining such a case in a negative way.

The case-studies cover the whole Mediterranean basin with occasional incursions into the Middle East and a few shallower comments on the New World. In so doing, the chapter draws on two of the more developed and better researched traditions of household archaeology worldwide: the Levant and circum-Mediterranean archaeology –encompassing small and intermediate-scale ranked organisations– and Mesoamerican and Andean archaeologies in the Western hemisphere –covering state-level and imperial polities. This way the sample encompasses a wide range of socio-political variability, demonstrating the relevance of kinship even in the most integrated and stratified ancient polities. Since the most robust evidential reasoning is by collating independent lines of evidence, the chapter draws case-studies meeting the following criteria: a) from open-area excavations with high-quality published information on the contents and functional characterisation of buildings and spaces, and b) featuring textual and iconographic references or mortuary evidence on the nature and composition of the social groupings inhabiting them. The available archaeological literature offers information of very disparate quality and resolution.

9.4.1 Southwestern Asian roots of the composite aggregate

From the earliest agglomerations of early aceramic Neolithic (PPNA) villages in Southwestern Asia are well-known cases of informal bilateral aggregates, such as a Jerf el Amar (Syria) *ca.* 9000 BCE (Stordeur 2019). There, small conjugal dwellings clustered around an open central space with several shared buildings, including a large subterranean silo and several kitchens for communal cereal milling and cooking (Fig. 9.2C). However, to my knowledge, the oldest virilocal and formal –compact and en-suite– courtyard edifice likely housing large sections of patrilineages is Building E within the Eanna precinct at Uruk (Irak) at the later Chalcolitic Uruk period of Mesopotamia (*ca.* 3100 BCE) (Fig. 9.3). This was a one-off

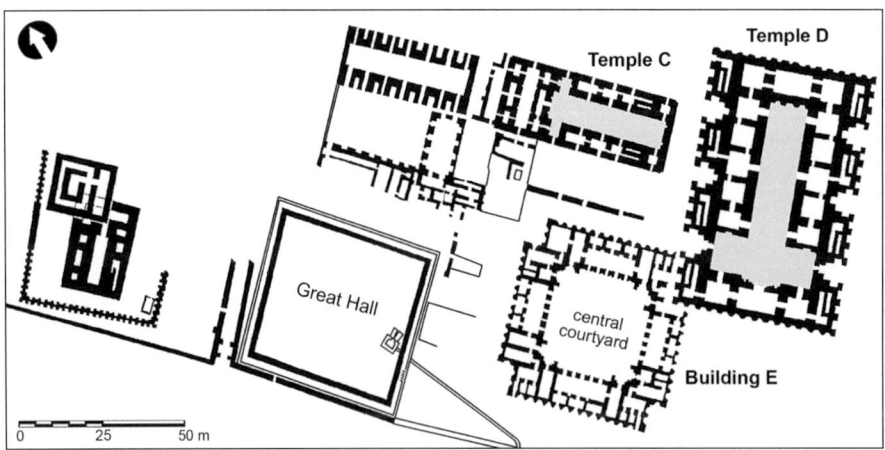

Figure 9.3. The earliest Chalcolithic virilocal extended households from the Eanna precinct at Uruk (Irak) (*ca.* 3100 BCE): the unique large courtyard edifice (Building E) and several tripartite or T-shaped buildings (Temples C and D) with their central hallways shaded in grey (modified from Ur 2014: fig. 3).

case among ubiquitous coeval examples of the so-called cruciform or T-shaped building: a monumental tripartite edifice with a roofed central hallway –for banqueting and social activities– connecting smaller adjoined rooms at both sides, whose earliest precedents can be traced back to the early Chalcolithic Ubaid period (5000-4000 BCE) (Ur 2014). These buildings have traditionally been interpreted either as sacred temples or secular residential palaces, following conventional reasoning. Ur (2014: 262) underlines that temples and houses were indistinguishable and echoed the tripartite template from the earlier Ubaid house. The difference between them was only in scale, and not from a qualitative standpoint, hosting either divine or lay joint residential groups. This researcher envisages such architecture within a patrimonial system of nested households that contributed to the emergence of the first cities at this dynamic hub, and accounts for this fractal model as the emic understanding of kinship itself (Ur 2014: 264).

9.4.2 Spreading the model in the Bronze Age

For the following millennia, the "courtyard-house" was a very successful architecture that typically accommodated sections of elite patrilineages. Urban districts were also important, and in Mesopotamia cuneiform documentation mentions the Akkadian word *bābtum* – derived from *bābum* meaning gate– a concept that, according to Schloen (2001: 287), conveys both a city district and an agnatically related corporate urban clan, since "each neighbourhood itself was enclosed and had its own gate." The Bronze Age (3000-1200 BCE) in this setting was the heyday of this kind of household, a process that went hand-in-hand with regional processes of urban agglomeration and the widespread emergence of socio-political ranking. During this period, we can find assorted virilocal compounds, following unilineal descent and bilateral networking and both as isolated rural estates or within the urban fabric of large cities, surrounding palatial quarters, some even representing palatial districts themselves such as at Middle Bronze Age Knossos (Greece) or Late Bronze Age Ugarit (Syria). However, the standard narrative only distinguishes between "nuclear" and "extended families" at most, with no reference to residence or descent (e.g., Apostolaki 2015).

Beyond the realm of Bronze Age large territorial states in Anatolia and the Upper Euphrates, archaeological research is increasingly acknowledging that Levantine palaces such as Tel Kabri (historic Palestine) echoed contemporary residential courtyard complexes and functioned as large households, replicating at an oversized scale their *modus operandi* (Yasur-Landau *et al.* 2015; Susnow and Goshen 2021). Both in the Levant and in the Aegean worlds these courtyard mega-compounds or palaces are undergoing a deep revision. Thus, part of academia is steadily regarding them as "great households" within heterarchical landscapes where Lévi-Straussian House Societies are often invoked (Chesson 2003; Ruiz-Gálvez 2013; González-Ruibal and Ruiz-Gálvez 2016; Relaki and Driessen 2020). The basic courtyard-focused design was reproduced in Cretan Late Minoan centres such as Quartier Nu at Malia (Greece) (Driessen and Fiasse 2011) (Fig. 9.4A). This is a three-winged aristocratic building made up of four multi-room houses –named here from A to D– each with an individual hearth and whose entrances face diverse directions, arranged around a pebbled courtyard and featuring three large ritual pits (*favissae*). Based on the distribution of findings and the presence of a communal stand-alone kitchen, its excavators conclude that the wings were complementary, and the large complex "was occupied by a single extended household with different family units" that they relate with the Archaic Cretan *startos* or patriclan made of several *oikoi* or households (Driessen and Fiasse 2011: 295).

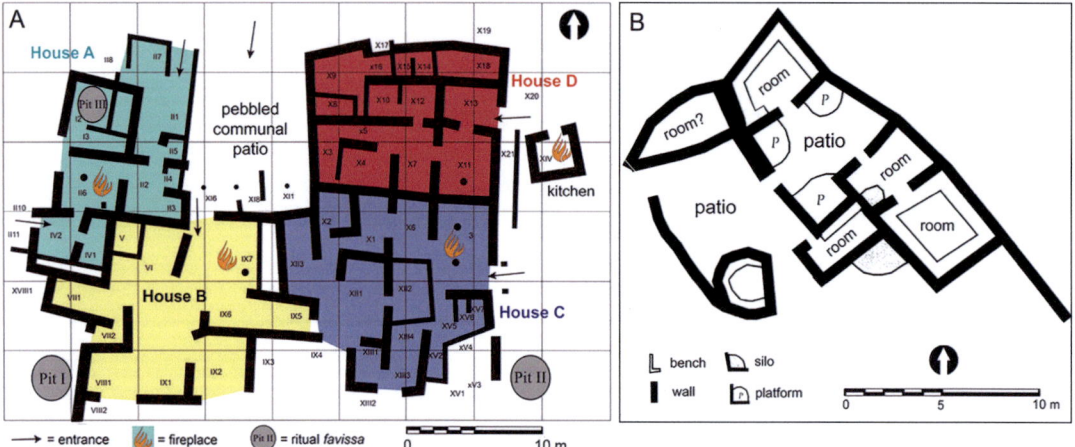

Figure 9.4. Reinterpreted Bronze Age courtyard compounds: A) Postpalatial Quartier Nu at Malia (Greece), 1330-1200 BCE, a likely bilocal and bilateral joint household (modified from Driessen and Fiasse 2011: fig. 25.1); B) Early Bronze Age compound 23 at Arad (historic Palestine), 2500 BCE, reinterpreted here as a patrilineal residential group (modified from Chesson 2003: fig. 1).

However, the relative independence of its oversized constituent dwellings coupled with their spatial subordination suggest a bilocal and bilateral composite complex instead, with no clear main house (Fig. 9.4A).

On other occasions, the kinship organisation of purported cases of House Societies might be discredited. Indeed, if we follow Ensor's (2013a: 47-49, 2021: 27-28) restrictive criterion, only the combination of bilateral and bilocal practices defines Lévi-Straussian *maisons* as *residential-household groups* (see above). Thus, household compound 23 at Arad (historic Palestine) (Chesson 2003) (Fig. 9.4B) or the Bronze-Iron Age Sardinian *nuraghi* (González-Ruibal and Ruiz-Gálvez 2016) have been posited as examples of House Societies. Yet these pretty formal walled aggregates of conjugal-sized dwellings whose individual entrances face open-air patios better match the cross-cultural pattern indicative of virilocal and unilineal descent as the residences of patrilineage groups (Ensor 2013a, 2021).

The Syrian site of Ras Shamra (ancient Ugarit) was the capital of a small city-based commercial and agrarian kingdom in the thirteenth century BCE (Schloen 2001: 320-335; González-Ruibal and Ruiz-Gálvez 2016: 395-401). It has been crucial in discussions on social organization at that time and deserves a closest insight. From the over 6 ha unearthed out of the 22 ha-tell, urbanism is arranged in tightly packed and irregularly shaped quarters separated by narrow winding alleys. Composite households of variable sizes, forms and wealth were common occurrences. Minor, humbler and simpler dwellings always agglomerated around the leading houses of the patriarchs/heads of extended groupings, as indicated by archaeology and cuneiform archival documentation. Schloen (2001: 317-347), drawing on M. Weber's patrimonial theorisation, interprets this layout in terms of his "house of the father" model of multiscalar or fractal hierarchization vehiculated by unilineal agnatic descent –i.e., populated by patrilineages, patriclans and patronymic associations. González-Ruibal and Ruiz-Gálvez (2016) argue for a House Society account with bilateral Houses as protagonists. Both proposals (Schloen 2001: 320-328; González-Ruibal and Ruiz-Gálvez 2016: 397-398, fig. 6) pay special attention to Insula 6 in Ville Sud,

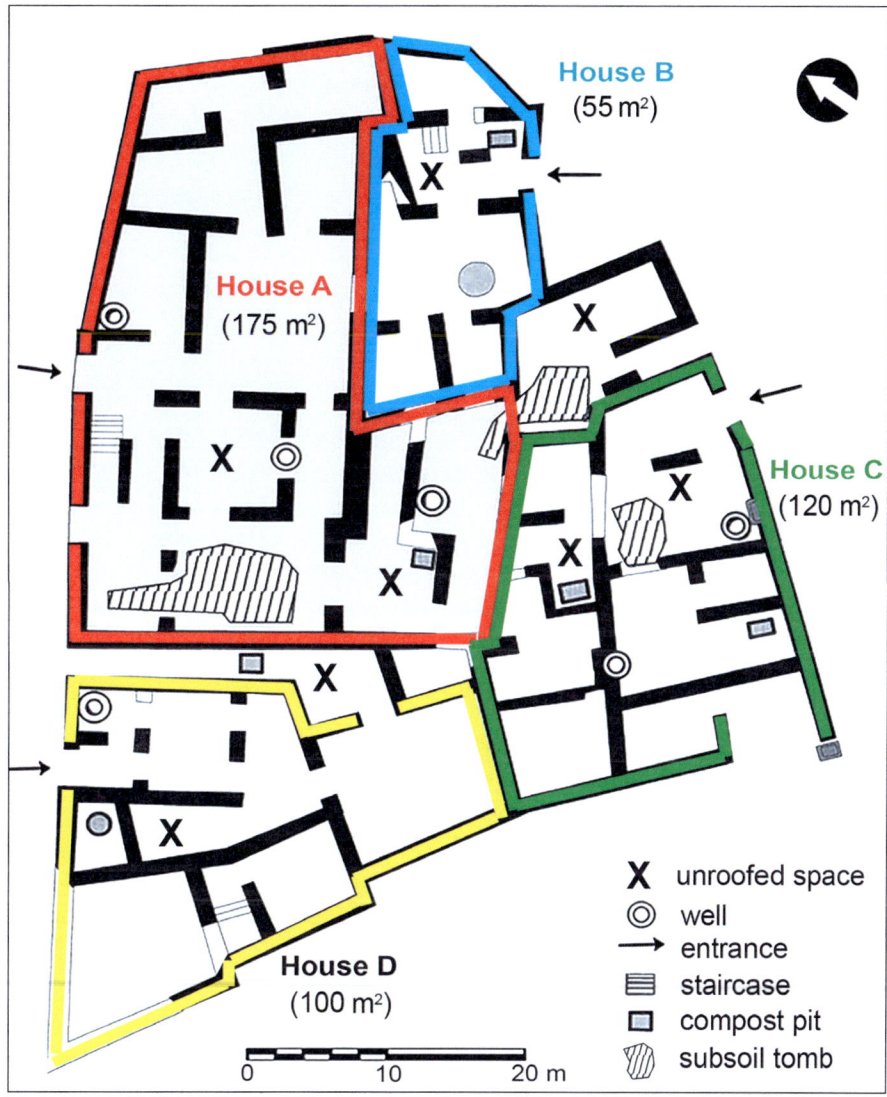

Figure 9.5. Insula 6 at Late Bronze Age Ugarit/Ras Shamra (Syria), 1200 BCE, a likely bilocal and bilateral joint residential group (modified from Schloen 2001: fig. 21, González-Ruibal and Ruiz-Gálvez 2016: fig. 4).

within a large excavation transect (5,700 m²) that uncovered the central-southern part of the tell. Insula 6 is a wide preconceived block irregular in plan, with four two-storey adjoining dwellings (Houses A to D) featuring shared maintenance and production facilities on the ground floor and living quarters on the upper storey (Fig. 9.5). Every individual (one-entry) multi-room house may be regarded as an oversized (i.e., noble) dwelling featuring at least one inner open-air patio, a compost pit and three of them also have wells and underground monumental tombs. House A is consensually regarded as the wealthiest and main dwelling within this urban district, hosting in its *ca.* 175 m² of roofed space either a polygamous group –as posited by its excavator and questioned by Schloen (2001: 320-328)– or the leader's (patriarch?) subunit including its unmarried offspring and retainers with their families. The other three noble dwellings –Houses B-D, ranging between 55-120 m²– are clearly minor or ancillary subunits (smaller, simpler) and plausibly accommodated the leader's married sons/cousins/clients with their families and dependents. Individual entrances facing multiple directions (Ensor 2013: 68), the occurrence of three small tombs –instead of one large unilineal collective mausoleum for a whole corporate group– and the absence of a household-wide focal point within this urban district suggest that this was a bilocal and bilateral insula. These observations provide support to the House Society reading (González-Ruibal and Ruiz-Gálvez 2016: 397-398) over other accounts in patrilineal terms.

9.4.3 The Iron Age momentum of the "courtyard compound"

This Old-World survey ends with the Iron Age, roughly in the first half of the first millennium BCE. This period was characterised by the Mediterranean-wide spread of the sociopolitical formula of the "courtyard compound" beyond the Levantine and Asian hubs where it had taken root over millennia prior. As a self-sufficient social cell, this form of conviviality fitted the requirements of workforce, mutual support and heritage transmission in ranked and increasingly expanding agrarian and commercial societies (Schloen 2001; González-Ruibal and Ruiz-Gálvez 2016). In the Levantine Iron Age (1200-600 BCE), an architectural design was very successful: the so-called pillared, central courtyard "hash-plan" or "four-room" compact two-storey building. This composite edifice featured a central hall or small roofed patio –hosting shared domestic chores and social life– agrarian facilities and stable on the ground floor and resting/living quarters on top (Schloen 2001; Gilboa *et al.* 2014; Gadot and Yasur-Landau 1998; Ruiz-Gálvez 2020). These buildings likely accommodated patrilineage residential groups of three-four conjugal subunits totalling some 10 kinspeople on average. Such a composite social grouping has been related by Schloen (2001: 136-183) with literary and biblical references to the "house of the father" (Heb. *bêt 'āb*), a corporate patrilineage or patronymic association nested into a higher-order patriclan (*mišpāhôt*).

There are several well excavated and published Iron Age I (1200-1100 BCE) cases, such as Area G House at Tel Dor (historic Palestine) (Gilboa *et al.* 2014), with a half-paved and half-roofed central patio (Fig. 9.6A). Another case is Building 00/K/10 from Megiddo (historic Palestine) (Gadot and Yasur-Landau 1998), a single-storey humble house with nine small rooms around an unroofed central patio, totalling 110 m² of roofed area and 65 m² of living space (Fig. 9.6B). The building contained all necessary facilities –workshops, looms, storerooms, *tabuns* and baking trays, etc.– as a self-sufficient unit. Its fortunate –for the archaeologists– sudden collapse by an earthquake, the careful documentation of all finds and their exhaustive publication allow a rare highly-detailed functional and social characterisation. Excavators estimate that it was inhabited by an extended group of six-seven co-residents (farmers). The fallen roof trapped a number of occupants very close to the estimated ones, based on the allocation of people per m² (Fig. 9.6B): seven/eight individuals –three children and five adults, four of them inside the edifice. This bioarchaeological sample would be most suitable for biomolecular analyses –aDNA for biological relatedness and stable isotope analyses for mobility and diet– to ascertain whether this grouping was unilineal or bilateral. In any case, the most likely scenario is that not all the co-residents will display close biological relatedness. Identifying genealogical "nuclear families" *per se* has no bearing at all in kinship characterisation or social organisation (Ensor 2013a; 2021), despite misleading and ethnocentric claims by geneticists to the contrary.

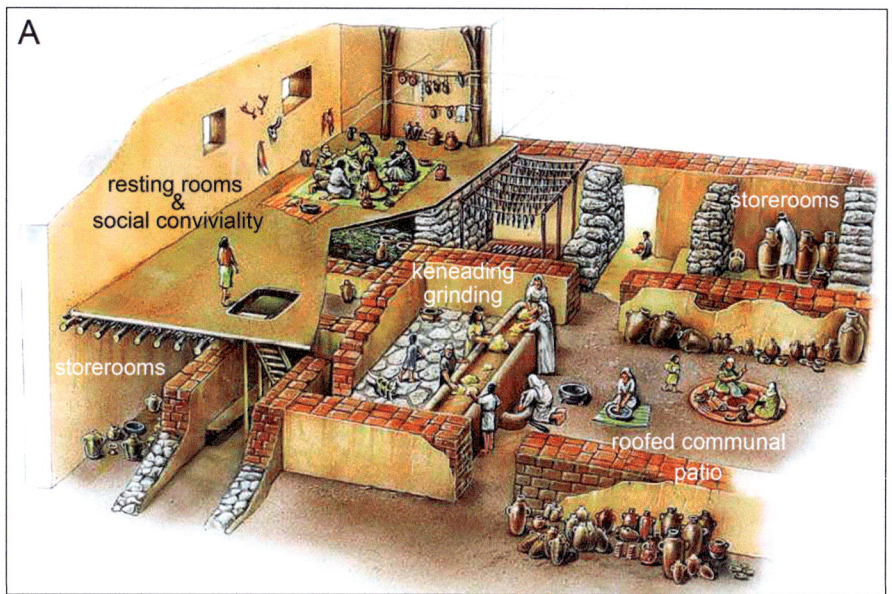

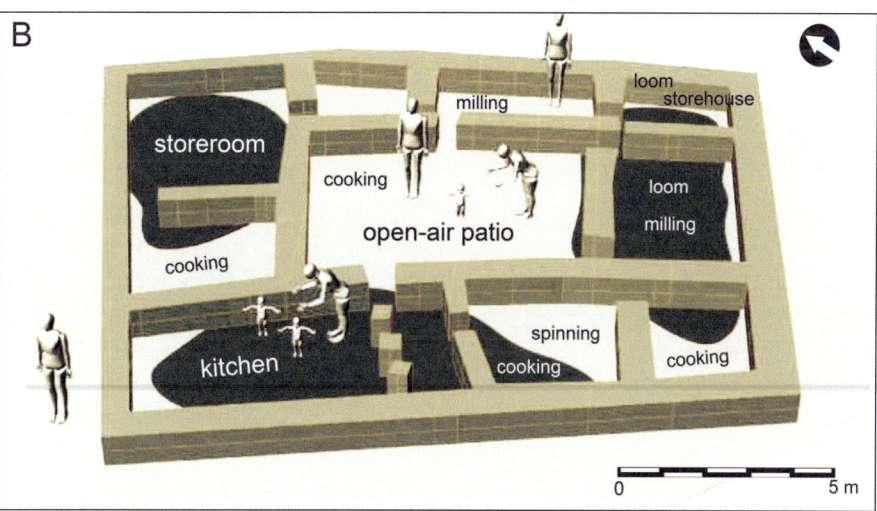

Figure 9.6. Levantine Iron Age I (1200-1000 BCE) courtyard compounds dwelled by patrilineage sections of diverse socioeconomic status, indicating activity areas and domestic chores: A) affluent two-storied Area G House at Tel Dor (Palestine) (modified from Gilboa *et al.* 2014: fig. 16); B) modest single-storied Building 00/K/10 from Megiddo (Palestine) (modified from Gadot and Yasur-Landau 1998: fig. 33.1; Ruiz-Gálvez 2020: fig. 2).

Ruiz-Galvez (2013: 102-106, 2020) has carefully studied the archaeology of these Levantine –first Canaanite and later Phoenician– commercial kingdoms, emphasizing the adoption of a comparable household organisation and its quick seaborne irradiation across the Mediterranean reaching Iberia. Beneath the apparent wide variability of buildings, her analysis identifies the same composite residential group as a self-sufficient and heterarchical socio-economic unit. She addresses it from the Lévi-Straussian corporate and bilateral institution of the House and under the light of the

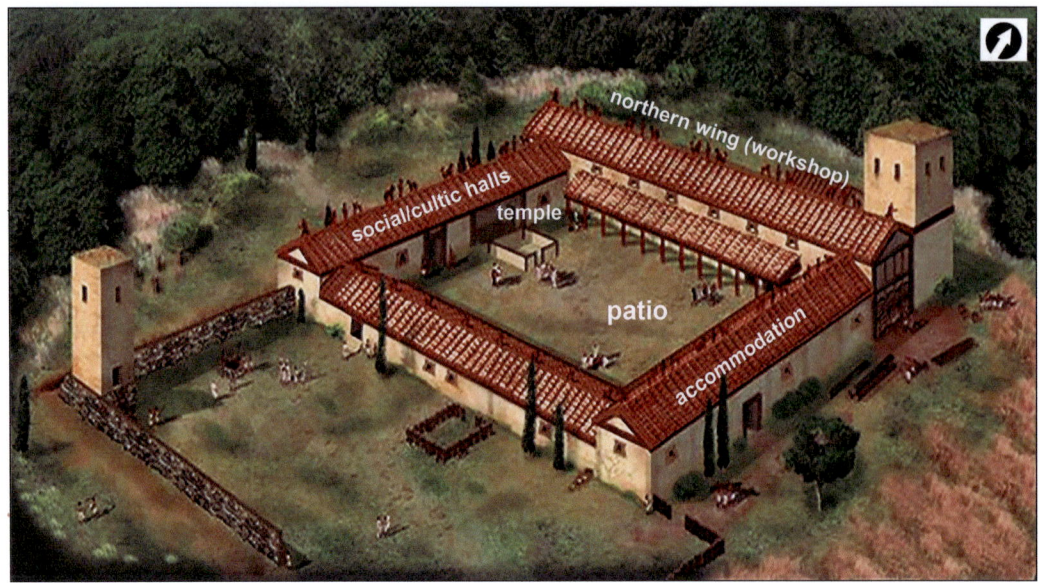

Figure 9.7. The Etruscan Iron Age rural courtyard compound of Poggio Civitate (Murlo, Italy) *ca.* 550 BCE, reinterpreted here as the seat of an aristocratic patrilineal and virilocal group (modified from Tuck 2021).

Weberian patrimonial model reclaimed by Schloen (2001). Household archaeology of Phoenician commercial outposts on the Iberian coasts is well developed. In the eighth and seventh centuries BCE such foundations proliferated, with multi-functional and joint-family households such as at Cerro del Villar (Málaga, Spain) (Delgado *et al.* 2014) or the small and ephemeral compact courtyard building at Abul (Setúbal, Portugal) (Mayet and Silva 2005). In 2008-2009, expansion works at Malaga airport prompted important excavations at La Rebanadilla (Málaga, Spain), which revealed a multi-building neighbourhood including a sanctuary and dwellings detected in 5,000 m², out of which 400 m² were unearthed. From the patchy information published (Sánchez Sánchez-Moreno *et al.* 2012) this old island setting hosted reticulate urban arrangements with several adobe courtyard buildings and a nearby cremation necropolis featuring Levantine rituals, all radiocarbon-dated to *ca.* 850-750 BCE. Therefore, La Rebanadilla can be regarded as the material correlate of an Oriental socio-economic formula of conviviality that reached the Iberian littoral at an early stage and was extremely successful later throughout Iberia (Grau Mira this volume).

Finally, the Archaic phase (*ca.* 600-550 BCE) of the "courtyard compound" of Poggio Civitate (Murlo, Italy) (Tuck 2021) (Fig. 9.7) is another well-known case of an elite courtyard household. It is a huge –*ca.* 360 m² including the patio– closed and compact quadrangular building with four wings –three of them colonnaded– around an open patio with a small dynastic shrine or archaic temple devoted to the ancestors (aka the *maiores*), also likely represented in the apotropaic *akroteria*. The northern wing hosted an enormous two-storied workshop, whereas other wings accommodated storerooms, kitchens, aristocratic dwellings and social halls devoted to ostentatious banqueting and social life. Such a building is far larger than anything known in the Mediterranean for its time and has prompted debate in typical exclusionary terms; interpreted either as a noble rural palace, a specialised workshop, a religious sanctuary, a political audience or a meeting hall. Poggio Civitate was all this in a compact and organic way. The complex worked as an aristocratic self-sufficient rural estate, housing a large residential grouping with a remarkable domestic retinue and occasional specialist craftspeople (Tuck 2021). Since its later design clearly adheres to a very formal and preplanned unitary building program, following Ensor's (2013a, 2021) criteria for his *residential-household group* this case would not meet the markers of a bilateral and bilocal House (contra González-Ruibal and Ruiz-Gálvez 2016: 424-426) and might tally better with a patrilineage faction instead.

Beyond these pre-Classical households in the ancient Mediterranean and Near East, historical archaeology and textual documentation are very eloquent sources on the nature of subsequent strictly patrilineal and virilocal groupings who live in "courtyard houses". I cannot help but mentioning the classical Greek patrilineage or *γένος* composed of several households (*οἶκοι*), the similar Roman *gens* led by the *paterfamilias* and inhabiting the *domus* (González-Ruibal and Ruiz-Gálvez 2016), the Palestinian *zaʾila* (Schloen 2001: 108) or the *qawm* or patronymic association of neighbourhoods (*ḥawāʾir*) sharing the patronym (*mansūba*) in medieval Iberian al-Andalus (Ortega Ortega 2024).

9.4.4 A New World historical epilogue: Mesoamerican and Andean cases

There are excellent prospects to assess comparable examples in the Western hemisphere, especially from multi-stranded approaches integrating household and funerary evidence. Indeed, in the Americas we can find the pioneer and far more developed tradition of Mesoamerican household archaeology. Beyond the groundbreaking study by Ensor (2013b) of Mayan sites in the Yucatán peninsula (Fig. 9.2C), there are potential cases as splendid as the over 2,000 mapped apartment compounds of the city of Teotihuacán (México) in the Early Classic period (*ca.* 200-550 CE), the largest in the continent at Pre-Columbian times. In its grid-like arrangements of rectangular residential quarters lived both aristocrats –in oversized dwellings– and commoners –fitting cross-cultural size standards for residential space. Such a wide residential complex spread along the Street of the Dead between the northern Pyramid of the Moon and the southern Pyramid of the Sun (Fig. 9.8A). This urban project has recently been characterised as social and anti-hierarchical in nature (Graeber and Wengrow 2021: 343-351). It hosted what might be regarded as typical virilocal compounds of subordinate and interdependent dwellings around patios, with shared facilities –some with courtyard shrines overlying the tombs of adult males with wealthy burial furnishings– and abundant intramural sub-floor burials (Manzanilla 1996, 2009; Murakami 2019; Smith *et al.* 2019) highly prone to kinship analyses after critical context-based assessment (Fig. 9.8B).

There is also a wide field of exploration of this approach in the emerging branch of Andean household archaeology, which delves into state-level and imperial kingdoms. A case in point is the World Heritage listed civic center of Chan Chan (Perú), the 6-km² large and densely packed capital of the empire of Chimor or Chimú (*ca.* 1000-1476 CE). Royal dynasties organised in patrilineages (*kuraka*) dwelled at least nine –surviving Spanish looting– huge districts (*ciudadelas* or citadels) (Fig. 9.9A). These were high-walled and single-entry adobe compounds, with impressive open-air courtyards and interior facilities, including storerooms, so-called *audiencias* and massive dynastic mausoleums

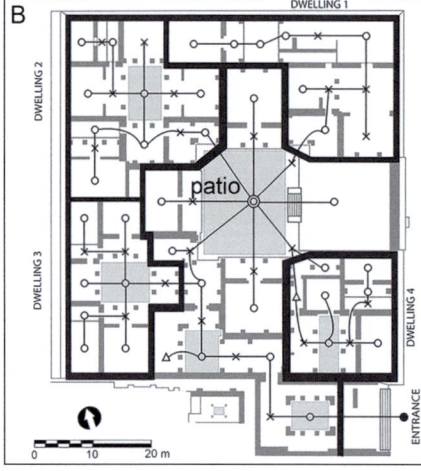

Figure 9.8. Courtyard compounds at Teotihuacán (México) *ca.* 450 CE: A) artistic rendering of the Atetelco palace (foreground) and apartment compounds on the west side of the Street of the Dead, with the Pyramid of the Sun (background) (modified from Wikimedia Commons); B) network analysis of the palace Zacuala, an enormous single-entrance complex made up of four oversized houses around a central patio (modified from Smith *et al.* 2019: fig. 9).

–multi-storey burial platforms (*huacas*)– (Kolata 1990; Colisi *et al.* 2009; Rengifo 2020; Moore 2021) (Fig. 9.9B). Finally, the metropolis of Cuzco (Perú), capital of the short-lived Inca empire or Tawantinsuyu (1476-1533 CE), also held noble and regal patrilineage groups –each part of a large patriclan (*ayllu*)– accommodated in virilocal walled districts with a single entrance called *kanchas*. These contained oversized subordinate dwellings (*wasi*) hosting residential subunits and arranged on three or four sides of an open-air patio, encapsulated by the famous Coricancha and its House of the Sun *wasi* (Bauer 2011; Morris and von Hagen 2011).

9.5 Final Remarks

This chapter has resorted to a social archaeology and household-focused approach to unravel the physical remains of collective ways of arranging domestic life vehiculated by kinship. It has supported the relevance of this pathway of evidential reasoning and the fruitfulness of studying past (including prehistoric) kinship through the household among illiterate societies. To do so, the chapter has avoided oversimplistic visions in terms of "extended families" or all-encompassing "household groups" –concepts too lax and archaeologically unobservable– to bring to the forefront the disregarded protagonists of these extinct societies: the co-resident, collaborating and practice-based residential groupings. This text has also attempted to transcend the mere classification and formal description –however sophisticated it may be– of residential architecture. By circumventing the usual exclusionary or dichotomous functional categories used when analysing the domestic sphere –by means of more flexible and inclusive notions– it has intended to delve deeper into the sociological meaning of household materiality. The revisitation of controversial and uncontested issues on household and kinship has aimed to shed new light over current confusion. It has set out to avoid the old culturalist categories of normativism and its conceptual watertight boxes, rehearsing as an alternative transversal

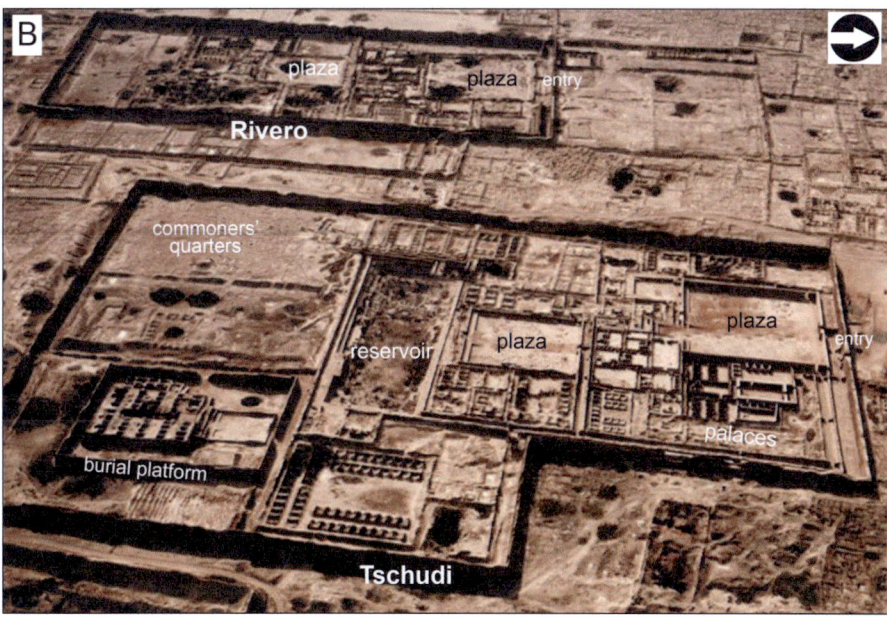

Figure 9.9. Andean high-walled adobe *ciudadelas* or royal patrilineal and patrilocal courtyard compounds within the Chimú imperial capital of Chan Chan (Perú) (*ca.* 1000-1476 CE): A) aerial view of *ciudadelas* Rivero (left) and Tschudi (right) (modified from *Google Earth*); B) oblique aerial view from the east of both walled compounds, showing the large entry courtyards in northern and central quarters, and massive adobe curtain walls subdividing the compounds (modified from Wikimedia Commons).

accounts cutting across various archaeological cultures, even independent and seemingly incomparable realities in different continents and timeframes.

The text has reviewed a selective suite of habitual and well-studied archaeological cases of joint residential groups. It has concentrated on multi-house compounds of small commoner conjugal dwellings and oversized noble individual subunits –virilocal and bilocal examples associated with patrilineal and bilateral patterns– over other commoner options –e.g., large matrilocal and medium-sized semi-matrilocal dwellings. It was not a chapter's goal to comprehensively characterize the whole kinship arrangements in the target ancient societies; it has tracked instead a concrete material and social configuration across diverse spatio-temporal coordinates in a comparative manner. In so doing, it has necessarily dispensed with further joint-family composite combinations –e.g., matrilocal and matrilineal or semi-matrilocal and bilocal customs– which should not be ruled out from the whole picture.

This survey has allowed to glimpse the occurrence of several kinship patterns of residence and descent practices featured in the ethnohistorical record with a distinctive physical signature. Beyond their apparent architectural disparity, in collating these material assemblages with case-based inductive abstractions the scope here has been experimental and generative of hypotheses instead of reconstructive. Such analogy-based generalisations are more critically constructed and better supported than a great deal of inductive reasoning commonly accepted in the archaeological literature (Ensor 2021: 124). Despite the lack of cross-cultural statistical testing of some of the proposed scenarios, there is enough room to broadly identify a handful of kinship practices combinations. Indeed, when excavations are large and recording fine-grained enough and their results are sufficiently published –crucially containing plans with indication of the entrances, hearths and the functional characterisation of all rooms– it is often possible to detect such patterns.

The joint residential group was a ubiquitous formula of conviviality in the protohistoric and early historic past, when it was a key social agent of small- and large-scale organisations and even within large territorial states and expansive empires. The chapter has revealed how since the most ancient examples in sixth-millennium BCE Mesopotamia (Ur 2014), the overlapping and multi-task nature of the target domestic arrangements has confused archaeologists, often operating from their own cultural notions of separate functional spheres (see criticism in Schloen 2001; Ruiz-Gálvez 2020). This has led in diverse scholarly traditions worldwide to empty nominal debates about the function or the role of such architecture. But such parochial disquisitions and the exclusionary labels used –e.g., temples, sanctuaries, palaces, residences, workshops, markets, storehouses, etc.– make little sense outside the culture-historical framework within which they were coined and have worked for decades.

Lastly, the essay has tried to advance in understanding the domestic realm as an active participant of social dynamics, rather than a mere deactivated container. Thus, the multi-family architectural compounds hosting composite social units can be envisaged as social and material assemblages –both aristocratic and subaltern– that entailed an entanglement of distinctive lifestyles, social practices and materiality. In the case of Mediterranean protohistory, its adoption and irradiation across the Mediterranean from precedents in Southwestern Asia was likely led by Eastern Mediterranean peoples –

Phoenician and Aegean settlers– at diverse times, and these pristine littoral occurrences contrast with previous more community-focused arrangements (Ruiz-Gálvez 2013). There is no need to keep resorting to ill-defined and vague cultural "influences" or using this materiality as an indicator for "acculturation". The lesson to be learnt here is the crucial role played by the extended residential group as a key historical actor leading the social transformations involved in these processes of change.

In sum, a comparative approach to household archaeology informed by kinship is capable of underlining commonalities in demographic, functional and social terms, offering a robust interpretive framework for understanding household variability, while helping to prompt the discussion of far more relevant issues in the sphere of humanities. This approach is proving its wide-range potential for interpreting an array of modalities of organising social life across time and space. A social archaeology of kinship is to pay more attention to the whole range of social actors instead of only progenitors or "pedigrees" simply because these are visible and traceable in bioarchaeological datasets. It should also focus on understanding how kinship is culturally defined, and the decisions made regarding it (Ensor 2013, 2021). Kinship has to do with the protagonists of history, either households or nested corporate groupings from a wide range of kin-based options: unilineal lineages or clans, ramages and bilateral networks, genealogical or kindred-related units, etc. From a social archaeology perspective, these should be our target historical protagonists, which keep being overlooked, elusive and liquid in grand macro-narratives. It is surprising how since the 1990s the pervasive and ubiquitous use of sociological theory on social agency or agent-based modelling in archaeology have hardly contributed to substantially mitigate the dismissing of these ancient –both individual yet foremost corporate– subjects.

We cannot be fully sure that the analysed residential groupings always adhered to kinship principles instead of other mechanisms. Yet applying the logic of Ockham's razor to humanistic reasoning, the simplest and most parsimonious option –strongly supported by ethnohistorical and textual evidence– is to assume that kin-based relatives were the social subjects of our account, instead of non-kin co-residents –e.g., organised in sodalities, craft-based groupings, secret societies, phratries, etc. The burgeoning burial-centred biomolecular approaches –aDNA, stable isotopes– are considered here independent bodies of evidence providing crucial information on consanguineal progeny, diet or mobility, but such science-based knowledge is to be collated with household archaeology research. Since our social interpretations will always be underdetermined and subject to equifinality, the kinship framework may serve as an effective and reliable scaffolding to sketch working hypotheses, learn from their collation with materiality and encourage robust cross-cultural and multi-stranded evidential reasoning.

Acknowledgements

An early draft version benefitted from critical insight by Brad Ensor, whom I am very grateful for his invaluable suggestions. I am solely responsible for any factual mistakes, misunderstanding or misinterpretation. Charles Bashore proofread and improved my English text. This contribution is part of research project ARQPARENT (PID2019-104349GA-I00, MCIN/AEI/10.13039/501100011033) funded by the Spanish Ministry of Science.

References

Albertz, R., Nakhai, B.A., Olyan, S.M. and Schmitt, R. (eds.) (2014): *Family and Household Religion. Toward a Synthesis of Old Testament Studies, Archaeology, Epigraphy, and Cultural Studies.* Winona Lake: Eisenbrauns.

Apostolaki, E. (2015): On the Household Structure of Neopalatial Society. In Cappel, S. Günkel-Maschek, U. and Panagiotopoulos, D. (eds.), *Minoan Archaeology. Perspectives for the 21st Century.* Louvain: Presses Universitaires du Louvain, pp. 223-239.

Bauer, B. (2011): *The Sacred Landscape of the Inca: The Cusco Ceque System.* Austin: University of Texas Press.

Bilge, E. (2019): *The 'courtyard house'. A spatial reading of domestic architecture in ancient Anatolia and Greece.* MA Thesis. Ankara: Middle East Technical University.

Blanco González, A. and Grau Mira, I. (in press), Kinship Practices and House Societies: Socio-political Entanglements in the Iberian Iron Age (450-300 BC). In Souvatzi, S., Bickle, P. and Cvecek, S. (eds.), *Prehistoric Kinship: Contemporary Perspectives in Archaeology and Bioarchaeology.* Cambridge: Cambridge University Press.

Colosi, F., Fangi, G., Gabrielli, R., Orazi, R., Angelini, A. and Bozzi, C.A. (2009): Planning the Archaeological Park of Chan Chan (Peru) by means of satellite images, GIS and photogrammetry. *Journal of Cultural Heritage* 10(1), pp. e27-e34, DOI: 10.1016/j.culher.2009.08.002.

Chesson, M.S. (2003): Households, Houses, Neighborhoods and Corporate Villages: Modeling the Early Bronze Age as a House Society. *Journal of Mediterranean Archaeology* 16(1), pp. 79-102. DOI: 10.1558/jmea.v16i1.79.

Delgado, A., Ferrer, M., García, A., López, M., Martorell, M. and Sciortino, G. (2014): Arquitectura doméstica en El Cerro del Villar: uso y función del espacio en el edificio 2. In Arruda, A.M. (ed.), *Fenicios e Punicos por terra e por mar. Actas do VI Congresso Internacional de Estudos Fenicios e Punicos.* Lisboa, pp. 338-344.

Driessen, J. and Fiasse, H. (2011): Burning down the house: Quartier NU at Malia. An ArcView analysis. In Glowacki, K.T. and Vogeikoff-Bogan, N. (eds.), *STEGA: The archaeology of houses and households in ancient Crete.* Hesperia Supplement 44. Princeton: American School of Classical Studies at Athens, pp. 285-296.

Ensor, B.E. (2013a): *The Archaeology of Kinship: Advancing Interpretation and Contributions to Theory.* Arizona: University of Arizona Press.

Ensor, B.E. (2013b): *Crafting Prehistoric Maya Kinship.* Tuscaloosa: University of Alabama Press.

Ensor, B.E. (2021): *The Not Very Patrilocal European Neolithic. Strontium, aDNA, and Archaeological Kinship Analyses.* Oxford: Archaeopress Archaeology.

Gadot, Y. and Yasur-Landau, A. (2006): Beyond finds: Reconstructing life in the courtyard Building of the Level K-4. In Finkelstein, I., Ussishkin, D. and Halpern, B. (eds.), *Meggido IV. The 1998-2002 Seasons.* Tel Aviv: Tel Aviv University, pp. 583-600.

Gilboa, A., Sharon, I. and Zorn, J.R. (2014): An Iron Age I Canaanite/Phoenician courtyard House at Tel Dor: A Comparative Architectural and Functional Analysis. *Bulletin of the American Schools of Oriental Research* 372, pp. 39-80. http://www.jstor.org/stable/10.5615/bullamerschoorie.372.0039.

Graeber, D and Wengrow, D. (2021): *The Dawn of Everything. A New History of Humanity.* New York: Farrar, Straus and Giroux.

Godelier, M. (2011): *The Metamorphoses of Kinship.* London: Verso.

González-Ruibal, A. (2006): House societies vs. kinship-based societies: An archaeological case from Iron Age Europe. *Journal of Anthropological Archaeology* 25, pp. 144-173. DOI: 10.1016/j.jaa.2005.09.002.

González-Ruibal, A. and Ruiz-Gálvez, M. (2016): House Societies in the Ancient Mediterranean (2000–500 BC). *Journal of World Prehistory* 29(2), pp. 383-437. DOI: 10.1007/s10963-016-9098-8.

Goody, J. (1956): *The Developmental Cycle in Domestic Groups*. Cambridge: Cambridge University Press.

Goody, J. (1976): *Production and Reproduction: A Comparative Study of the Domestic Domain*. Cambridge: Cambridge University Press.

Grau Mira, I. (2013): Unidad doméstica, linaje y comunidad: estructura social y su espacio en el mundo ibérico (ss. VII-I a.C.). In Gutiérrez Lloret, S. and Grau Mira, I. (eds.), *De la estructura doméstica al espacio social. Lecturas arqueológicas del uso social del espacio*. Alicante: Universidad de Alicante, pp. 59-76.

Harris, O. (1982): Households and their boundaries. *History Workshop Journal* 13, pp. 143-152.

Harris, O. (2018): Extended Family. In *The New Palgrave Dictionary of Economics*. London: Palgrave Macmillan. DOI: 10.1057/978-1-349-95189-5_425.

Hayden, B. and Cannon, A. (1982): The corporate group as an archaeological unit. *Journal of Anthropological Archaeology* 1(2), pp. 132-158. DOI: 10.1016/0278-4165(82)90018-6.

Hernando, A. (2017): *The fantasy of individuality. On the Sociohistorical Construction of the Modern Subject*. Cham: Springer.

Huebner, S.R. (2017): A Mediterranean Family? A Comparative Approach to the Ancient World. In Huebner, S.R. and Nathan, G. (eds.), *Mediterranean Families in Antiquity: Households, Extended Families, and Domestic Space*. Oxford: Wiley-Blackwell, pp. 3-26.

Kertzer, D.I. (1991): Household History and Sociological Theory. *Annual Review of Sociology* 17, pp. 155-179. DOI: 10.1146/annurev.so.17.080191.001103.

Kolata, A. (1990): The Urban Concept of Chan Chan. In Moseley, M. and Collins, A.C. (eds.), *The Northern Dynasties: Kingship and statecraft in Chimor*. Washington: Dumbarton Oaks Research Library and Collection, pp. 107-145.

Manzanilla, L.R. (1996): Corporate Groups and Domestic Activities at Teotihuacán. *Latin American Antiquity* 7(3), pp. 228-246. DOI: 10.2307/971576.

Manzanilla, L.R. (2009): Corporate life in apartment and barrio compounds at Teotihuacan, Central Mexico: Craft specialization, hierarchy and ethnicity. In Manzanilla, L.R. and Chapdelaine, C. (eds.), *Domestic Life in Prehispanic Capitals. A Study of Specialization, Hierarchy and Ethnicity*. Ann Arbor: University of Michigan, pp. 21-42.

Mayet, F. and Silva, C.T. (2005): *Abul: fenícios e romanos no vale do Sado. Phéniciens et romains dans la vallée du Sado*, Setúbal.

Moore, J.D. (2021): *Ancient Andean Houses: Making, Inhabiting, Studying*. Gainesville: University Press of Florida.

Morris, C. and von Hagen, A. (2011): *The Incas. Lords of the Four Quarters*. London and New York: Thames and Hudson.

Murakami, T. (2019): Labor Mobilization and Cooperation for Urban Construction. Apartment Compounds at Teotihuacan. *Latin American Antiquity* 30(4), pp. 741-759. DOI: 10.1017/laq.2019.78.

Nash, D.J. (2009): Household Archaeology in the Andes. *Journal of Archaeological Research* 17(3), pp. 205-261. DOI: 10.1007/s10814-009-9029-7.

Netting, R., Wilk, R. and Arnould, E. (eds.) (1984): *Households. Comparative and Historical Studies of the Domestic Group.* Berkeley: University of California Press.

Nevett, L.C. (2023): *Ancient Greek Housing.* Cambridge: Cambridge University Press. DOI: 10.1017/9780511979262.

Pacifico, D. and Truex, L.A. (2019): Why Neighborhoods? The Neighborhood in Archaeological Theory and Practice. *Archaeological Papers of the American Anthropological Association* 30(1), pp. 5-19. DOI: 10.1111/apaa.12110.

Relaki, M. and Driessen, J. (2020): *Oikos. Archaeological Approaches to House Societies in the Bronze Age Aegean.* Louvain: Presses Universitaires de Louvain.

Rengifo, C. (ed.) (2020): *Chan Chan: esplendor y legado. Redescubriendo la antigua capital del Chimor.* Trujillo: Ministerio de Cultura.

Risch, R., Haak, W., Krause, J. and Meller, H. (2023): Kinship, sex, and biological relatedness. The contribution of archaeogenetics to the understanding of social and biological relations. In Meller, H., Krause, J., Haak, W. and Risch, R. (eds.), *Kinship, sex, and biological relatedness. The contribution of archaeogenetics to the understanding of social and biological relations.* Heidelberg: Propylaeum, pp. 9-25. DOI: 10.11588/propylaeum.1280.c18044.

Ruiz-Gálvez M. (2013): *Con el fenicio en los talones. Los inicios de la Edad del Hierro en la cuenca del Mediterráneo.* Barcelona: Bellaterra.

Ruiz-Gálvez, M. (2020): Colonias fenicias, casas y la 'casa' como institución. In Celestino, S. and Rodríguez, E. (eds.), *Un viaje entre el Oriente y el Occidente del Mediterráneo. IX Congreso Internacional de Estudios Fenicios y Púnicos*, vol. I. Mérida: Instituto de Arqueología de Mérida, pp. 461-470.

Sánchez Sánchez-Moreno, V., Galindo San José, L., Juzgado Navarro, M. and Dumas Peñuelas, M. (2012): El asentamiento fenicio de La Rebanadilla a finales del siglo IX AC. In García Alfonso, E. (ed.), *Diez años de arqueología fenicia en la provincia de Málaga (2001-2010).* Sevilla: Junta de Andalucía, pp. 67-86.

Schloen, J.D. (2001): *The House of the Father as Fact and Symbol. Patrimonialism in Ugarit and the Ancient Near East.* Leiden: Brill.

Smith, R. (ed.) (1984): *Land, Kinship and Life Cycle.* Cambridge: Cambridge University Press.

Smith, M.E. (2010): The archaeological study of neighborhoods and districts in ancient cities. *Journal of Anthropological Archae*ology 29(2), pp. 137-154. DOI: 10.1016/j.jaa.2010.01.001.

Smith, M.E., Chatterjee, A., Huster, A.C., Stewart, S. and Forest, M. (2019): Apartment compounds, households, and population in the ancient city of Teotihuacán, México. *Ancient Mesoamerica* 30(3), pp. 399-418. DOI: 10.1017/S0956536118000573.

Souvatzi, S. (2008): *A social archaeology of households in Neolithic Greece: An anthropological approach.* Cambridge: Cambridge University Press.

Souvatzi, S. (2017): Kinship and Social Archaeology. *Cross-Cultural Research* 51(2), pp. 172-195. DOI: 10.1177/1069397117691028.

Souvatzi, S., Bickle, P. and Cvecek, S. (eds.) (in press): *Prehistoric Kinship: Contemporary Perspectives in Archaeology and Bioarchaeology.* Cambridge: Cambridge University Press.

Stordeur, D. (2019): *Le village de Jerf el Ahmar (Syrie, 9500-8700 av. J.-C.). L'architecture, miroir d'une société néolithique complexe.* Paris: CNRS Éditions.

Susnow, M. and Goshen, N. (2021): House of a king, house of a god? Situating and distinguishing palaces and temples within the architectonic landscape of the Middle and Late Bronze Age southern Levant. *Levant* 53(1), pp. 69-91. DOI: 10.1080/00758914.202 1.1935097.

Ortega Ortega, J. (2024): Viviendas y parientes. Observando la sociedad andalusí desde abajo. In Bermejo Tirado, J. and Blanco González, A. (eds.), *Arqueología de los espacios domésticos en la península ibérica. De la Prehistoria reciente a la Edad Media.* Vitoria, Universidad del País Vasco, pp. 195-214.

Tuck, A. (2021): *Poggio Civitate (Murlo).* Austin: University of Texas Press.

Ur, J.A. (2014): Households and the Emergence of Cities in Ancient Mesopotamia. *Cambridge Archaeological Journal* 24(2), pp. 249-268. DOI: 10.1017/S095977431400047X.

Yasur-Landau, A., Cline, E.H., Koh, A.J., Ben-Shlomo, D., Marom, N., Ratzlaff, A. and Samet, I. (2015): Rethinking Canaanite Palaces? The Palatial Economy of Tel Kabri during the Middle Bronze Age. *Journal of Field Archaeology* 40(6), pp. 607-625. DOI: 10.1080/0093 4690.2015.1103628.

Spatial Patterns, Households and Kinship Practices in the Iron Age of Eastern Iberia (450-200 BCE)

Ignasi Grau Mira

University Institute for Research in Archaeology and
Historical Heritage (INAPH), University of Alicante, Spain,
ignacio.grau@ua.es

Abstract

The aim of this paper is to present a new proposal on the social structure of South-eastern Iberian communities during the Late Iron Age, between the fifth and third centuries BCE of the current Southern Valencian Country. Based on the analysis of settlement structure, build environment and the domestic spatial organisation in a series of urban centres, we propose and describe two different urban models. The first one is based on organic and strictly pre-planned layouts; the second one is based on more malleable and fluid spatial arrangements. These architectural patterns can be explained by a social organisation based on unilineal and bilateral kinship, respectively. The social structure and kinship practices are also related to more hierarchical or heterarchical political models among the Iberian groups of the region.

Keywords: *Iberian Iron Age, urban layout, household, social organization, kinship practices.*

10.1 Urbanization in Iberia: from general processes to local dynamics

Iberians is the term assigned by ancient Greco-Latin texts to peoples that inhabited the Mediterranean area of the Iberian Peninsula, and it is usually employed in academic literature to refer to a series of Iron Age societies in this region. These groups developed mainly between the sixth and the first centuries BCE in a context of economic growth marked by an increase in connectivity with the Mediterranean Basin. Iberian communities shared several linguistic and cultural traits but were organized into various small-scale territorial districts (Ruiz and Molinos 1998; Aranegui 2012). They were characterized by the development of social complexity, the emergence of urbanism, early states and economic sophistication with interregional exchange (Ruiz Rodríguez 2008; Sinner and Grau Mira in press). These dynamics are expressed by centralization and urbanization

In Blanco-González, A. and Alarcón-García, E. (eds.) 2025, *A Social Archaeology of Kinship in Iberia and Beyond. Recent Multistranded Approaches from aDNA to Household Archaeology.* Leiden: Sidestone Press, pp. 209-228.

processes, in which a series of medium-sized towns, generally between 1 and 10 ha, were remarkably important in shaping landscapes. These sites are locally named *oppida* –sing. *oppidum*, a Latin term adopted to describe these urban fortified nuclei– generally built on prominent locations, that facilitated visual control over the surrounding countryside and nearby subordinate rural settlements. *Oppida* had many quadrangular houses that were adjoined to form dense urban centres with straight streets. The main public buildings of these sites were defensive structures, as walls, towers and fortified gates. The development of these *oppida* resulted in the configuration of an Iberian fortified landscape typical of this culture and, to a certain extent, typical of other areas on the Peninsula (Armada and Grau Mira 2023).

In this general framework of urbanization, notable differences can be found in the layout of the settlements and the sizes of the urban centres. The processes of urbanization resulted in different nuclei: from large centres to multiple smaller towns related to different political structures, which range from hierarchical forms to heterarchical models. Hierarchical centralized structures are evident in Southern Iberia, where strong processes of nucleation in fortified urban centres, the ostentatious forms of elite representation such as prominent noble residences, and conspicuous displays of power in the funerary rituals are found (Ruiz Rodríguez 2008). This hierarchical society is identified as a class-based gentilitate organization, pivoting on agnatic segments akin to the Roman patrilineage (*gens*). According to this model, such a patriarchal and unilineal-descent system was led by a patron aristocracy that practiced softened forms of servitude, channelled by the institutions of the *fides* and the clientele (Ruiz Rodríguez and Molinos 1998).

However, in other regions, as Eastern Iberia, the settlement patterns, funerary practices and consumption behaviour involved more heterarchical structures, characterized by various social groups accessing different sources of economic, military, or religious power. These groups deployed different social strategies to gain and enlarge their power and command. The control over their peasant local communities most likely overlapped, balancing each other to some extent and they lack powerful and stable hierarchies and socio-political centralization so characteristic of other regions of Iberia (e.g., Vives-Ferrándiz 2013; Ruiz-Gálvez 2018; Grau Mira 2019: 345-346). Individualized forms of authority –typical of ranked patrilineages headed by paramount aristocrats– are elusive and the archaeological record points to the political power of decentralized elite groups scattered in the landscape (Bonet *et al.* 2015; Grau Mira 2019).

In sum, within the Iberian societies there were striking differences among social and political structures, and were probably related to diverse historical dynamics, economic strategies, and correspondingly assorted kinship practices (Blanco-González and Grau Mira in press). Focusing on this later feature, in the following pages I will review archaeological evidence that allows us to characterize kinship practices and socio-political entanglements by Iberian peoples in Eastern Iberia. This analysis focus on the built environment and urban planning as a way of approaching the existence of different social structures within different territories. The purpose of this work is precisely to point out the occurrence of urban schemes related to different social models that must have existed even in nearby and neighbouring areas of Iberia. I will analyse a group of Iberian urban centres that extend in the current territory of Southern Valencian Country (Fig. 10.1). The chronological framework covers between the fifth and fourth centuries BCE, during the time of formation and consolidation of the Iberian urban settlements in the region. That process occurred in

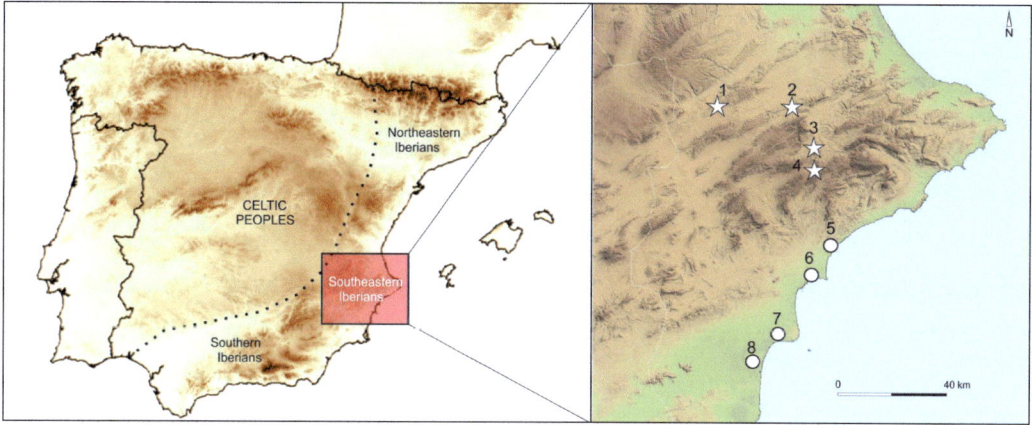

Figure 10.1. Map of the study area with the sites mentioned in the text: 1. La Bastida de les Alcusses; 2. La Covalta; 3. La Serreta; 4. El Puig d'Alcoi; 5. La Illeta dels Banyets de Campello; 6. El Tossal de les Basses; 7. La Picola; 8. El Oral.

a framework of intense relationships of the local population with Mediterranean peoples –Punic and Greek populations– possibly inhabiting the coast in mixed communities.

10.2 Social structures and built environment: Urban settlements with different principles of organization

To be able to research the different models of social organization in Iberian communities in our area of study, *Household Archaeology* is crucial. This approach centres on the detailed study of groups and domestic spaces and their interconnectedness, which lead to the configuration of settlements and communities. As we will see throughout these pages, this perspective will allow us to reach a better understanding of how different basic social units interact when building different corporate groups, and therefore define different relations regarding production, consumption, trade and reproduction (Souvatzi 2008; Vives-Ferrándiz 2013: 96).

Based on the archaeological record available, we can recognize constructive patterns that repeat; thus, we can better understand the social logic that modelled the urban layouts that have been documented. To this end, we centre our study around two different levels of observation. On one hand, the aggregation of domestic structures to build neighbourhoods, and on the other, the urban layout of the settlements. Spatial aggregations at a settlement level allow us to recognize habitation strategies and the formulization of operative schemes of different communities. We will follow by describing the two models that can be recognized in our study area. Further, we will explore the configuration of the houses in and of themselves, as residences for basic domestic groups that unit to create neighbourhoods. In the structuration of buildings, we can contrast a series of transcultural indicators that allow us to trace fluctuations in both domestic cycles and processes of familial relationships (Ensor 2013, 2021; Souvatzi 2008, 2017). To this end, we centre on the household, as the social cell based on cooperation practices that can be traceable throughout the archaeological record (Harris 1985; Souvatzi 2008). The different types of domestic units and their modifications can be interpreted in terms of familial relationship processes based on spatial and contextual information (Grau Mira and Vives-Ferrándiz 2018; Blanco-González and Grau Mira in press).

10.2.1 Organic model and planned layouts

The settlement plans in the coastal region of study are clear examples of organic and strictly pre-planned layouts. These are the settlements of La Picola, El Tossal de les Basses, Illeta del Campello or El Oral (Fig. 10.2). These settlements with organic layouts have originally planned configuration of walls, streets and houses that follow long rectilinear shapes. All the houses have common aligned facades and shared walls between houses, showing a simultaneous construction of several domestic units to conform the blocks. The dense occupation of space, without empty interstices, makes it difficult to expand the habitat. Only on very few occasions are enlargements of some houses identified from the occupation of open spaces, such as streets. The last common feature of these urban sites is the clear differentiation of homes in size and complexity, expressing marked differences in household composition. The most excavated and best-known site fitting this model is El Oral (Fig. 10.2B) dated to the fifth century BCE, whose

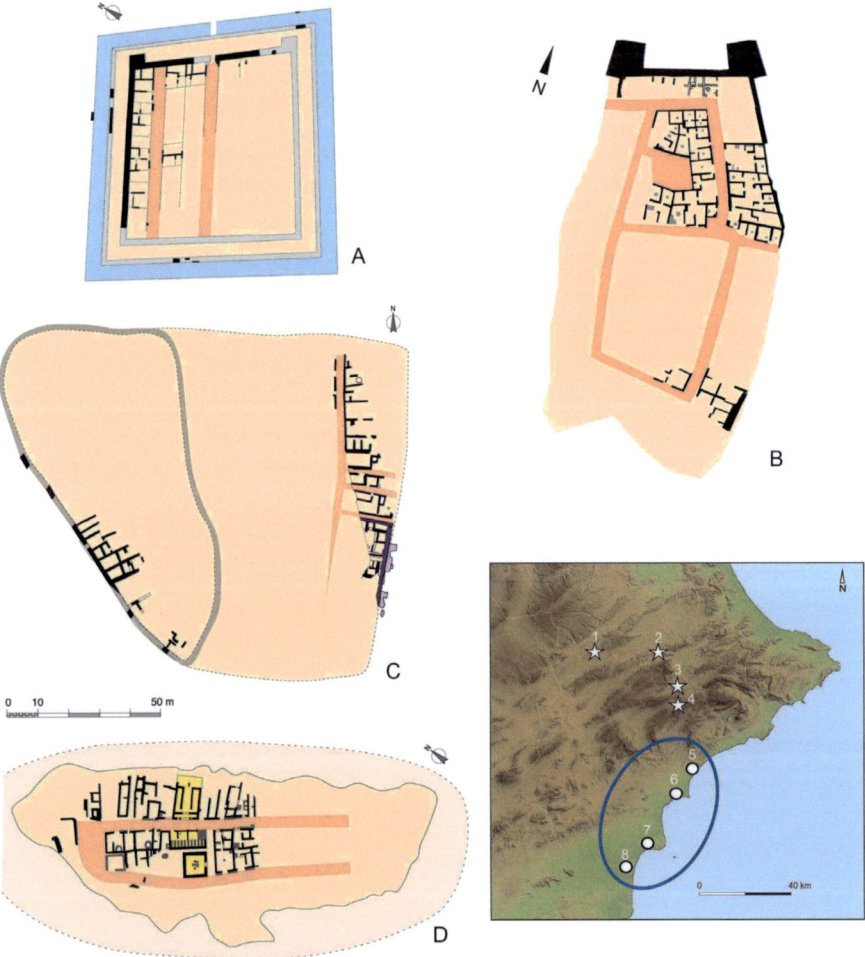

Figure 10.2. Plan of the sites with organic layout: A) La Picola; B) El Oral; C) El Tossal de les Basses; D) La Illeta dels Banyets de Campello.

plateau was densely occupied by multi-building and tightly packed blocks separated by orthogonal streets.

These pre-planned designs show a rigidity that may be related to clearly established intergenerational residence patterns and related to unilineal hierarchical descent structures. The interpretation of these spatial and architectural arrangements in terms of patrilocal and patrilineal practices is backed up by abundant ethnohistorical Mediterranean evidence tested with written sources from pre-classical times (Schloen 2001; González-Ruibal and Ruiz-Gálvez 2016; Huebner and Nathan 2017; Relaki and Driessen 2020). Such elongated quarters plausibly housed patrilocal descent groups of relatives or more loose patronymic associations of neighbouring households or quasi-kin clients. In any case, within these formal neighbourhoods, buildings were diverse in size, construction quality, domestic contents and even specialized spaces. This patterning goes hand in hand with a patriarchal and vertically stratified or pyramidal social organization, not necessarily divided into social classes (Schloen 2001). It also exhibits clear-cut social categories, and a relative invisibility of improvisation and deviant occurrences.

10.2.2 Amalgamation model and fluid spatial arrangements

By contrast, ways of nucleation in the inland region were based on more malleable and fluid spatial arrangements. Here, hilltop settlements as La Covalta, El Puig d'Alcoi, La Bastida de les Alcusses or La Serreta (Fig. 10.3) adhere to middle-density urbanism, characterized by miscellaneous dwelling compounds (Grau Mira 2013). In this model, only the construction of the walled precinct and fortification buildings are the evidence of common operations, but the construction of houses is due to individual initiatives following basic planning guidelines. None of the houses are attached to the fortified precinct and built at the same time as the wall, unlike the previous model that shows the simultaneous construction of the houses and fortifications. Open areas and straight streets are respected but the facades on common fronts of several houses are not aligned. As a result, the completed layout of these settlements in the result of sequential amalgamation of houses until space saturation.

The peculiar small size of houses in these hilly settings is worth noting, since most are between 10-35 m^2. This had to do with the partial allocation of social life and everyday chores in open-air exterior spaces, and the need to create smaller spaces easier to heat during the harsh winters (Grau Mira 2013). Several wards often juxtaposed freely in multi-room aggregates. These wide residential groups were composed of multiple kin-related conjugal dwellings, who often shared collective outbuildings: stockyards, ovens, storehouses, workshops, etc. All these residential clusters materialized diverse architectural designs which testify the processes of expansion/accretion and reduction/fission cycles of these residential groups.

In sum, settlement layouts consist of scattered and widely spaced random aggregates of buildings forming free-standing neighbourhoods separated by open spaces or winding lanes, and this scheme clearly contrasts with that observed in the planned model (Fig. 10.4). This is very much in keeping with cognatic principles, lacking any stringent and long-lasting social agreement, and responds to boom-and-bust cycling via construction/demolition dynamics. The common foundation of isolated neolocal dwellings, the wide variety of their designs, and their informal arrangements are all indicative of bilocality (Ensor 2013, 2021; Blanco-González and Grau Mira in press).

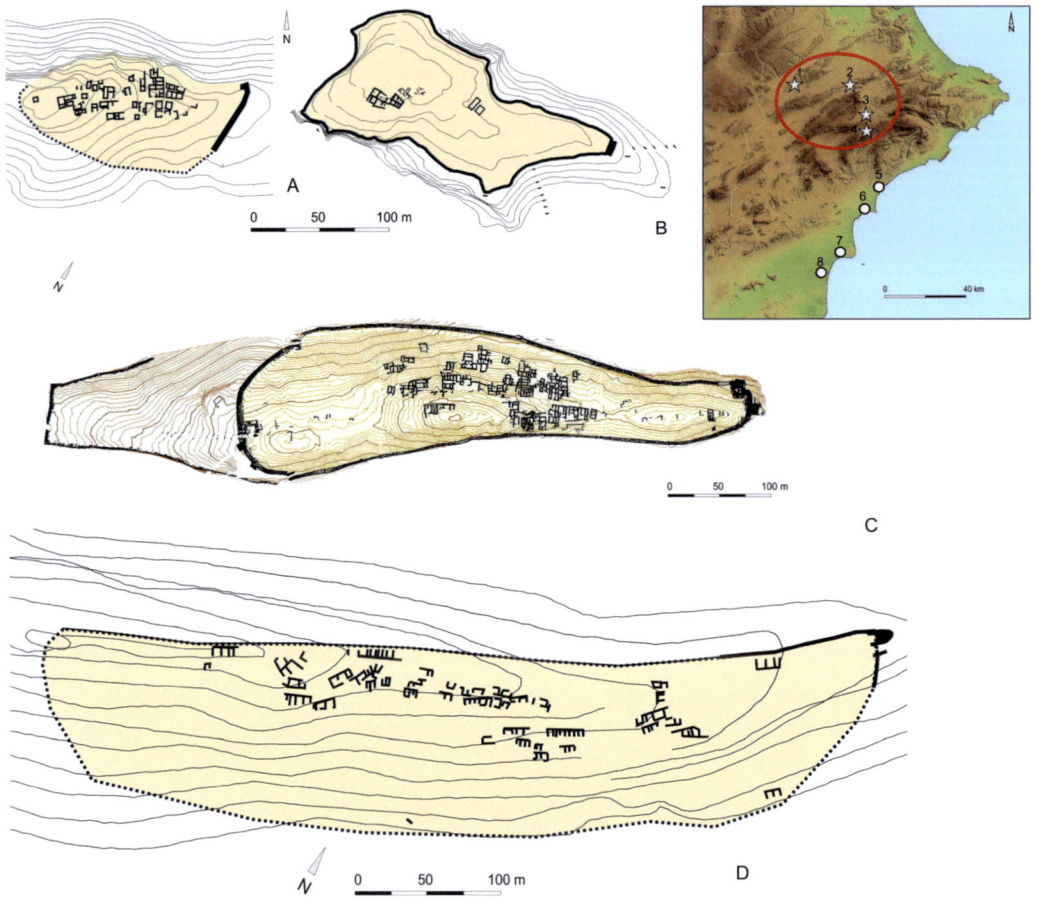

Figure 10.3. Plan of the sites with a flexible amalgamated layout: A) La Covalta; B) El Puig d'Alcoi; C) La Bastida de les Alcusses; D) La Serreta.

10.3 Diversity in household composition

When considering the *household* as the analytical basis, we can consider its size as an indicator of both the residential group, and post-marital residential practices. We can also observe the architectonic organization and aggregation patterns that are drawn by the social practices as indicators of the organization of descendants (Ensor 2013, 2021; Souvatzi 2017; Pacifico and Truex 2019). Regarding this matter, recently, Blanco-González (2024: 79) has synthesised the basic sociological variants associated to the configuration of households. Basically, we can establish two possibilities:

a. The single-family, nuclear or marital household, which is inhabited by two generations (the progenitors and their adopted or biological descendants). This is the most common configuration within the ethnohistorical record, as the most repeated stage throughout any domestic structure since it represents a young domestic unit (with immature descendants until their coming-of-age or they reach the age in which they can marry) as well as the final, senile, divorced or widowed stage. The bilateral nature would impose

Figure 10.4. Comparative plans at the same scale of the sites with organic and amalgamated layouts:
A) El Oral; B) La Covalta.

an early emancipation of male descendants with the objective of founding their own
household far from their natal one (neolocality).

b.   The grouping of marital units in a multiple, complex, extended or aggregated household
     that would conform multigenerational and multifamily corporate units (of up to three
     generations), based on the cohabitation of different marital cells that share sustenance,
     land ownership and everyday activities.

These two realities of simple and complex units can be found in different urban models. In
the following section we will describe these households and their functions based on some
examples form our Iberian study case.

10.3.1 Simple households

Households IIIC and IIID of El Oral (Fig. 10.5, 1) serve as good examples of basic households in
settlements that present organic and planed layouts, but these houses are also very frequent
in the amalgamated model. They are almost identical houses with a rectangular layout that are
divided into two spaces: one main space that contains the hearth and would be main area of
the household. The domestic assemblages, mostly pottery sets, within the area present varied
functions, while of other equipment endorse the multifunctional character of this space and
resting area. The second space is slightly smaller and is located to the back of the building,
acting as a storage unit (Abad and Sala 1993: 60). Both houses present similar dimensions,
household IIIC occupies 17.7 m², while IIID occupies 17.3 m² (Abad and Sala 1993: 165).

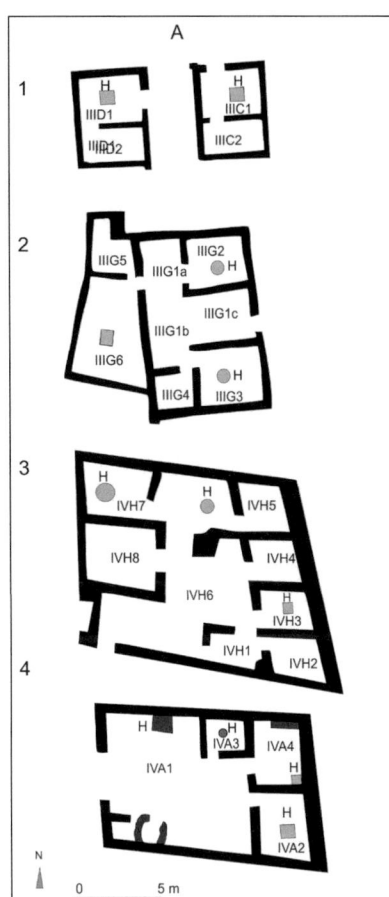
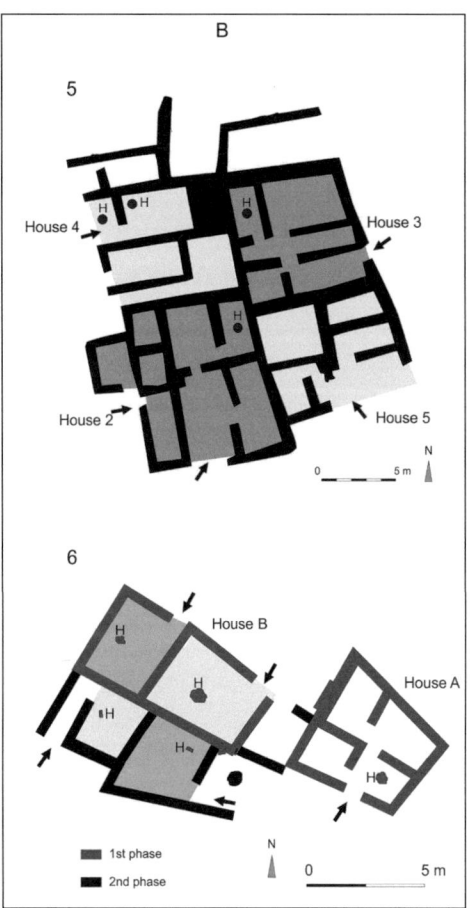

Figure 10.5. Plants of diverse houses of the organic (A) and amalgamated (B) layouts. 1, 2, 3 and 4 from El Oral, 5 from La Bastida de les Alcusses, and 6 from El Puig d'Alcoi.

The layout, disposition and size represent a module that repeats in other households documented at the site, such as household IIB, which is slightly larger with 21.10 m² (Abad and Sala 1993: 165). In all of them, the chamber that acts as the dwelling area is similar: IIIC1 having 10.85 m², IIID1 with 10.85 m² and IIB1 with 15.75 m² (Abad and Sala 1993: 165). The replication of this formal pattern and size reenforces the organic character of the constructions found throughout the *oppidum*, designating similar plots for these domestic units.

These households are dissociated and open to the exterior, without intermediate spaces between the street and the nucleus of the household with the hearth. The simplicity and reduced size are related –as previously mentioned– with marital households that would carry out most of their day-to-day activities within these spaces.

10.3.2 Complex households

Together with the double-spaced basic households we find other structures with similar constructive and formal characteristics, similar equipment and goods, but with a more complex design. One of these structures is household IIIG which initially was a household inscribed within a rectangle that repeats and duplicates the layout previously described, presenting two symmetrical simple houses that compose one large house. The design incorporates a semi-covered spaces in the shape of a 'T', that acts as a small redistribution area (IIIG1), that separates the exterior of the house from the interior (Abad *et al.* 2001: 159). The final configuration of the house presents three housing units: IIIG2 with 10.75 m^2, IIIG3 with 13.55 m^2 and the largest of them, IIIG6 with 24.65 m^2, which size can be explained due to an expansion towards an open space without the constrictions that the other units had.

Household IIIG presents an aggregated layout, with the nuclear hearth in the deeper parts of the house while the access would be materialized through a transitional space in the form of a vestibule. The existence of this space, located between the street and the living areas clearly indicates the intention of creating a separation between the exterior and the interior.

Other households with larger dimensions and more rooms present a more complex layout, but still follow the same basic modules of replicated households that are organized around a central patio. One of these houses is IVH (Sala and Abad 2006: 30-31). It is a large household occupying 209 m^2 inscribed within a trapezoid attached to the wall surrounding the settlement. It presents an elbowed entrance at the southeast side that gives access to a large patio (IVH6) that distributes circulation to different areas of the household. To the north of the patio there is room IVH7 with a large central hearth that evidences a space dedicated to the transformation a preparation of food. Next to this spaces there is another hearth in a non-individualized space of the patio. The main room of the complex can be found to the west of the central patio (room IVH8), which due to its architectonic finishes, can be identified as a noble house (Sala and Abad 2006: 31) and has a surface of 16.30 m^2. On the opposite side of the patio there are a series of rooms that form the east wing of the household. These are rooms IVH5 with 8 m^2, IVH4 with 7 m^2, IVH3 with 8 m^2 and IVH2 with 7 m^2, this last one with a vestibule that precedes room IVH1. These small spaces present pavements or plastered walls, as well as similar domestic materials, and have been interpreted as resting areas with a certain degree of care and presence (Sala and Abad 2006: 31).

Another house that presents the same layout is the named IVA (Sala and Abad 2006: 31-32). Again, it is a household with a large entrance that leads to a large patio (IVA1) that distributes the circulation and acts as the main area for carrying out different activities within the household, based on the equipment and ceramic repertoire. From this patio one can access three similar rooms: IVA2 with 11 m^2, IVA3 with 11 m^2 and IVA4 with 11.7 m^2. These rooms offer limited remains of activity, though hearths have been found in two of them (IVA2 and IVA3), with equipment and domestic features. These spaces have been interpreted as resting areas or meeting spaces (Sala and Abad 2006: 32).

These households, independently from an assessment regarding their status, are characterized by the coherent replication of identical rooms, many of them with hearths, found in larger architectonic units, which are related to multi-family compounds. This model is compatible with one of three different households that share the same homes, possibly virilocal homes, based on the total habitable surface, different from the large matrilocal households. In these houses, married sons would remain in the household of the father-patriarch, sharing equal rights of inheritance, either practicing strict patrilocality –nuclear integrated household development– or the virilocal cohabitation of male brothers and their families after the death of the patriarch (Blanco-González 2024: 79). In the case of IVH, the main room (IVH8) possibly belongs to the senior couple, who are accompanied by their descendants and their respective families.

As we can see, these households had liminal or transitional spaces in the form of patios or corridors. This space has a dual nature, since it generates a place of encounter between the exterior and the interior, but also has the potential of dividing spaces, activities and people. From the social point of view, the use of this liminal space has the function of avoiding non-desired intrusions in the main area of the household, as well as acting as an effective symbol that sends a message of division of outside and indoor people and activities. The existence of these transitional spaces evidence two facts that must have been interconnected. Firstly, the relations between members from within and outside the domestic unit must have been sufficiently frequent that it created the need to regulate them by means of architectonic devices. The second aspect, connected to the first, is that the type of relationships must have necessarily been submitted to specific rules that were channelled through the transitional space. Therefore, there was a spatial configuration that frames, distinguishes and articulates the circulation, categories of people, and activities in a pattern of controlled encounters (Grau Mira 2013). This transitional space between the exterior and interior, allowed for the division as well as integration of the different members of the domestic unit, and these members with regards to outside visitors. This space suggests the necessity and willpower to control the encounters and social dynamics that required situations of co-presence. When building the house with a liminal space, it generates a pattern of movement and occupation that clearly conditioned the encounters between members of the household and visitors: favouring them, controlling them or avoiding them. These extended families would choose a model of closed-of households in which the experimentation of space favoured the development of social distance between residents and the outside world. In other words, since the first moments of socialization during infancy, there was a clear difference between the space belonging to the family and the space for individuals outside the family. These social norms were acquired through a physical experience and interaction with the space (Bourdieu 1972) that facilitate the codification of social norms and creates a clear sense of the public-exposed social sphere and the privet-closed social sphere.

Contrary to this model, simple households are defined by the fact that their spaces were never integrated based on the presence of transitional spaces. This recurring pattern regarding the organization of spaces with few possibilities of segregation of space, and the dominance of exposed life, suggest an intense contact between different families and groups. The favouring of this interconnectedness must have enhanced the possibility of forming groups at a higher level than that of the conjugal family.

10.3.3 Amalgamated compounds

Complex compounds are typical of the amalgamated model of Iberian urbanism. Good examples are found at La Bastida de Les Alcusses (Moixent, Valencia) (Fig. 10.3C) a four-hectare hilltop *oppidum* occupied between the late fifth century BCE until the third quarter of the fourth century BCE when was suddenly destroyed by a violent attack. This *oppidum* has been extensively excavated since the early twentieth century (Díes *et al.* 1997; Bonet and Vives-Ferrándiz 2011; Vives-Ferrándiz 2013) providing a good plan to study these urban forms. In this site edifices arrange loosely and haphazardly in about 20 stand-alone clusters, separated by lanes practicable to oxen-drawn vehicles. This intra-site layout fits the cross-cultural pattern for bilateral kinship (Ensor 2013, 2021). The so-called complexes systematically produce evidence for at least some basic domestic activities: residence, storage of staple and culinary activities around the hearths. Nonetheless, they vary widely in size, architectonic sophistication, functions, and contents. A few cases display one or two quadrangular wards, totalling between 13-45 m^2 of living area, and can be confidently envisaged as the home of small conjugal units or –given the overall small living floors– semi-matrilocal households (e.g., complexes 13, 15, 16, 17, 18, 19 or 20). But most were larger multi-room blocks made of several conjugal dwellings that also hosted antechambers, stables and workshops, totalling 100 150 m^2 or 67 100 m^2 of living floor (e.g., complexes 2, 3, 4, 6, 9, 10 or 11).

Complex 2 (Fig. 10.5, 5) is an eloquent exemplar of the most successful households (Díes *et al.* 1997; Bonet and Vives-Ferrándiz 2011; Blanco-González and Grau Mira in press). It is an aggregate of four multi-roomed dwellings of diverse living sizes; every dwelling is identified by a series of wards sharing a hearth: house 2 with seven departments, including a stable, and 80 m^2 (120 m^2 in total); house 3 with five rooms and 56 m^2 out of 84 m^2; house 4, divided into four wards, including a stable and 72 m^2 (48 m^2 of living size); and house 5 with five wards and 47 m^2. Every dwelling hosted a conjugal unit or a semi-matrilocal group, and the four households were somehow related as kindred without losing their independence (Blanco-González and Grau Mira in press). In this regard, it is very telling that within bilateral and bilocal patterns "there is usually no focal point, and individual entrances face multiple directions" (Ensor 2013: 68), as in the case at hand.

Not every complex represented an autonomous production and consumption unit (Vives-Ferrándiz 2013). In this regard, agrarian iron tools (ploughs, sickles) and culinary transforming implements (quernstones, ovens) were unevenly distributed. Thus, not all compounds had a plough or had one or two at most yet complexes 7 and 10 had five each. As for querns, they were abundant within complex 2 but absent from nearby complex 3. More exclusive activities were undertaken by complex-family households living in multi-dwelling aggregates. Thus, weighing devices (balance plates and weights) were exclusive of complexes 3 and 11; textile activities (inferred from loom weights and scissors) were represented in complex 2 with six looms and complex 10 with three looms, whereas other neighbourhoods had none or two looms at most. This patterning suggests a ranked, non-equitable organization incompatible with a strict isonomic ethic (Blanco-González and Grau Mira in press). This seems akin to a horizontal distribution of wealth and social control, involving several nodes of social power and diverse sources for achieving it (Vives-Ferrándiz 2013).

Finally, it is very eloquent that some compounds hosted supra-household functions. Thus complex 7 was a large barn or silo likely shared by a handful of corporate groups. Likewise, complex 5, isolated on the highest hillock, was a monumental meeting hall featuring storerooms and an open-air courtyard where conspicuous consumption of luxury foodstuffs was assiduously overtaken. Moreover, social cohesion practices are also attested, such as the ritualized deposition in the rampart's principal door –a liminal spot of community-wide importance– of abundant burnt remains, including the wooden doors and five warrior panoply assemblages. These depositions have been interpreted as the burial of the door, accompanied by inalienable offerings of several high-ranking corporate factions (Bonet and Vives-Ferrándiz 2011; Vives-Ferrándiz 2013).

Another good example of this type of amalgamated compounds is provided by El Puig d'Alcoi (Alcoi, Alacant) (Fig. 10.3B), a hilltop settlement around 1.5 hectares in area, whose urbanism was loosely arranged. The site was occupied from the seventh century to the end of the fourth century BCE and its excavation has allowed to know the sequence of creation of the households and the urban layout (Grau Mira and Segura 2013). The initial phase of every such neighbourhood is represented by a neolocal dwelling occupied by the founding conjugal unit –typically a newly married couple with their unmarried offspring. Subsequently, all known domestic biographies entailed their undergoing demographic and landholding growth (Blanco-González and Grau Mira in press).

On the top of El Puig one such compound has been exhaustively studied: houses A and B (Fig. 10.5, 6). In an earlier phase (late fifth century BCE) house A was a small four-room building totalling 35 m^2 with internal partitions hosting a conjugal household. Its entrance led to a kitchen and storeroom and two inner and more private restrooms, which produced a remarkable ceramic set of imported Attic vessels. Meanwhile house B was made of two adjoined dwellings hosting independent conjugal units. This might be related to natolocality: two brothers that decided to remain in their bilocal natal group (Ensor 2021: 162-163). In a later phase (fourth century BCE), house B attracted more inhabitants, and the construction expanded southward with two additional dwellings and two exterior outhouses: a cellar or pantry, and a semi-roofed space for grinding grain and culinary activities. During this period, the haphazardly arranged aggregate matches the cross-cultural pattern indicative of bilateral descent, lacking any unitary focal point, with its entrances facing two opposite directions as if stressing their independence. By then the compound was likely inhabited by five conjugal families forming an extended household: house A as an enduring conjugal unit, and house B hosting a semi-matrilocal group (ca. 44 m^2), likely inhabited by sisters that remained in their bilocal group and shared utilities (Blanco-González and Grau Mira in press).

After collating architecture and domestic equipment, it is apparent that occupants of house A were richer or higher-ranked than those of house B, perhaps dependent clientele-like kindred. The occurrence of intra-household subgroupings with varying residence strategies strongly suggests bilocality (Ensor 2021). Finally, it is worth mentioning the ritualized subsoil deposit of three ovicaprids and a human neonate under house B, carried out as a founding ritual by the corporate group reinforcing the links of the group (Blanco-González and Grau Mira in press).

The formation of aggregated households documented at El Puig could have been due to the rapid growth of the domestic group which lead to the expansion from two conjugal families, at the end of the fifth century BCE, to four units during the beginning of the fourth century BCE (Grau Mira and Segura 2013: 177-190). This duplication of families could have occurred due to a natural demographic growth, with the consequent increase of population. In that case, all descendants of the original family would occupy newly built spaces, indicating a pattern of bilateral descendance.

Other examples that can attest to this rapid process of aggregation lead us to scenarios in which bonds are established through bilateral descendance. For example, at La Covalta we find processes of adjacency which, from a structural point of view, indicate a rapid growth of the inhabited zone. New houses are physically linked to one another, making use of previously constructed structures, therefore deducing a strong link between households. If it were the opposite, a block of independent houses with independent walls that isolate each domestic unit would have appeared, as can be seen in other areas of the site. Once again, this layout was a result of a rapid demographic growth and multiplication of the original families in a short timeframe of the second half of the fourth century BCE. Nonetheless, the rapidness of these processes of expansion can also respond to social processes of aggregation, with families connecting together due to political affiliations, as clientele links. In this sense, amalgamation of households could be justified based on social relations, work cooperation, the shared use of means of production, etc.

It seems clear that during this period, Iberian society was characterized by a complex form of interaction between different types of households. It is possible that these transformations in which the growth of some domestic units have been documented, could encompass large biological groups, as well as incorporate other affiliate groups with no blood ties. In this sense, there would be a reconfiguration of social groups from the base that would lead to the creation of the lineage structure, faction or kindreds. Therefore, we can recognize social processes based on the transformation of households through the expansion or decrease according to spatial reconfiguration. Functional diversity and accumulation of wealth seems to be reflected in the blocks within the settlement, reinforced by a sense of supra-familial groupings (Bonet and Vives-Ferrándiz 2011: 253). In any case, we place agency at the level of the wider domestic group and recognize the importance of amalgamation of basic households that are connected in a complex network of economic and social relations. The analysis of these social and spatial dynamics is key to understand the formation of Iberian groups, communities and landscapes.

10.4 Discussion: Spatial patterns and kinship practices

The study of Iberian Iron Age settlements and the different proposals of social organization have had great importance regarding the conceptualization of the *oppidum* as a defining and particular model of the Iberian urbanization process. Nonetheless, this important milestone has consequently led to a generalizing effect of the different urban realities, emphasising similarities among *oppida* while lacking to explaining the important differences documented between them. In other words, different urban models have been compared based merely on the differences in size and the processes of political centralization, emphasizing different manners of power structuration (Ruiz Rodríguez 2008: 814-822; Bonet *et al.* 2015). In this paper, our objective was to complement these perspectives by focusing on social organization and how urban layouts and domestic units are structured, while relating them to social aspects such as kinship.

In premodern societies, kinship –real or metaphorical– structures social relations, as well as indicates a sense of belonging to different corporative groups, while presenting the motive for the organization of production and redistribution models (Peregrine 2001: 44; Souvatzi 2008; Ensor 2013b: 56-120). Under these assumptions, generally, societies whose descendance is unilinear and based on patrilineage or matrilineage, present more rigid structures, since belonging to a line of descendants is non-negotiable and is acquired at birth (Ensor 2013a). Theses unilineal schemes are well represented in the organic urban structures and the complex household models documented at El Oral, in the coastal region of our area of study. We have interpreted these houses as virilocal households articulated around a patio or distribution corridor. The existence of houses with larger or smaller spaces that were more or less well constructed would indicate the social status of the household which, in spite of this, would share the same organizational scheme independent of their social level.

These structures are common among Iron Age Iberian groups with organic urban schemes which, on the other hand, are more frequent in the larger region of Mediterranean Iberia. Coetaneous examples to those alluded to in the eastern area can also be found at other sites such as Puente Tablas (Jaén) among Southern Iberians, or El Castellet de Banyoles (Tarragona) and Ullastret (Girona), among the Northern Iberians, just to quote few examples.

In contrast to this rigid scheme, we also find bilateral descendance where members can negotiate whether they belong to one group or another or change their membership throughout time. They can also belong to one or more groups simultaneously and therefore have the right to use both lines of ascendants to transmit rights and goods, which lead to much more flexible structures. Furthermore, both spouses can belong to the same corporative group that compete with the rest through the demonstration of wealth and power through ceremonial dances and festivities (Ensor 2013a: 152; Ensor 2013b: 32-33). These kinship structures originate malleable and flexible urban models, configurated through the amalgamation of different units building architectonic complexes. It is possible to associate this type of model to bilateral descendance and a malleable political structure characteristic of a heterarchical organization, characterized by the competition between different power groups. The examples presented, belonging to the settlements of El Puig and La Bastida de les Alcusses, are located in the inland area of the study region, and together with other well-known examples such as La Covalta or La Serreta, present a recurring pattern in the region.

The differentiating aspects that we can find between the diverse models of spatial structuration also remit to different socioeconomic models. Scholars defend that, in traditional societies, kinship is the means that motivate the organization of different modes of production and distribution, as well as social reproduction, through the creation of alliances, marriage strategies or ceremonies. In other words, far from being an autonomous structure, kinship coordinates social and economic relations, fomenting societal cohesion (Ensor 2013b; Souvatzi 2017).

The flexibility of power structures that is present in heterarchical models is expressed in different power groups that compete within the frame of a decentralized economies in which existed different strategies of production and distribution operating simultaneously that were controlled by different groups. The spatial expression of this model would be the existence of different aggregations in which diverse productive strategies are present. In the well-known case of La Bastida there is a diversity in the function and control of certain resources on behalf of the different households found throughout the city. The detailed study regarding the sources of economic power carried out by Vives-Ferrándiz have highlighted the control of different resources, from agricultural products to the transformation of metal, which are then distributed and complement each other within the urban community (Vives-Ferrándiz 2013). We therefore can identify a structure based on socio-spatial economic and power networks that are superimposed and intersect among each other (Mann 1991: 14), and it is around these networks that society is structured.

In the case of more hierarchical settlements, as is the case of El Oral, studied in this paper, it is probable that power would be held by only a few members, who would accumulate the forms of economic and political power. In this coastal settlement, research has highlighted the importance of superregional trade, especially with the Punic sphere to the South and the Greek areas of influence to the North (Abad *et al.* 2003). The administration of trade networks depends on the control of interpersonal relationships within a framework characterized by the absence of institutions such as markets, which opens the door for trade to be controlled by a few families that have and maintain these networks. This economic structure could explain the capacity of some lineages to develop more hierarchical systems, maintained through a restrictive unilineal lineage. In any case, based on the intensity of each source of power and control that any lineage has, social structures develop differently and acquire varied forms. Therefore, the control of different sources of power would allow for stronger social control. This would be the case of more hierarchical structures, where the elite control these sources, in contrast to the diversification of power sources among different groups form the inlands which present a more heterarchical nature. Finally, we can observe that different models of sociopolitical organization are dependent on the generation of resources, their exploitation, production, transformation and distribution (Mann 1991: 45). The mode of production and distribution would therefore condition social, political and symbolic processes (Harris 1985: 13-71).

10.5 Final remarks

For different reasons, fundamentally due to the increment in production and the development of interregional trade networks within an expansive context, protohistoric societies of Southeastern Iberia would begin to concentrate in aggregated nucleuses located in strategic positions of the region. The sum of social units offers new stimuli for the creation and development of organizational structures through the aggregation of information, capital, people and resources, consequently contributing to the formation of new forms of community and the development of complexity. This dynamic began during the Late Bronze Age, supported by trade relations between different regions of the Peninsula and the increase in contacts with Phoenician traders from the end of the eighth century BCE, which would lead to the crystallization of the Iberian society between the fifth and fourth centuries BCE.

It is inevitable to think in the different manners in which these sites were structured and planed, especially when comparing them to other nearby and coetaneous settlements with cultural and morphological differences. In this sense, the perspective of household archaeology and kinship offer heuristic tools that allow for the identification of differences among social structures and evaluate them. The characteristics defined throughout this paper would justify certain formulas of differential social organization, based on spatial and functional differences among neighbourhoods, and the flexibility or rigidity regarding the aggregation of households. As opposed to rigid structures, we found flexibility; in contrast to compact aggregations of households, we found separate compounds in different aggregations. The amalgamated urban model, that we have proposed here, must be related to the establishment of social units that would be a reflection of a diverse social network in which economic and political power is structured in different sources and is in continuous negotiation. This would make it possible for the existence of patterns of neolocality based on the development of new households that would opt to aggregate themselves to one of the multiple compounds, indistinctly opting for either a mother or father lineages, showing bilateral kinship. On the other hand, the organic model with a more rigid structure would be regimented by unilineal lineages which would reenforce hierarchical structures.

Similarly, heterarchical and flexible organization with a bilateral pattern of descendance would be evidenced by burial spaces. Though it is not our objective in this paper to analyse these contexts, we want to highlight that the structure of necropolises contemporaneous to the settlements mentioned in this text are characterized by a polycentric organization, with different tombs that are organized in different spaces, like what has been observed at the settlements. In all of them tumular structures of larger or smaller dimensions have been identified and are the focal points that organize the funerary space, acting as the central point where other family and affiliated members are buried. This would connect with the pattern of bilateral descendance in which both woman and men are transmitters of rights and goods, and both develop symbolic discourses of power (Grau Mira and Comino 2021).

Associated to hierarchical urban structures we can similarly identify accentuated forms of ideological representation and funerary ostentation expressed thorough funerary sculptures. Certainly, in the coastal area of our study region one of the largest concentrations of Iberian sculptures can be found, concentrated around the important area of L'Alcúdia d'Elx (Chapa 1985: 163-164; León 1998: 37-38). Some important examples would be La Dama d'Elx or the warrior sculpture named the *Cardiophylax* of l'Alcúdia, to which we must add dozens of examples of animal sculptures and, to a lesser degree, human ones. Traditionally, these archaeological elements have been analysed as an artistic and cultural expression, characterized by the development of an original Iberian art with important Hellenistic influences, framed by an intense process of cultural interaction. But the aspect that is now being highlighted is that they were an ostentatious representation of power associated to enclaves characterised by hierarchical models of organization, as is proven by the specific geographic distribution of these elements.

Finally, we would like to highlight that having chosen a comparative framework we have emphasized the differences between two organizational models. Nevertheless, when trying to characterize the forms of social organization we must consider that these structures would be placed upon a variable gradient that oscillates between hierarchization –or vertical structures of power– and corporativism –or horizontal strategies. Therefore, we can identify different forms of both heterarchy and hierarchy among the different groups and communities that have been studied, without restricting the interpretations to a basic polarity. Future studies must take into account this variability and characterize the different social strategies that accompany kinship in the creation of social structures throughout the Iberian Iron Age.

Acknowledgements

I would like to thank the editors of this book, Eva Alarcón and Antonio Blanco, for the opportunity to present my ideas in this collective initiative. This work was supported by the Conselleria d'Innovació, Universitats, Ciència i Societat Digital -GVA under grant CIAICO/2022/049 RURITANIA.

References

Abad, L. and Sala, F. (1993): *El poblado ibérico de El Oral (San Fulgencio, Alicante)*. Valencia: Diputación de Valencia.

Abad, L., Sala, F., Grau, I., Moratalla, J., Pastor, A. and Tendero, M. (eds.) (2001): *Poblamiento ibérico en el Bajo Segura: El Oral (II) y La Escuera*. Madrid: Real Academia de la Historia.

Abad, L., Grau, I., Sala, F. and Moratalla, J. (2003): Ancient trade in south-eastern Iberia: the lower Segura River as focus of exchange activities. *Ancient West and East* 2(2), pp. 265-282. DOI:10.1163/9789004495432_006.

Aranegui, C. (2012): *Los iberos ayer y hoy. Arqueologías y culturas.* Madrid: Marcial Pons.

Armada Pita, X.L. and Grau Mira, I. (2023): The Iberian Peninsula. In Haselgrove, C., Wells, P.S. and Rebay-Salisbury, K. (eds.), *Oxford Handbook of European Iron Age.* Oxford: Oxford University Press, pp. 305-344.

Belarte, M.C. (2008): Domestic Architecture and Social Differences in North-Eastern Iberia during the Iron Age (*c.* 525-200 BC). *Oxford Journal of Archaeology* 27(2), pp. 175-199. DOI:10.1111/j.1468-0092.2008.00303.x.

Belarte, M.C., Bonet, H. and Sala, F. (2009): L'espai domèstic i l'organització de la societat ibèrica: els territoris de la franja mediterrània. In Belarte, M.C. (ed.), *L'espai domèstic i l'organització de la societat a la protohistòria de la Mediterrània occidental (Ier mil·lenni).* Barcelona: Universitat de Barcelona and Institut Catalá d'Arqueologia Clàssica, pp. 93-124.

Blanco-González, A. (2024): Arqueología del hogar extenso virilocal en el suroeste peninsular (siglos VII-V a.C.). In Bermejo Tirado, J. and Blanco González, A. (eds.), *Arqueología de los espacios domésticos en la península ibérica. De la Prehistoria reciente a la Edad Media.* Vitoria: Universidad del País Vasco, pp. 75-96.

Blanco-González, A. and Grau Mira, I. (in press): Kinship Practices and House Societies: Socio-political Entanglements in the Iberian Iron Age (450-300 BC). In Souvatzi, S., Bickle, P. and Cvecek, S. (eds.), *Prehistoric Kinship: Contemporary Perspectives in Archaeology and Bioarchaeology.* Cambridge: Cambridge University Press.

Bonet, H. and Vives-Ferrándiz, J. (2011): *La Bastida de les Alcusses. 1928-2010.* Valencia: Diputación Provincial de Valencia.

Bonet Rosado, H., Grau Mira, I. and Vives-Ferrándiz Sánchez, J. (2016): Estructura social y poder en las comunidades ibéricas de la franja central mediterránea. In Belarte, M.C., García, D. and Sanmartí, J. (eds.), *Les estructures socials protohistòriques a la Gàl.lia i a Ibèria. Homenatge a Aurora Martín i Enriqueta Pons. VII Reunió Internacional d'Arqueologia de Calafell (Calafell, del 7 al 9 de març de 2013).* Tarragona: Institut Catalá d'Arqueologia Clàssica, pp. 251-272.

Bourdieu, P. (1972): *Esquisse d'une théorie de la pratique.* Genève: Droz.

Chapa Brunet, T. (1985): *La escultura ibérica zoomorfa.* Madrid: Ministerio de Cultura.

Díes, E., Bonet, H., Álvarez, N. and Pérez Jordà, G. (1997): La Bastida de les Alcuses (Moixent): resultados de los trabajos de excavación y restauración. Años 1990-1995. *Archivo de Prehistoria Levantina* 22, pp. 215-295.

Ensor, B.E. (2013a): *The Archaeology of Kinship: Advancing Interpretation and Contributions to Theory.* Arizona: University of Arizona Press.

Ensor, B.E. (2013b): *Crafting Prehispanic Maya Kinship.* Tuscaloosa: The University of Alabama Press.

Ensor, B.E. (2021): *The Not Very Patrilocal European Neolithic. Strontium, aDNA, and Archaeological Kinship Analyses.* Oxford: Archaeopress Archaeology.

Glowacki, K.T. and Klein, N.L. (2011): The analysis of 'Dark Age' domestic architecture: The LM IIIC settlement at Kavousi Vronda. In Mazarakis Ainian, A. (ed.), *The 'Dark Ages' Revisited: Acta of an International Symposium in Memory of William D.E. Coulson.* Volos, pp. 407-418.

González-Ruibal, A. and Ruiz-Gálvez, M. (2016): House Societies in the Ancient Mediterranean (2000–500 BC). *Journal of World Prehistory* 29(2), pp. 383-437. DOI: 10.1007/s10963-016-9098-8.

Grau Mira, I. and Segura, J.M. (2013): *El* oppidum *ibérico de El Puig d'Alcoi. Asentamiento y paisaje en las montañas de la Contestania.* Alcoi: Ajuntament d'Alcoi.

Grau Mira, I. (2013): Unidad doméstica, linaje y comunidad: estructura social y su espacio en el mundo ibérico (ss. VI-I a.C.). In Gutiérrez Lloret, S. and Grau Mira, I. (eds.), *De la estructura doméstica al espacio social. Lecturas arqueológicas del uso social del espacio.* Alacant: Universitat d'Alacant, pp. 57-76.

Grau Mira, I. (2019): Social dynamics in eastern Iberia Iron Age: Between inclusive and exclusionary strategies. In Currás, B.X. and Sastre, I. (eds.), *Alternative Iron Ages. Social Theory from Archaeological Analysis.* London: Routledge, pp. 337-358. DOI: 10.4324/9781351012119-18.

Grau Mira, I. and Comino Comino, A. (2021): Mujeres en los modelos sociales y las estructuras de poder del sureste de iberia (siglos V-IV a.n.e.): una lectura desde los espacios funerarios. *Trabajos de Prehistoria* 78(2), pp. 309-324. DOI: 10.3989/tp.2021.12278.

Harris, M. (1985): *El materialismo cultural.* Madrid: Alianza Editorial.

Huebner, S.R. and Nathan, G. (eds.) (2017): *Mediterranean Families in Antiquity: Households, Extended Families, and Domestic Space.* Oxford: Wiley-Blackwell.

León, P. (1998): *La sculpture des Ibères.* París: Ed. Harmattan.

Llobregat Conesa, E.A., Cortell Pérez, E., Juan Moltó, J. and Segura Martí, J.M. (1992): El urbanismo ibérico en La Serreta, *Recerques del Museu d'Alcoi* 1, pp. 37-70.

Mann, M. (1991): *Las fuentes del poder social, II.* Madrid: Alianza Editorial.

Pacifico, D. and Truex, L.A. (2019): Why Neighborhoods? The Neighborhood in Archaeological Theory and Practice. *Archaeological Papers of the American Anthropological Association* 30(1), pp. 5-19. DOI: 10.1111/apaa.12110.

Peregrine, P. (2001): Matrilocality, corporate strategy, and the organization of production in the Chacoan World. *American Antiquity* 66(1), pp. 36-46.

Relaki, M. and Driessen, J. (2020): *Oikos. Archaeological Approaches to House Societies in the Bronze Age Aegean.* Louvain: Presses Universitaires de Louvain.

Ruiz-Gálvez Priego, M. (2018): ¿Sociedad de clase o... Sociedad de Casa? Reflexiones sobre la estructura social de los pueblos de la Edad del Hierro en la Península Ibérica. In Rodríguez Díaz, A., Pavón Soldevila, I. and Duque Espino, D.M. (eds.), *Más allá de las casas. Familias, linajes y comunidades en la protohistoria peninsular.* Cáceres: Universidad de Extremadura, pp. 13-40.

Ruiz Rodríguez, A. (2008): Iberos. In Gracia, F. (ed.), *De Iberia a Hispania.* Barcelona: Ariel, pp. 734-844.

Ruiz Rodríguez, A. and Molinos, M. (1998): *The Archaeology of the Iberians.* Cambridge: Cambridge University Press.

Sala, F. and Abad, L. (2006): Arquitectura monumental y arquitectura doméstica en la Contestania. *Lucentum,* 25, pp. 23-46.

Schloen, J.D. (2001): *The House of the Father as Fact and Symbol. Patrimonialism in Ugarit and the Ancient Near East.* Leiden: Brill.

Sinner, A.G, and Grau Mira, I. (in press): The Iberians: A Mosaic of Mediterranean Peoples. In Seth, B. and Murray, S. (eds.), *Oxford Handbook of Mediterranean Iron Age,* Oxford: Oxford University Press.

Souvatzi, S. (2008): *A Social Archaeology of Households in Neolithic Greece. An Anthropological Approach*. Cambridge: Cambridge University Press.

Souvatzi, S. (2017): Kinship and Social Archaeology. *Cross-Cultural Research* 51(2), pp. 172-195. DOI: 10.1177/1069397117691028.

Vives-Ferrándiz, J. (2013): Del espacio doméstico a la estructura social en un *oppidum* ibérico. Reflexiones a partir de la Bastida de Les Alcusses. In Gutiérrez Lloret, S. and Grau Mira, I. (eds.), *De la estructura doméstica al espacio social. Lecturas arqueológicas del uso social del espacio.* Alacant: Universitat d'Alacant, pp. 95-110.

Against Citizenship: Kinship and Heterarchy in Late Roman Celtiberia (Spain)

Jesús Bermejo Tirado

Department of Humanities, History, Geography
and Art, Carlos III University of Madrid, Spain,
jbtirado@hum.uc3m.es

Abstract

This chapter challenges traditional perspectives on Late Roman Celtiberia by applying a kinship focused approach to archaeological evidence from sites such as the *villa* of Dehesa de Cuevas de Soria (Spain). We argue that following the decline of classical Roman municipal structures in the third century CE, a heterarchical society emerged where kinship groups became central to political and social organization. Our analysis focuses on bilateral kinship, the symbolic significance of houses and heraldic artefacts. We aim to demonstrate how kinship-based groups, such as the Irrics, filled the void left by declining municipal models and citizenship status. Spatial configurations of rural estates (*villae*), votive inscriptions, and the use of anagrams and *signacula* support this interpretation. This approach moves beyond narratives of decline, revealing a complex process of social transformation and resilience. By employing anthropological models and microhistorical perspectives, we offer a nuanced understanding of Late Roman society in the Iberian Peninsula, with implications for studying other areas of the late antique world.

Keywords: *Late Roman Celtiberia, heterarchy, household archaeology, kinship structures, Citizenship.*

11.1 Introduction

The study of social and political structures in Late Antiquity has long been dominated by traditional historiographical perspectives, often framed through the lens of decline and crisis (e.g., Heather 2005; Ward-Perkins 2005). However, recent archaeological discoveries and theoretical advancements offer opportunities to reassess these interpretations, particularly in peripheral regions of the Roman Empire. This paper explores the social and political landscape of Late Roman Celtiberia in Central Iberia (Spain), proposing a new

In Blanco-González, A. and Alarcón-García, E. (eds) 2025, *A Social Archaeology of Kinship in Iberia and Beyond. Recent Multistranded Approaches from aDNA to Household Archaeology.* Leiden: Sidestone Press, pp. 229-248.

framework for understanding the complex dynamics at play during this transformative period based on the analysis of kinship from a cross-cultural perspective (Godelier 2011; Ensor 2013a; Sahlins 2013). An important point to our analysis is the application of some concepts taken from Claude Lévi-Strauss's House Society model (Lévi-Strauss 1987; Carten and Hugh-Jones 1995; González-Ruibal 2006). Originally developed with a strong influence from his contemporary conceptions on medieval feudal France (Cohen 2004), this conceptual framework has proven remarkably versatile in recent archaeological studies. We argue that its application to Late Roman contexts can provide fresh insights and challenge long-held assumptions regarding social organization during this period.

By re-examining archaeological evidence from key sites from Late Roman Celtiberia, we aim to demonstrate how the decline of classical Roman state structures in the third century CE led to the emergence of a heterarchical society. In this new social landscape, Houses –understood as both physical structures and social units– became central to political and social organization. This perspective offers a more nuanced understanding of Late Roman society, moving beyond simplistic narratives of decline to reveal a complex world of adaptation and resilience. Our analysis will focus on several key aspects of the House Society model, including the importance of bilateral kinship, the symbolic and material significance of the house itself, and the presence of heraldic artefacts. By applying these concepts to the archaeological record of Late Roman Celtiberia, we hope to shed new light on the social dynamics of this period and region. This approach not only challenges traditional historiographical perspectives but also demonstrates the potential for anthropological models to enrich our understanding of historical periods, providing a valuable framework for re-examining the Late Roman world in broad terms.

11.1.1 Kin-based societies and states

At the risk of oversimplification and reiterating some conceptual discussions, we can state that many prehistoric archaeologists have traditionally been interested in analyzing kinship structures as a key feature in understanding the social organization of past human groups. By "kinship structures" we refer to the organizational patterns of kinship relationships within a cultural group, encompassing descent systems, postmarital residence arrangements, and all familial ties established within a cultural framework, both in cognatic and agnatic terms. Indeed, numerous publications have used the term "kin-based societies" to refer to the social and political structures of many different prehistoric communities around the world. In many cases, the concept of a kin-based society has been used in opposition to the state –and its political, ideological, and legal apparatuses– which typically characterizes many cultural formations in historical periods. From an evolutionary perspective which often influences our thinking even when we try to abstract ourselves from it, these two concepts have been used as opposite poles within the same process of social evolution (Flannery 1972; reviewed and nuanced in 2002).

For many reasons, I oppose this evolutionary conception of history. The importance of kinship structures must also be considered fundamental for understanding historical societies, even within the framework of post-industrial societies of liquid capitalism, and, of course, their importance must be noted in the social archaeology of historical communities. From my perspective as a historical archaeologist, a dialectical perspective that situates both concepts within the same fluctuating axis is much more constructive than a dichotomous evolutionary stance. Attempting to reduce history to a classification based on abstract categories is an exercise in reductionism (Wolf 1982: 6). It is futile to analyze any historical period by rigidly fitting it into one of these two generic conceptual axes: kin-based society or state (in line with these criticisms see Yoffee 2005; Pauketat 2007; Ensor 2013a, 2013b; Feinman 2023). More than as mutually exclusive possibilities, in this work we propose that both concepts –kinship and state– can be considered as two scales of political analysis that operate with different intensities within various prehistoric and historic societies. Despite being against this dichotomous or evolutionary view of both concepts, their application as a discursive axis can be very useful for guiding the social analysis of past communities. In this way, a discussion on which of these two scales –kinship and state– has greater importance in articulating the social relations of human groups can be introduced. This will allow to produce new interpretations regarding the evolution of the structures and political relationships that operated throughout any given historical period.

A key point in this matter is how we can distinguish which of these two axes has more prominence at each historical moment. Indeed, I believe that the differences between kinship society and the state become more evident when we consider the relationship between kinship relationships and citizenship. Citizenship is understood here as the political status leading to access to a set of political rights –especially those related to property and inheritance (see Earle 2017)– and duties by an individual based on belonging to a particular social community. Applying the concept of citizenship as a kind of social thermometer, if I were to define both terms as extremes of a dialectical axis, I would say that a kin-based society is one in which the social structure of citizenship is subordinate to kinship, while a state is one in which kinship is subordinate to citizenship. To illustrate this dialectic approach, we can use the historical example of the Roman state and its Mediterranean expansion policy. This conquest process entailed –among many other things– a permanent change in the way kinship was structured in various regions such as the Iberian Peninsula. Contrasting with an evolutionary conception that tries to understand this change process as part of a "civilizing" process whereby local pre-Roman communities obtained municipal status, my view is much more pessimistic. In my opinion, the relationship between the transmission of citizenship and the adoption of canonical kinship relations within Roman private law, at least in legal terms and social projection, opens a horizon of nonconformity with these family models. If a legally sanctioned marriage between two citizens was the legal requirement for the transmission of citizenship rights, those who did not or could not attain this privileged status of citizenship were exempt from the obligation to adopt these types of kinship (Wallace-Hadrill 1981). Roman funerary epigraphy has allowed us to document not only the marginal survival of certain pre-Roman kinship forms in various regions of the Iberian Peninsula (Edmondson 2005; Gorrochategui *et al.* 2011) but also the emergence of hybrid models that highlight the diversity of kinship structures documented in the communities that inhabited the Iberian Peninsula throughout Roman rule (Bermejo Tirado 2018). Even in the Early Imperial period –when various political and legal apparatuses established very detailed regulations on family relationships– various social and political factors exerted a great influence, allowing for the emergence of alternative kinship forms to those that the Roman rulers sought to impose.

Despite lacking a deep understanding of the social structure of kinship relations in pre-Roman communities, most data from the so-called Palaeohispanic epigraphy seem to confirm the importance of various supra-conjugal kinship groups (Ramírez Sánchez 2002, 2003, 2007). These corporate groups (Hayden and Cannon 1982; Schloen 2001) –which include different patrilocal/virilocal models of extended family (Huebner and Nathan 2017)–were predominant in articulating the social relations of these communities. The situation changed substantially with the Roman conquest. From this point on –as various studies have repeatedly shown– the model of the nuclear family established through a legally sanctioned marriage between two Roman citizens became predominantly represented in the extensive funerary epigraphy recorded in all the cities and territories of the Iberian Peninsula. This change in trend was proposed for the case of the Italian Peninsula in the seminal study published in the mid-1980s by Brent Shaw and Richard P. Saller. It has traditionally been interpreted as a reflection of a process of acculturation through which pre-Roman kinship traditions and structures were progressively replaced by those established by Roman private law (Shaw and Saller 1984). Conversely, this cultural reading of the changes in the kinship structure of the ancient communities of the Iberian Peninsula has been subject to various criticisms. Some of these criticisms focus on the celebratory nature of many funerary inscriptions in which these conjugal kinship relationships are recorded (Martin 1996; Curchin 2000; Huebner 2013: 31-57). These works argue that funerary epigraphy does not show so much the true structure of kinship relations of the Roman communities but rather indicates those close relatives who would take charge of covering the expenses of the funerary rituals.

Against this cultural view of the changes in kinship models in the Iberian Peninsula following the Roman conquest, I propose an interpretation of this process based on sociopolitical issues. The spread of the Roman nuclear family model as the predominant framework within the kinship structures of this period is mainly derived from legal issues rather than cultural or behavioral aspects. Returning to the discussion mentioned above, the key factor for understanding this change in kinship structures during the Roman period is related to the normative framework in which the reproduction of citizenship rights within the juridical-administrative structures of the Roman state occurs. It is well known that various Roman governments used access to citizenship as a political integration tool comparable in importance to conventional military campaigns (Lavan 2016, 2019a). However, many scholars of Roman funerary epigraphy have not sufficiently focused on the fact that, according to Roman private law –especially from the legislation linked to the social policy of the Augustan regime– one of the main requirements for the proper transmission of Roman citizenship rights was the sexual reproduction of heirs within the framework of a marriage legally sanctioned by the state itself. Those individuals who –regardless of their cultural background– attained the status of Roman or Latin *cives* had to adopt the kinship structure model determined by Roman private law, and more specifically, the marriage regulations imposed by laws such as the *Lex Iulia de maritandis ordinibus* (Treggiari 1996; Osgood 2019).

11.1.2 Kinship and heterarchy

Since the early years of the twenty-first century, within part of the archaeological literature, we have witnessed the decline of kinship analysis as a central element for understanding the social structures of past communities. In parallel, various archaeologists have openly criticized the use of the concept of kinship for the social analysis of human groups in various prehistoric contexts around the world (Gillespie and Joyce 2000). Many of these criticisms are based on a functionalist interpretation of some concepts developed in traditional ethnographic studies on kinship (Gillespie 2000). As an alternative to these paradigms, some critics (Gillespie 2000; González-Ruibal 2006: 144-146; Beck 2007a: 4-6) have advocated applying the House Society model (Carsten and Hugh-Jones 1995), based on the concept of *maison* coined by Claude Lévi-Strauss (1987: 152).

The implementation of this interpretive model in archaeology has been particularly successful, and in recent decades we have witnessed its utilization for the analysis of multiple contexts (e.g., Beck 2007b; Thomas 2015; Wiersma 2020; Moreno García 2022). One of the key factors in understanding the success of these approaches in recent studies lies in the significance of the House, both in its architectural and metaphorical sense, as a material correlate of the corporate group itself in archaeological contexts. Nevertheless, as Ensor (2013a: 10-27, Ensor this volume) has aptly pointed out, this emphasis on the material aspect –something logically valued by archaeologists– has served to eclipse or undervalue (e.g., Currás 2019: 21-23 and 61-73) the importance of other bonds and immaterial elements that are extremely significant for any society –and especially for prehistoric communities– such as kinship and familial relations.

Echoing many of the arguments put forth by Ensor, the House Society model is a form of kinship-based organization, specifically a bilateral and bilocal residential group, referred to as a *residential-household group* (Ensor 2013a: 23-45). In advocating for the importance of kinship structures in preindustrial societies, González-Ruibal and Ruiz-Gálvez (2016) have reviewed previous approaches to propose a new synthesis to solve this discussion. They argue that, indeed, kinship relations are not opposing and mutually exclusive models but rather complementary facets. As these authors themselves have emphasized, Lévi-Strauss referred to the complementary nature of both the material and kinship components of the House when formulating his definition of *maison*: "Moral person, keeper of a domain composed simultaneously of material and immaterial possessions, which perpetuates itself by the transmission of its name, its fortune, and its titles in a real or fictive line considered legitimate on the sole condition that this continuity can express itself in the language of kinship or alliance, and most often, both" (Lévi-Strauss 1987: 152). Based on this hybrid definition of *maison*, the House Society would function as a model of political structuring of communities articulated through kinship relations. Even though I agree with Ensor in considering that the House Society is fundamentally a type of kinship model based on bilateral descent groups, I do not agree with limiting its application to bilocal postmarital residence strategies, and specific nuances will be later discussed in detail. For this reason, I consider that it could be situated on a different conceptual plane than kin-based societies and states in their modern conception. In this way, we can find Houses operating as main social agents within political contexts with marked rank differences, heterarchical systems and complex or decentralized administration apparatuses (González-Ruibal and Ruiz-Gálvez 2016). In other words, the application of this analytical category allows us to transcend the evolutionist conception of traditional political theory, escaping the conventional dichotomy between kin-based societies and states.

11.2 Kinship and heterarchy in Late Roman Celtiberia

This chapter focuses on a region, Roman Celtiberia, which never constituted a Roman administrative demarcation. Nevertheless, its common geographical and demographic characteristics have shaped its historical trajectory under the Roman rule. Located on the fringes of the Iberian System (Fig. 11.1), with a core area situated between the upper valleys

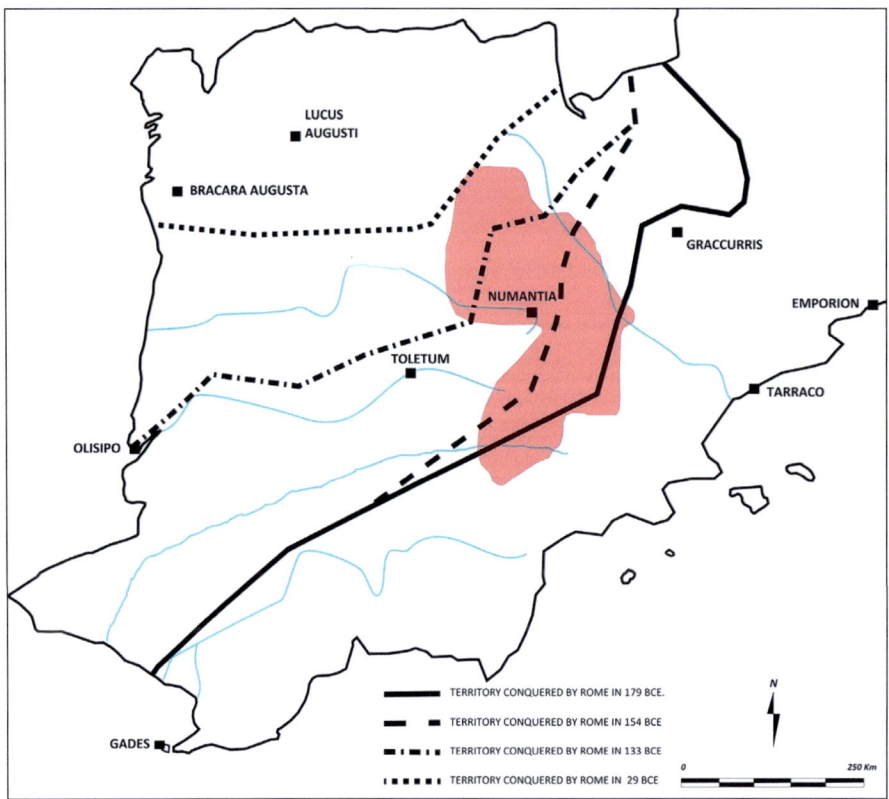

Figure 11.1. Map of the phases of the Roman conquest of the Iberian Peninsula using a model extracted from the late medieval recreation of Ptolemy's *Geographia*, with the territory of Celtiberi, Arevaci, Pelendones, and Turmogos in red (drawn by the author, after the fifteenth century CE Florentine codex curated at the Biblioteca Nacional de España).

of the Duero and Ebro rivers, this region remained under Roman domination from the mid-second century BCE to the fifth century CE. Generally, we can consider the Celtiberian region as a peripheral area within the Roman Empire. This is partly due to communication difficulties related to its geopolitical location, making it hard to access maritime routes, and partly due to the scarce economic resources that could be massively exploited by the metropolis (Salinas de Frías 1986; Burillo 2007).

Some of my previous research has focused on the study of Roman houses and households in this region (Bermejo Tirado 2014a). To synthesize and avoid repetitions, my research has discovered the existence of a great diversity of situations through comparative analysis of domestic architecture, family relationships, and recorded onomastics documented in the extensive corpus of funerary epigraphy (Bermejo Tirado 2018). These range from Roman kinship canonical models to others practically indistinguishable from those recorded in the Iron Age, with hybrid models (Bermejo Tirado 2020). However, this situation underwent a major turning point at the dawn of Late Antiquity, in the third century CE, when the decline of most of the ideological apparatuses of the classical Roman state, and especially the promulgation of the *Constitutio Antoniniana* during the reign of Emperor Caracalla marked a significant change (Lavan 2019b). As is well known, this decree granted Roman citizenship universally to all freeborn individuals regardless of the civic and marital status of their parents (Garnsey 2004). In other words, it meant the disappearance of the subordination relationship between citizenship status and kinship structures.

These elements, among other things, marked a shift in the development of political and social structures in this and other regions of the Late Roman world. Such a shift has not been analyzed with sufficient anthropological perspective. Conventional historiography has applied a Gibbonian perspective to this period, marked by the idea of crisis and the emergence of an aristocratic caste of *honestiores* that would come to occupy the gap left by the old Roman municipal structures and an increase in the importance of the rural world in the framework of Late Roman political networks. In archaeological terms, it is an era marked by the rise of several aristocratic rural residences and *villae*. In a way, it can be asserted, at least in much of the Iberian Peninsula –and the Celtiberian area is no exception– that the archaeology of the Late Roman period has been fundamentally an archaeology of *villae* (Grau Mira and Bermejo Tirado in press). Despite the undeniable importance of these architectural and archaeological models in previous historiography, their application to the interpretation of most of these aristocratic rural states and *villae* of the Late Roman period and the social and political structures of these communities is simply an exercise in anachronism.

As an alternative to these traditional historiographical perspectives, we believe that applying an interpretive model based on the intriguing reinterpretation of the concept of House Society proposed by González-Ruibal and Ruiz-Gálvez (2016) can help generate new frameworks for archaeological inquiry that may overcome old historicist models. The political and social structure of Late Roman Celtiberia cannot be equated with a kin-based society like those of the later prehistoric period. However, the historical changes that occurred in the Roman Empire from the third century CE onward prevent us from speaking of a classical state in the sense that we have outlined. Therefore, a heterarchical situation arises in which the municipal and citizenship structures that were hegemonic during the previous centuries of Roman domination are displaced by the emergence of Houses as moral persons –using the literal expression coined by Lévi-Strauss– and key political agents in the region.

11.2.1 The Irrics as an example of a Late Roman House

To support the validity or at least the heuristic utility of this model for rethinking the social and political structure of this region in the Late Roman period, we will use as a guiding thread the list of common traits of House Societies proposed by González-Ruibal and Ruiz-Gálvez (2016: 220-231) to discuss our case study. In addition, we will focus a significant portion of this discussion on the archaeological record of the *villa* of Dehesa de Cuevas de Soria (Soria, Spain) (Taracena 1930; Mariné 2007; Bermejo Tirado 2014b: 281-324), a site excavated on three different occasions (Fernández Galiano 2011), as well as other Late Roman examples in the region.

The discovery of a votive inscription in the surroundings of the *villa* at Dehesa de Cuevas de Soria (Sanz Aragonés *et al.* 2011) (Fig. 11.2) has allowed us to document a personal name, *Titus Irricus*, whose onomastic structure is certainly non-canonical and contains the cognominal reference to "the Irrics", which also appears in a considerable number of funerary inscriptions (Jimeno 1980: 74, 86; for a recent contextual study of

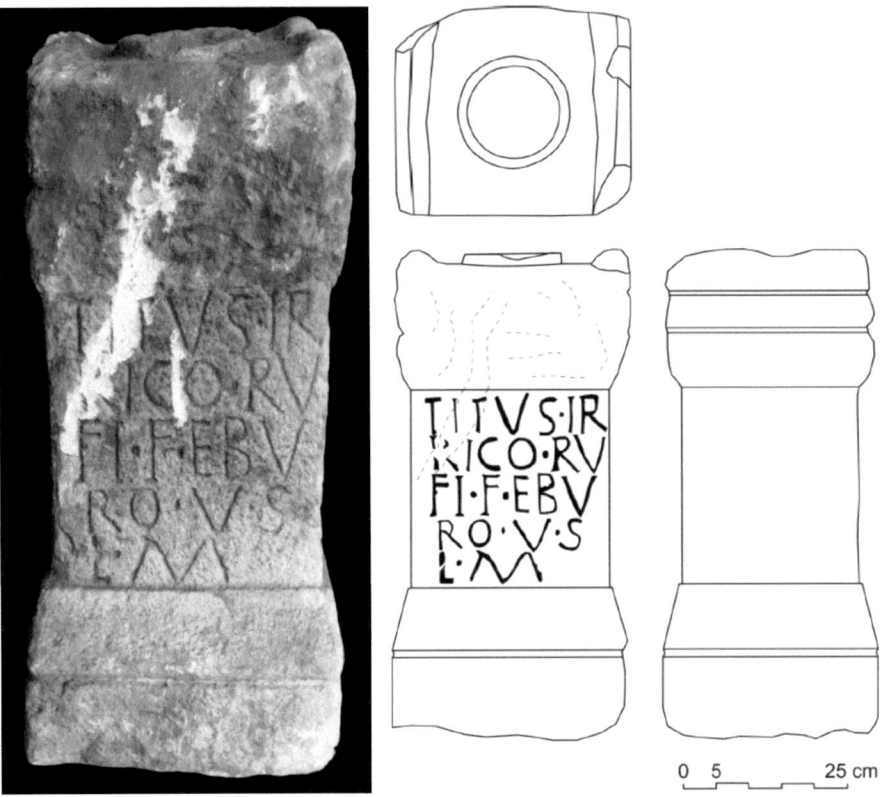

Figure 11.2. Photograph and drawing of the votive altar devoted to *Eburos* (after Sanz Aragonés *et al.* 2011: fig. 4).

the inscriptions, see Abascal Palazón 2017). The *ductus* and style of the inscription can reasonably be assigned to the initial third century CE. The fact that the onomastic structure of the name is non-canonical indicates that, regardless of their civic status, they wanted to emphasize this so-called *cognomen* –Latin onomastic reference to an extended cognatic group– with probable pre-Roman origins as a reference for personal identity in this public message. These Irrics may serve as an example of one of these Houses, that competed with one another as political agents within a heterarchical context. The votive inscription of *Titus Irricus*, for example, highlights how individuals and groups might assert their identity and status through public monuments, reinforcing their role as central actors in the local heterarchical social structure.

11.2.2 The *villa* of Cuevas de Soria as a patri-/virilocal extended household

Another common trait related to the analysis of House Societies is the importance of bilateral kinship, cognatic systems, and other kinship forms in which the wife's patrilineage is relevant. The significance of such kinship relations can be inferred directly from texts or indirectly through material –especially epigraphic– evidence (Edmondson 2015: 565-568). Examples of this include women inheriting property in the absence of males or the presence of women in rituals and kin-making strategies.

Figure 11.3. Mosaic of the *Dominus Iulius*. Bardo Archaeological Museum (Tunisia) (source: Wikipedia).

For our period, multiple examples illustrate the validity of this model. One notable case is Melania the Younger, whose biography was recorded by, among others, Gerontius, a late antique author (White 2010). Born into a wealthy aristocratic family of Hispanic origin, she was the daughter of senator *Valerius Publicola*. Her lands and *villae* extended from Roman Hispania to Rome, Campania, Sicily, Roman Africa, Mauretania, and Britain. In more material terms, perhaps the most explicit representation of the importance of women in these contexts is found in the *Dominus Iulius* Mosaic discovered in Carthage (Tunisia). This mosaic depicts the *domina* participating as a protagonist alongside her husband in a ritual of offering the fruits of the *domus* (Fig. 11.3).

Returning to the archaeological record of Roman Celtiberia, we believe that the spatial configuration of the *pars urbana* of the *villa* at Cuevas de Soria –like many others in the context of the Duero Valley– can be interpreted as an example of an extended patri/virilocal household. This interpretation follows the proposals put forward by Blanco González in the context of the Early Iron Age period in the southwestern Iberian Peninsula and beyond, as recently published in several works (Blanco González *et al.* 2022; Blanco González 2024). While acknowledging historical differences, the space syntax analysis proposed for a building like Cancho Roano by Jiménez Ávila in an article on the privatization of the complex, which must be nuanced by small methodological errors in the application of some of these indices, offers results surprisingly like those obtained from my own application

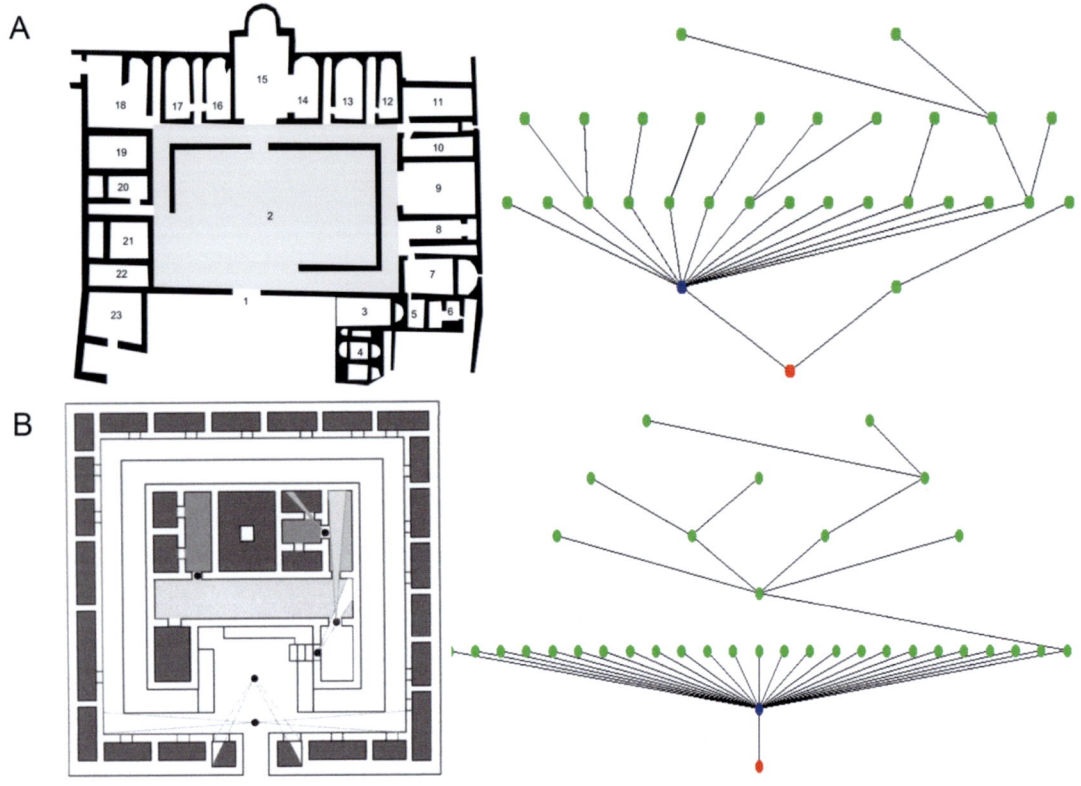

Figure 11.4. Comparison of the schematic plans of Cuevas de Soria (A) and Cancho Roano (B) and their access graphs (after Bermejo Tirado 2018: fig. 3 and Jiménez Ávila 2005: fig. 4).

of such space syntax tools for the case of Cuevas de Soria (Jiménez Ávila 2005) (Fig. 11.4). Although our archaeological knowledge of other similar settlements documented in other regions of Roman Celtiberia, such as San Pedro de Valdanzo (Soria, Spain) (Jimeno *et al.* 1988-89) is much more incomplete than in the case of the *villa* at La Dehesa, what we know about the configuration of these spaces indicates that we are likely dealing with structures of great similarity both in size and spatial configuration around an extended central courtyard. In broad terms, this could indicate the widespread adoption of this spatial configuration model among Late Roman *villae* built in this region of Celtiberia.

The application of space syntax analysis, as demonstrated in our case study of Cuevas de Soria, provides quantitative support for the virilocal extended household hypothesis. The striking similarities between the spatial configurations of Late Roman *villae* and earlier Iron Age buildings, such as Cancho Roano (Badajoz, Spain) and other later prehistoric buildings discussed by other colleagues (Grau Mira this volume; Blanco González this volume), suggest a degree of cross-cultural similarity in the social organization of different periods.

At this point, we consider it appropriate to introduce a series of clarifications to refine our interpretation. The first issue concerns the search for cross-cultural patterns. We wish to emphasize that our comparison in no way seeks to suggest a possible cultural or historical continuity between the later prehistory and the Late Roman period. The aim of our comparison is analytical, intended to better understand which spatial configuration models may reflect both the organization of descent groups and postmarital residence patterns (Ensor 2013a, 2013b) during this period. In this regard, our interpretation of Cuevas de Soria as a settlement linked to a community with virilocal or patrilocal postmarital residence patterns is developed within a broader framework, characterized by the articulation of bilaterally organized descent groups. We cannot overlook that our interpretation differs slightly from Ensor's (2013a: 160, this volume) recurring concept of the House Society model, which he limits to a single framework characterized by bilateral descent organization and bilocal postmarital residence patterns.

While we agree with linking the concept of House Society to bilateral descent models in complex social contexts, we differ in limiting its application to bilocal postmarital residence patterns. For us, the primary characteristic of these Houses is their capacity to use and modify different forms of kinship articulation from an intergenerational perspective. Although we fundamentally agree with Ensor in emphasizing kinship as a key organizing principle of political relationships in many past societies, we believe this cannot be understood rigidly or restricted to specific postmarital residence models. Instead, these patterns may have shifted over various chronological periods, as we suggest occurred in Roman Celtiberia during the transition from the Early Roman Empire to Late Antiquity.

11.2.3 The House as a key symbolic and material component

Material remains reveal that Houses are imbued with religious or cosmic symbolism, including ritual offerings in their vicinity. In the case of Cuevas de Soria, this function is related to the discovery of the inscription we mentioned earlier, which refers to the onomastic reference of the Irrics. This inscription is a votive altar dedicated to a pre-Roman deity, [EBUROS] (Fig. 11.2), which, as a theonym, has not been documented in the Celtic area, but is clearly linked to another anthroponym in genitive plural (*Eburanco*) from the nearby Celtiberian city of *Uxama Argaela* (Soria) (*CIL* II, 2828; Ramírez 2007: 1162) that could be interpreted as a reference to another House or corporate group in the region. The maintenance of a cult and the performance of rituals –as evidenced by the presence of this altar– serve to assert the value of local cults as social articulators of these groups during the Late Roman era.

A strong material investment in the House is another obvious element for the identification of this model from an archaeological point of view. Most of the settlements that we currently label as *villae* in the interior of the Iberian Peninsula belong to a phase of monumentalization that occurred within a specific chronological period between the late third and the fourth centuries CE, lasting just over a century. This is something that is now more systematically documented thanks to the application of stratigraphic documentation criteria in more recent interventions. Hence, analyzing this type of settlement according to the classical perspective of the *Re Rustica* treatises is, as I just exposed, an exercise in anachronism.

In the case of Cuevas de Soria, thanks to excavations in 2011, we can conclude that the monumentalization phase, which includes the paving of the complex with a series of magnificent mosaics, the application of wall painting decoration to simulate the use of marble materials, and the construction of a small but very complex thermal feature that includes a heating system within the walls of the enclosure (Mariné 1985), was equipped during this timespan of the Late Roman period. These features, among many other things, represent a clear investment that goes beyond the survival margins of the community and can only be explained concerning the social needs of their inhabitants during this period.

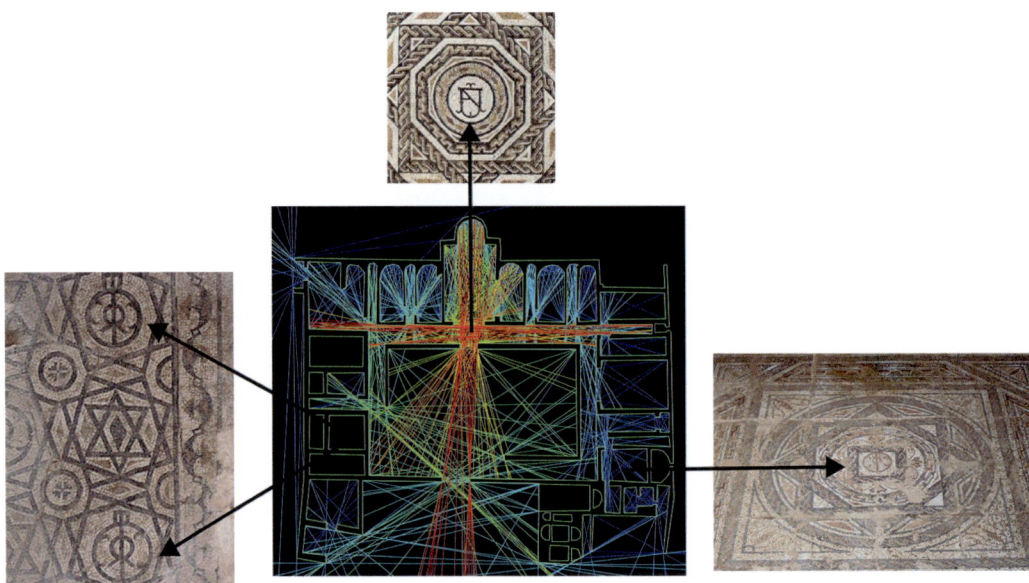

Figure 11.5. Axial map of the *villa* of La Dehesa at Cuevas de Soria (Soria, Spain) and indication of the anagrams/*signacula* recorded in its mosaics.

The monumentalization of the *villa* at Cuevas de Soria reflects a broader trend in the Roman Empire during the Late Antique period, when significant investments in domestic architecture symbolized the social status and political influence of the inhabitants. This substantial material investment can be interpreted as a means of asserting and reinforcing the social hierarchy and the power dynamics within the local community.

11.2.4 Heraldic artefacts

Another common trait in this social model is the presence of titles of nobility, rank, or office (González-Ruibal and Ruiz-Gálvez 2016). Even if we cannot infer these aspects from specific texts, the material culture offers examples that can help visualize these aspects: sceptres, staffs, diadems, or signet rings are obvious examples. One of the mosaic pavements of Cuevas de Soria (Fig. 11.5) –currently on display at the National Archaeological Museum in Madrid– was located in a clearly highlighted axial position within the ensemble. Among the motifs reproduced, it depicts what can be interpreted as an anagram, most likely with a certain heraldic significance. More significantly for this discussion, the shape of this anagram has been rightly related by Gómez Pantoja (2001) –among others– to a *signaculum*, that is, a signet metal tool for marking livestock (Gómez Pantoja 2001; Bermejo Tirado 2016). This can be interpreted as a symbolic reference to the origin of the wealth of the house. Although we have not found in the material culture of this villa any artefact that can be linked to a *signaculum*, in the not-too-distant *villa* of Arellano, in Navarra, an almost complete specimen has been documented within the context of an archaeological level sealed by a sudden fire (Mezquíriz Irujo 2007-2008: 99) (Fig. 11.6).

Much less certain, but much more significant for our regional study, is the interpretative hypothesis for the identification of a metallic fragment from excavations carried out in San Pedro de Valdanzo (Soria, Spain) in the late 1980s, another Late Roman *villa* with a

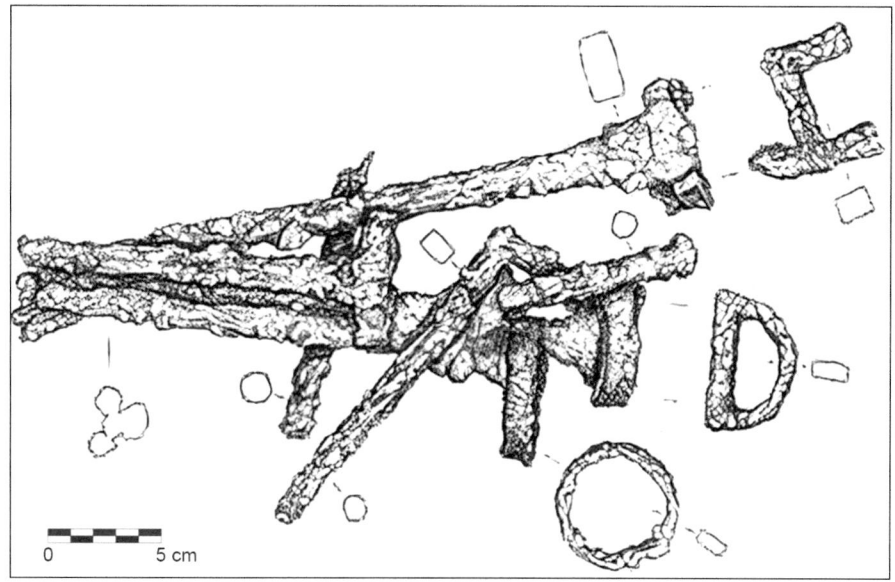

Figure 11.6. *Signaculum* from the Late Roman *villa* of Arellano (after Mezquíriz-Irujo 2007-08, fig. 28).

Figure 11.7. Comparison between the geometric composition of the central *emblemata* from the Late Roman villa of Cuevas de Soria (A) and San Pedro de Valdanzo (B) (after Bermejo Tirado 2016: fig. 4).

central courtyard located not far from Cuevas de Soria (Fig. 11.6). The significance of this interpretation lies in the fact that San Pedro de Valdanzo is a settlement that, despite being much less archaeologically known, presents a similar morphological configuration (Jimeno *et al.* 1988-89). Furthermore, the mosaics documented at this site show an astonishing similarity, unique in the Iberian Peninsula, to the pavements documented at Cuevas de Soria (Torres 1990). In this case, there is also a central geometric figure framing another motif, which has disappeared due to breakage, but which would likely reproduce a similar anagram/ *signaculum* to the one displayed at Cuevas de Soria (Bermejo Tirado 2016) (Fig. 11.7).

One last common trait to which we will refer, according to the proposal of González-Ruibal and Ruiz-Gálvez (2016), is the presence of explicit references to the house as something that includes the material building, its material wealth, and the extended family

Figure 11.8. Detail of the Irrics anagram in a mosaic from Cuevas de Soria (courtesy of Diputación Provincial de Soria).

that inhabits it. In this case, also thanks to the interventions carried out in 2011, new pavements were documented with a much larger number of anagrams/*signacula*, in this case much more alluding to the lineage of the Irrics (Abascal Palazón 2017) (Fig. 11.8), whose reference we have seen documented in the aforementioned votive altar. This is certainly an explicit reference that serves to relate the kinship of the Irrics to the material source of wealth of the house, which is repeated in multiple rooms of the complex. Therefore, if we accept this interpretation of the complex as a virilocal or patrilocal extended household, it will serve to confirm the identification of this corporate group inhabiting the *villa* with the lineage of the Irrics.

11.3 Discussion: Citizenship, kinship, and heterarchy in Late Roman Celtiberia

At the beginning of this chapter, we referred to the importance of the concept of citizenship as a thermometer of the relevance of the state as the primary political framework. Following this same logic, we will use the example of the Irrics to frame a discussion on the interpretation of the political structures that characterized these rural sectors of the Celtiberia region during the Late Roman period. First, we must make an obvious reference to the pre-Roman etymological origins of this anthroponym. But even more important, if possible, is the onomastic structure in which this name is recorded. Unlike other more canonical formulations that in previous centuries served to unequivocally express a privileged civic status, in these descriptions we find a heterodox formulation that again emphasizes belonging to a specific lineage at the expense of the civic status.

This onomastic phenomenon that occurred from the third century CE onwards is now widely known in Roman epigraphy studies and can be interpreted as a clear symptom of the decline of municipal administration as a political reference for local communities, even for their elites. Parallel to this onomastic process, the association of this House with local pre-Roman cults –indicated by the votive altar dedicated to the god *Eburos*– also points to a certain decline of canonical cults more conventionally linked to the administration of municipal centres, in favour of others associated with a tradition predating the arrival of the Romans.

In more purely archaeological terms, we have noted the similarity in the spatial configuration of multi-family dwelling and productive complexes such as in Cuevas de Soria and probably San Pedro de Valdanzo (Figs. 11.4), as well as many other so-called *villae* of the Duero Valley (Bermejo Tirado 2018), to models linked to patrilocal structures associated with extended families that we find in different points of the Mediterranean. This hypothesis regarding the interpretation of the *partes urbanae* of many *villae* as extended households is not entirely novel, but it deserves renewed attention considering recent archaeological discoveries and theoretical developments. Indeed, the architect J.T. Smith (1997) openly discussed this possibility, laying the groundwork for further exploration. Smith's insights, while groundbreaking, were primarily based on architectural analysis. Our current research builds upon this foundation by integrating a wider range of archaeological evidence, including spatial analysis, material culture, and epigraphic data. This multifaceted approach allows for a more nuanced understanding of the social dynamics within these Late Roman *villae*.

The coincidence of conventional space syntax patterns and other proxemic indicators within these housing arrangements is an analysis that must be undertaken on a more solid quantitative basis than the preliminary one advanced in this chapter. Pending more systematic sampling, the overall similarity of the analyzed cases seems to support the hypothesis that establishes a recurring pattern regarding the existence of a common spatial layout for these extensive virilocal residences on a cross-cultural comparison. The importance of this traditional kinship system represents a shift in trend compared to the frameworks considered canonical in Roman private law, which prioritized the nuclear family sanctioned through a legally recognized marriage as the fundamental guarantor of the sexual reproduction of citizens and the property rights over their assets (Wallace-Hadrill 1981; Saller and Shaw 1984; Edmondson 2015). This decline in kinship forms linked to canonical citizenship frameworks also coincides with the total disappearance in Roman Celtiberia of those domestic architectural types that best adapted to the social needs of the nuclear families of privileged citizens, which were the majority in the cities of Roman Celtiberia until the third century CE (Bermejo Tirado 2020: fig. 7).

Someone might think that our interpretation of this period as a House Society could be read as a process of involution. This would imply accepting the interpretation of the classical historiographical paradigm of the end of the Roman Empire as a period of decline and general crisis. On the contrary, we believe that the development of these Houses is the result of a process of formation of local rural communities (Grey 2011) with their own political agency. The material culture and epigraphic texts generated by these communities refer to forms of kinship, religious practices, and markers of social rank that are very different from those that defined the classical municipal model in the Roman world. All these elements serve to characterize political communities of a heterarchical nature in which the influence of the social policy of the Roman imperial administration seems to become marginal in the face of the strength of Houses such as that of the Irrics.

11.4 Final remarks

The presence of religious symbols and the practice of local cults –such as the votive altar dedicated to *Eburos*– highlights the integration of spiritual and religious elements into the domestic sphere. This integration underscores the role of Houses not just as living spaces but as centers of religious and cultural identity. It reinforces the idea that the House Society model, which emphasizes the House as a key social and political unit, is particularly useful for understanding the social structures of Late Roman Celtiberia.

The review of archaeological evidence from Cuevas de Soria and other similar sites through the lens of other methodological (space syntax) and theoretical (House Society, heterarchy) models provides a robust framework for interpreting the social and political organization of this region during the Late Roman period. The material culture, spatial configurations, and religious practices all point towards a complex and nuanced social structure where Houses functioned as central units of social, economic, and political life. This perspective allows us to move beyond traditional historiographical models and offers a more comprehensive understanding of the social dynamics in Late Roman Celtiberia.

This perspective, firstly articulated by authors like Ensor (2013a) or González-Ruibal and Ruiz-Gálvez (2016), can be applied to the Late Antiquity to offer a deeper understanding of the region's complex social dynamics during a period of significant transition. Our analysis suggests that kinship groups –here exemplified by the Irrics– played a pivotal role in shaping the political and social structures of the region. These Houses effectively filled the gap left by the declining municipal model and the waning importance of classical citizenship status in the Late Roman period. This approach may face scepticism from scholars accustomed to more traditional historiographical frameworks. However, the application of anthropological models based on concepts such as kinship, House Society, and heterarchy provides a more locally grounded perspective than conventional narratives. Traditional concepts like the *villa schiavistica* or the "crisis of the third century CE" often impose broad, generalized interpretations that may obscure the complexities of local experiences and adaptations.

By employing a heterarchical framework, we can open new avenues for historiographical review, encouraging a shift towards microhistorical approaches that can capture the subtleties and variations within Late Roman society. This methodology allows us to move beyond simplistic narratives of decline or crisis, revealing instead a dynamic landscape of social innovation and resilience. Moreover, this perspective encourages us to reconsider the material culture of Late Roman *villae* not merely as expressions of elite status, but as integral components of a complex social system. The symbolic and practical importance of these structures in maintaining and reproducing social relations becomes clearer when viewed through the lens of House Society.

In conclusion, while our approach may contribute to challenge established historiographical traditions, we believe it offers a more holistic and contextually sensitive understanding of Late Roman Celtiberia. By embracing anthropological models and micro-scale perspectives, we can develop a richer, more nuanced appreciation of the diverse ways in which communities navigated the changing social and political landscape of Late Antiquity.

Acknowledgments

I would like to express my gratitude to Dr. Eva Alarcón and Dr. Antonio Blanco for inviting me to join the ARQPARENT project, as well as for the invitation to participate in the seminar held in 2024 at the University of Salamanca. I am also deeply thankful for Dr. Antonio Blanco's comments and revisions, which have significantly contributed to improving the text.

References

Abascal Palazón, J.M. (2017): Notas de epigrafía soriana y salmantina (Santervás del Burgo, Las Cuevas de Soria y Espino de los Doctores). *Oppidum* 13, pp. 81-104.

Beck, R.A. (ed.) (2007a): *The Durable House: House Society Models in Archaeology*. Carbondale: Southern Illinois University.

Beck, R.A. (2007b): The durable house: Material, metaphor, and structure. In Beck, R.A. (ed.), *The Durable House: House Society Models in Archaeology*. Carbondale: Southern Illinois University, pp. 3-25.

Bermejo Tirado, J. (2014a): *Arqueología biopolítica. La sintaxis espacial de la arquitectura doméstica romana en la Meseta oriental*. Madrid: La Ergástula.

Bermejo Tirado, J. (2014b): *Arqueología de los espacios domésticos romanos. Condiciones de vida y sociedad en la Meseta nordeste durante el periodo imperial*. Soria: Diputación Provincial de Soria.

Bermejo Tirado, J. (2018): Houses and society in Roman Celtiberia. *Mouseion* 15(2), pp. 173-218.

Bermejo Tirado. J. (2020): House form and household structure: The social analysis of urban domestic architecture in Roman Celtiberia. In Dardenay, A. and Laubry, N. (eds.), *The Anthropology of Roman Housing*. Turnhout: Brepols, pp. 259-292. DOI: 10.1484/m.ash-eb.5.119589.

Blanco González, A. (2024): Arqueología del hogar extenso virilocal en el suroeste peninsular (siglos VII-V a.C.). In Bermejo Tirado, J. and Blanco González, A. (eds.), *Arqueología de los espacios domésticos en la Península Ibérica. De la Prehistoria reciente a la Edad Media*. Vitoria: Universidad del País Vasco, pp. 75-96.

Blanco González, A., Padilla Fernández, J.J., Alario García, C., Macarro Alcalde, C., Alarcón García, E. *et al.* (2022): Un singular ambiente doméstico del Hierro I en el interior de la península ibérica: la casa 1 del Cerro de San Vicente (Salamanca, España). *Trabajos de Prehistoria* 79(2), pp. 346-361. DOI: 10.3989/tp.2022.12303.

Burillo, F. (2007): *Los celtíberos. Etnias y estados*. Barcelona: Crítica.

Carsten, J. and Hugh-Jones, S. (eds.) (1995): *About the House: Lévi-Strauss and Beyond*. Cambridge: Cambridge University Press. DOI: 10.1017/CBO9780511607653.

Cohen, A. (2004): The historical roots of French medievalism: Lévi-Strauss and the structuralist approach. In Cohen, A. (ed.), *Medievalism in Europe*. New York: Palgrave Macmillan, pp. 45-67. DOI: 10.1057/9780230505996_3.

Curchin, L. (2000): The Roman family: Recent interpretations. *Zephyrus* 53-54, pp. 535-550.

Currás, B. (2019): *Las sociedades de los castros entre la Edad del Hierro y la dominación de Roma. Estudio del paisaje del Baixo Miño*. Madrid: CSIC.

Earle, T. (2017): Property in prehistory. In Graziadei, M. and Smith, L. (eds.), *Comparative Property Law*. Cheltenham: Edward Elgar Publishing, pp. 3-37. DOI: 10.4337/97817853 69162.00008.

Edmondson, J. (2005): Family relations in Roman Lusitania: Social change in a Roman province? In George, M. (ed.), *The Roman Family in the Empire: Rome, Italy, and Beyond.* Oxford: Oxford University Press, pp. 183-230. DOI: 10.1093/acprof:oso/9780199268412.003.07.

Edmondson, J. (2015): Roman Family History. In Bruun, C. and Edmondson, J. (eds.), *The Oxford Handbook of Roman Epigraphy.* Oxford: Oxford University Press, pp. 559-581.

Ensor, B.E. (2013a): *The Archaeology of Kinship: Advancing Interpretation and Contributions to Theory.* Tucson: University of Arizona Press.

Ensor, B.E. (2013b): *Crafting Prehispanic Maya Kinship.* Tuscaloosa: University of Alabama Press.

Feinman, G.M. (2023): Reconceptualizing archaeological perspectives on long-term political change. *Annual Review of Anthropology* 52, pp. 347-364. DOI: 10.1146/annurev-anthro-060221-114205.

Fernández-Galiano Ruiz, D. (2011): *Los monasterios paganos. La huida de la ciudad en el mundo antiguo.* Córdoba: El Almendro.

Flannery, K.V. (1972): The cultural evolution of civilizations. *Annual Review of Ecology and Systematics* 3(1), pp. 399-426. DOI: 10.1146/annurev.es.03.110172.002151.

Flannery, K.V. (2002): The cultural evolution of civilizations. *Annual Review of Ecology, Evolution, and Systematics* 33, pp. 1-23. DOI: 10.1146/annurev.ecolsys.33.010802.150249.

García Moreno, J.C. (ed.) (2022): *From House Societies to States. Early Political Organisation, from Antiquity to the Middle Age*s. Oxford: Oxbow Books.

Garnsey, P. (2004): Roman citizenship and Roman law in the late Empire. In Swain, S. and Edwards, M. (eds.), *Approaching Late Antiquity: The Transformation from Early to Late Empire.* Oxford: Oxford University Press, pp. 133-155. DOI: 10.1093/acprof:oso/9780199297375.003.0005.

Gillespie, S.D. (2000): Introduction. In Gillespie, S.D. and Joyce, R.A. (eds.), *Beyond Kinship: Social and Material Reproduction in House Societies.* Philadelphia: University of Pennsylvania Press, pp.1-21.

Gillespie, S.D. and Joyce, R.A. (eds.) (2000): *Beyond Kinship: Social and Material Reproduction in House Societies.* Philadelphia: University of Pennsylvania Press.

Gómez Pantoja, J. (2001): *Pastio agrestis.* Pastoralismo en Hispania. In Gómez Pantoja, J. (ed.), *Los rebaños de Gerión. Pastores y trashumancia en Iberia antigua y medieval.* Madrid: Casa de Velázquez, pp. 177-213. DOI: 10.4000/mefra.3521.

González-Ruibal, A. (2006): House societies vs. kinship-based societies: An archaeological case from Iron Age Europe. *Journal of Anthropological Archaeology* 25(2), pp. 144-173. DOI: 10.1016/j.jaa.2005.09.002.

González-Ruibal, A. and Ruiz-Gálvez, M. (2016): House societies in the ancient Mediterranean (2000-500 BC). *Journal of World Prehistory* 29(3), pp. 217-336. DOI: 10.1007/s10963-016-9098-8.

Gorrochategui, J., Navarro, M. and Vallejo, J.M. (2011): Reflexiones sobre la historia social del valle del Duero: las denominaciones personales. In Navarro, M., Palao, J.J. and Magallón, M.A. (eds.), *Villes et territoires dans le bassin du Douro à l'époque romaine.* Bordeaux: Ausonius Éditions. DOI: 10.4000/books.ausonius.996.

Grau Mira, I. and Bermejo Tirado, J. (in press): Roman Iberia: An overview of the archaeology of rural landscapes and societies. In Tol, G. and Van Oyen, A. (eds.), *Cambridge Handbook of Roman Rural Archaeology*. Cambridge: Cambridge University Press.

Godelier, M. (2011): *The Metamorphoses of Kinship*. London: Verso.

Grey, C. (2011): *Constructing Communities in the Late Roman Countryside*. Cambridge: Cambridge University Press. DOI: 10.1017/CBO9780511994739.013.

Hayden, B. and Cannon, A. (1982): The corporate group as an archaeological unit. *Journal of Anthropological Archaeology* 1(2), pp. 132-158. DOI: 10.1016/0278-4165(82)90018-6.

Heather, P. (2005): *The Fall of the Roman Empire: A New History of Rome and the Barbarians*. Oxford: Oxford University Press.

Huebner, S.R. (2013): *The Family in Roman Egypt: A Comparative Approach to Intergenerational Solidarity and Conflict*. Cambridge: Cambridge University Press. DOI: 10.1017/CBO9780511894558.

Huebner, S.R. and Nathan, G. (eds.) (2017): *Mediterranean Families in Antiquity: Households, Extended Families, and Domestic Space*. Oxford: Wiley-Blackwell.

Jiménez Ávila, J. (2005): Cancho Roano: El proceso de privatización de un espacio ideológico. *Trabajos de Prehistoria* 62(2), pp. 105-124. DOI: 10.3989/tp.2005.v62.i2.71.

Jimeno Martínez, A. (1980): *Epigrafía romana de Soria*. Soria: Diputación Provincial de Soria.

Jimeno Martínez, A., Argente Oliver, J.L. and Gómez Santa-Cruz, J. (1988-1989): La villa romana de San Pedro de Valdanzo (Soria). *Zephyrus* 41-42, pp. 419-454.

Lavan, M. (2016): The spread of Roman citizenship (14-212 CE): Quantification in the face of high uncertainty. *Past and Present* 230(1), pp. 3-46. DOI: 10.1093/pastj/gtv043.

Lavan, M. (2019a): The army and the spread of Roman citizenship. *Journal of Roman Studies* 109, pp. 27-69. DOI: 10.1017/S0075435819000662.

Lavan, M. (2019b): Constitutio Antoniniana. In *Oxford Classical Dictionary*. Fifth edition. Oxford: Oxford University Press. DOI: 10.1093/acrefore/9780199381135.013.1794.

Lévi-Strauss, C. (1987): *The Way of the Masks*. New Haven: Yale University Press.

Mariné, M. (1985): Las termas de la villa de Cuevas de Soria. In *Actas del I Symposium de Arqueología Soriana*. Soria: Diputación Provincial de Soria, pp. 403-415.

Mariné, M. (2007): La villa romana de Cuevas de Soria, relato de una investigación sincopada. *Arevacon* 27, pp. 17-24.

Martin, D.B. (1996): The construction of the ancient family: Methodological considerations. *Journal of Roman Studies* 86, pp. 40-60. DOI: 10.2307/300422.

Mezquíriz Irujo, M.A. (2007-2008): Instrumentos de hierro para la explotación agropecuaria en época romana. *Trabajos de Arqueología Navarra* 20, pp. 197-228.

Osgood, J. (2019): Family history in Augustan Rome. In Whitton C., Gildenhard, I., Gotter, U., Havener, W. and Hodgson, L. (eds.), *Augustus and the Destruction of History: The Politics of the Past in Early Imperial Rome*. Cambridge: Cambridge Philological Society, pp. 135-156. DOI: 10.2307/j.ctv10kmc9n.12.

Pauketat, T.R. (2007): *Chiefdoms and Other Archaeological Delusions*. Lanham: AltaMira Press.

Ramírez Sánchez, M. (2002): Estelas funerarias y grupos de parentesco en la región Celtibérica. In *Actas del VII Congreso Internacional de Estelas Funerarias*. Santander: Fundación Marcelino Botín, pp. 139-156.

Ramírez Sánchez, M. (2003): Epigrafía latina y relaciones de parentesco en la región celtibérica: nuevas propuestas. In Armani, S., Stylow, A. and Hurlet-Martineau, B. (eds.),
   *Epigrafía y sociedad en Hispania durante el Alto Imperio: estructuras y relaciones
   sociales.* Madrid: Universidad de Alcalá de Henares and Casa de Velázquez, pp. 13-32.

Ramírez Sánchez, M. (2007): Los grupos de parentesco en la epigrafía latina hispánica.
   Genitivos de plural en -om/ - om. In Mayer, M., Baratta, G. and Guzmán, A. (eds.),
   Provinciae Imperii Romani inscriptionibus descriptae. Barcelona: Institut d'Estudis
   Catalans, pp. 1161-1168.

Sahlins, M. (2013): *What Kinship Is - And Is Not.* Chicago: University of Chicago Press.

Salinas de Frías, M. (1986): *Conquista y romanización de Celtiberia.* Salamanca: Universidad de Salamanca.

Saller, R.P. and Shaw, B.D. (1984): Tombstones and Roman family relations in the Prin-
   cipate: Civilians, soldiers and slaves. *The Journal of Roman Studies* 74, pp. 124-156.
   DOI:10.2307/299012.

Sanz Aragonés, A., Tabernero Galán, C., Benito Batanero, J.P. and Bernardo Stempel, P.
   (2011): Nueva divinidad céltica en un ara de Cuevas de Soria. *Madrider Mitteilun-
   gen* 52, pp. 440-452.

Schloen, J.D. (2001): *The House of the Father as Fact and Symbol: Patrimonialism in Ugarit
   and the Ancient Near East.* Leiden: Brill.

Smith, J.T. (1997): *Roman Villas: A Study in Social Structure.* London: Routledge.

Thomas, J. (2015): House societies and founding ancestors in early Neolithic Britain.
   In Renfrew, C., Boyd, M.J. and Morley, I. (eds.), *Death Rituals, Social Order and
   the Archaeology of Immortality in the Ancient World: "Death Shall Have No
   Dominion".* Cambridge: Cambridge University Press, pp. 138-150. DOI:10.1017/
   CBO9781316014509.

Torres, M. (1990): Los mosaicos en la Meseta Norte. *Boletín del Seminario de Estudios de
   Arte y Arqueología* 56, pp. 223-243.

Treggiari, S. (1996): Social status and social legislation. In Bowman, A.K., Champlin, E.
   and Lintott, A. (eds.), *The Cambridge Ancient History.* Cambridge: Cambridge University Press, pp. 873-904. DOI:10.1017/CHOL9780521264303.031.

Wallace-Hadrill, A. (1981): Family and inheritance in the Augustan marriage laws.
   Proceedings of the Cambridge Philological Society 27(207), pp. 58-80. DOI:10.1017/
   S0068673500004326.

Ward-Perkins, B. (2005): *The Fall of Rome and the End of Civilizatio*n. Oxford: Oxford
   University Press.

White, C. (2010): *Lives of Roman Christian Women.* London: Penguin Books.

Wiersma, C.W. (2020): House (centric) societies on the prehistoric Greek mainland. *Oxford
   Journal of Archaeology* 39, pp. 141-158. DOI:10.1111/ojoa.12190.

Wolf, E.R. (1982): *Europe and the People Without History.* Berkeley: University of California Press.

Yoffee, N. (2005): *Myths of the Archaic State: Evolution of the Earliest Cities and States.*
   Cambridge: Cambridge University Press.